Berger Automating with STEP 7 in STL

# Automating with STEP 7 in STL

SIMATIC S 7-300/400
Programmable Controllers

by Hans Berger

Publicis MCD Verlag

Die Deutsche Bibliothek – CIP-Einheitsaufnahme

**Automating with STEP 7 in STL :** SIMATIC S7-300/400
programmable controllers / issued by Siemens-Aktiengesellschaft,
Berlin and Munich. By Hans Berger. – Erlangen ; Munich : Publicis-
MCD-Verl.
Einheitssacht.: Automatisieren mit STEP 7 in AWL <dt.>

ISBN 3-89578-093-6

Buch. 1998
Gb.

Diskette. Programming examples. – 1998

The programming examples concentrate on describing the STL functions and providing SIMATIC S7 users with programming tips for solving specific tasks with this controller.

The programming examples given in the book do not pretend to be complete solutions or to be executable on future STEP 7 releases or S7-300/400 versions. Additional care must be taken in order to comply with the relevant safety regulations.

The author and publisher have taken great care with all texts and illustrations in this book. Nevertheless, errors can never be completely avoided. The publisher and the author accept no liability, regardless of legal basis, for any damage resulting from the use of the programming examples.

ISBN 3-89578-093-6

Issued by Siemens Aktiengesellschaft, Berlin and Munich
Published by Publicis MCD Verlag, Erlangen and Munich


Printed in Germany

# Preface

The new SIMATIC basic automation system unites all the subsystems of an automation solution under a uniform system architecture into a homogeneous whole from the field level right up to process control. This is achieved with integrated configuring and programming, data management and communications with programmable controllers (SIMATIC S7), automation computers (SIMATIC M7) and control systems (SIMATIC C7). With the programmable controllers, three series cover the entire area of process and production automation: S7-200 as compact controllers ("Micro PLCs"), S7-300 and S7-400 as modularly expandable controllers for the low-end and high-end performance ranges.

STEP 7, a further development of STEP 5, is the programming software for the new SIMATIC. Microsoft Windows 95 or Microsoft Windows NT has been chosen as the operating system, to take advantage of the familiar user interface of standard PCs (windows, mouse operation). STEP 7 complies with DIN EN 6.1131-3, the standard for PLC programming languages. For block programming you can choose between STL (statement list; an Assembler-like language), LAD (ladder logic; a representation similar to relay logic diagrams), FBD (function block diagram) and SCL (a Pascal-like high-level language). Several optional packages supplement these languages: S7-GRAPH (sequential control), S7-HiGraph (programming with state-transition diagrams) and CFC (connecting blocks; similar to function block diagram). The various methods of representation allow every user to select the suitable control function description. This broad adaptability in representing the control task to be solved significantly simplifies working with STEP 7.

This book contains the description of the LAD programming language for S7-300/400. In the first section, the book introduces the S7-300/400 automation system and explains the basic handling of STEP 7. The next section addresses first-time users or users changing from relay contactor controls; the "Basic Functions" of a binary control are described here. The digital functions explain how digital values are combined; for example, basic calculations, comparisons, data type conversion. With LAD, you can control program processing (program flow) and design structured programs. As well as a cyclically processed main program, you can also incorporate event-driven program sections as well as influence the behavior of the controller at startup and in the event of errors/faults. The book concludes with a general overview of the system functions and the function set.

The contents of this book describe Version 4.02 of the STEP 7 programming language.

Erlangen, March 1998

Publicis MCD Verlag

# The Contents of the Book at a Glance


Overview of the S7-300/400 programmable logic controller

**Introduction**

**1 S7-300/400 Programmable Controller**
- Structure of the Programmable Controller
- Configuring Stations
- Addressing Modules
- Memory Areas
- Address Areas

**2 STEP 7 Programming Software**
- Editing Projects
- Hardware Configuration
- Developing Programs, Symbol Table
- Program Test, Online Functions

**3 STL Programming Language**
- Program Processing
- Blocks
- Incremental and Source-Oriented Programming

PLC functions comparable to a contactor control system

**Basic Functions**

**4 Binary Logic Operations**
- AND, OR and Exclusive OR Functions
- Nesting Functions

**5 Memory Functions**
- Assign, Set and Reset
- Edge Evaluation

**6 Move Functions**
- Load and Transfer
- Accumulator Functions

**7 Timer Functions**
- Start Five Different Types of Timers

**8 Counter Functions**
- Count Up, Count Down, Set, Scan

Numbers, manipulating the contents of the accumulators

**Digital Functions**

**9 Comparison Functions**

**10 Arithmetic Functions**
- Four-function Math with INT, DINT and REAL Numbers
- Adding Constants
- Decrement/Increment

**11 Math Functions**
- Trigonometric Functions, Powers, Logarithms

**12 Conversion Functions**
- Data Type Conversion, Complement Formation

**13 Shift Functions**
- Shifting and Rotating

**14 Word Logic**
- AND, OR, Exclusive OR

Program run control, block functions

**Program Flow Control**

**15 Status Bits**
- Set and Evaluate
- Status Word

**16 Jump Functions**
- Branches, Jump Distributors, Loops

**17 Master Control Relay**

**18 Block Functions**
- Block Call, Block End
- Static and Temporary Local Data
- Data Addresses

**19 Block Parameters**
- Formal Parameters, Actual Parameters
- Declarations and Assignments, "Parameter Passing"

Processing the user program

## Program Processing

**20 Main Program**
Program Structure
Scan Cycle Time, Response Time
Program Functions
Data Exchange using System Functions
Start Information

**21 Interrupt Handling**
Hardware Interrupts, Watchdog Interrupts, Time Interrupts, Time-Delay Interrupts, Multiprocessor Interrupts

**22 Start-up Characteristics**
Complete Restart, Warm Restart
STOP, Memory Reset
Initializing Modules

**23 Error Handling**
Synchronous Errors
Asynchronous Errors
Diagnostics

Working with complex variables, indirect addressing

## Variable Handling

**24 Data Types**
Structure
Declaration and Use of Complex Data Types
User Data Types

**25 Indirect Addressing**
Area Pointers
DB Pointer
ANY Pointer
Indirect Addressing via Memory or Register
Working with an Address Register

**26 Direct Variable Access**
Load Variable Address
Storing Variables and Parameters in Memory

S5/S7 converter, block libraries, overview of STL operations

## Appendix

**27 S5/S7 Converter**
Preparations for Conversion
Converting STEP 5 Programs
Postprocessing

**28 Block Libraries**
Organization Blocks
System Function Blocks
IEC Converting Blocks
S5–S7 Converting Blocks
TI–S7 Converting Blocks

**29 Operation Set**
Basic Functions
Digital Functions
Program Flow Control
Indirect Addressing

# The Contents of the Diskette at a Glance

This book contains many illustrations and diagrams pertaining to the use of the STL programming language. All program sections shown in the book, as well as a number of additional examples, are also on the diskette which accompanies the book.

After installing the examples, you can open, view and print out all blocks; you can also modify them or copy them to other projects or libraries. Alll examples contain symbols and comments.

You will find information on installing the diskette-resident library on the introductory page of the book section entitled "Appendix".

| | |
|---|---|
| **Basic Functions**<br>Examples for STL Representation | **Program Processing**<br>Examples for SFC Calls |
| FB 104 Chapter 4: Series and Parallel Connection<br>FB 105 Chapter 5: Set/Reset Functions<br>FB 106 Chapter 6: Move Functions<br>FB 107 Chapter 7: Timer Functions<br>FB 108 Chapter 8: Counter Functions | FB 120 Chapter 20: Main Program<br>FB 121 Chapter 21: Interrupt Handling<br>FB 122 Chapter 22: Start-up Characteristics<br>FB 123 Chapter 23: Error Handling |
| **Digital Functions**<br>Examples for STL Representation | **Variable Handling**<br>Examples for Data Types and Variable Processing |
| FB 109 Chapter 9: Comparison Functions<br>FB 110 Chapter 10: Arithmetic Functions<br>FB 111 Chapter 11: Math Functions<br>FB 112 Chapter 12: Conversion Functions<br>FB 113 Chapter 13: Shift Functions<br>FB 114 Chapter 14: Word Logic | FB 124 Chapter 24: Data Types<br>FB 125 Chapter 25: Indirect Addressing<br>FB 126 Chapter 26: Direct Variable Access<br>FB 101 Elementary Data Types<br>FB 102 Complex Data Types<br>FB 103 Parameter Types |
| **Program Flow Control**<br>Examples for STL Representation | **Conveyor Example**<br>Examples for Basic Functions and Local Instances |
| FB 115 Chapter 15: Status Bits<br>FB 116 Chapter 16: Jump Functions<br>FB 117 Chapter 17: Master Control Relay<br>FB 118 Chapter 18: Block Functions<br>FB 119 Chapter 19: Block Parameters<br>Source Program for Block Programming (Chapter 3) | FC 11 Conveyor Belt Control System<br>FC 12 Goods Counter<br>FB 20 Feeding Conveyor<br>FB 21 Conveyor Belt<br>FB 22 Goods Counter |
| **Message Frame Example**<br>Examples for Working with Data | **General Examples** |
| UDT 51 Data Structure for Header<br>UDT 52 Data Structure: Message<br>FB 51 Generate Message<br>FB 52 Store Message<br>FC 61 Query Time-of-Day<br>FC 62 Checksum Generation<br>FC 63 Convert Date | FC 41 Limit Monitoring<br>FC 42 Out-of-limit alarm<br>FC 43 Computing Compound Interest<br>FC 44 Doubleword Edge Evaluation<br>FC 45 Converting S5 Floating-Point to S7 REAL<br>FC 46 Converting S7 REAL to S5 Floating-Point<br>FC 47 Copy Data Area (ANY Pointer) |

# Contents

**Introduction** . . . . . . . . . . . . . . . . . . . . . . **16**

**1 SIMATIC S7-300/400 Programmable Controller** . . . . . 17

1.1 Structure of the Programmable Controller . . . . . . . . . . . . . . . . . . 17

1.1.1 Components . . . . . . . . . . . . . . . . . . 17
1.1.2 S7-300 Station . . . . . . . . . . . . . . . 17
1.1.3 S7-400 Station . . . . . . . . . . . . . . . 17
1.1.4 Distributed I/O . . . . . . . . . . . . . . . 19
1.1.5 Communications . . . . . . . . . . . . . . 19

1.2 Configuring Modules . . . . . . . . . . 21

1.2.1 Signal Path . . . . . . . . . . . . . . . . . . 21
1.2.2 Slot Address . . . . . . . . . . . . . . . . . 21
1.2.3 Logical Module Address . . . . . . . . 21
1.2.4 User Data Area . . . . . . . . . . . . . . . 22

1.3 CPU Memory Areas . . . . . . . . . . . 24

1.3.1 User Memory . . . . . . . . . . . . . . . . 24
1.3.2 Memory Card . . . . . . . . . . . . . . . . 24
1.3.3 System Memory . . . . . . . . . . . . . . 25
1.3.4 Address Areas in System Memory . 25

**2 STEP 7 Programming Software** . . . . . . . . . . . . . . . . . . . **27**

2.1 STEP 7 Basic Package . . . . . . . . . 27

2.1.1 Installation . . . . . . . . . . . . . . . . . . 27
2.1.2 Authorization . . . . . . . . . . . . . . . . 27
2.1.3 SIMATIC Manager . . . . . . . . . . . . 27
2.1.4 Projects and Libraries . . . . . . . . . . 29
2.1.5 On-line Help . . . . . . . . . . . . . . . . 30

2.2 Editing Projects . . . . . . . . . . . . . . 31

2.2.1 Creating Projects . . . . . . . . . . . . . 31
2.2.2 Rearranging, Managing and Archiving . . . . . . . . . . . . . . . . . . 32
2.2.3 Project Versions . . . . . . . . . . . . . . 32

2.3 Configuring Stations . . . . . . . . . . . 33

2.3.1 Arranging Modules . . . . . . . . . . . . 34
2.3.2 Addressing Modules . . . . . . . . . . . 35
2.3.3 Parameterizing Modules . . . . . . . . 35

2.4 Configuring Communication Connections . . . . . . . . . . . . . . . . . 35

2.4.1 Networking Modules with MPI . . . 36
2.4.2 Configuring DP Communication . . 36
2.4.3 Configuring GD Communication . . 37
2.4.4 Configuring SFB Communication . 38
2.4.5 Network Configuration . . . . . . . . . 38

2.5 Developing Programs . . . . . . . . . . 39

2.5.1 Symbol Table . . . . . . . . . . . . . . . . 39
2.5.2 Editor . . . . . . . . . . . . . . . . . . . . . . 41
2.5.3 Reference Data . . . . . . . . . . . . . . . 42
2.5.4 Rewiring . . . . . . . . . . . . . . . . . . . 43

2.6 Debugging Programs . . . . . . . . . . 44

2.6.1 Connecting a PLC . . . . . . . . . . . . 44
2.6.2 CPU Information . . . . . . . . . . . . . 44
2.6.3 Loading the User Program into the CPU . . . . . . . . . . . . . . . . . . . . . . 45
2.6.4 Block Handling . . . . . . . . . . . . . . 45
2.6.5 Determining the Cause of a STOP 46
2.6.6 Monitoring, Modifying and Forcing Variables . . . . . . . . . . . . . . . . . . . 47
2.6.7 Program Status . . . . . . . . . . . . . . . 48

**3 The STL Programming Language** . . . . . . . . . . . . . . . . . **50**

3.1 Program Processing . . . . . . . . . . . 50

3.1.1 Program Processing Methods . . . . 50
3.1.2 Priority Classes . . . . . . . . . . . . . . . 51
3.1.3 Specifications for Program Processing . . . . . . . . . . . . . . . . . 52

3.2 Blocks . . . . . . . . . . . . . . . . . . . . . 53

3.2.1 Block Types . . . . . . . . . . . . . . . . . 53
3.2.2 Block Structure . . . . . . . . . . . . . . . 54
3.2.3 Block Properties . . . . . . . . . . . . . . 54

3.3 Programming Blocks . . . . . . . . . . 56

3.3.1 Structure of an STL Statement . . . . 56
3.3.2 Addressing Variables . . . . . . . . . . . 57
3.3.3 Programming Blocks with the Incremental Editor . . . . . . . . . . . . 59
3.3.4 Programming Source-File-Oriented Blocks . . . . . . . . . . . . . . . . . . . . . 62
3.3.5 Example of a Function Block with Instance Data Block . . . . . . . . . . . 67

3.4 Variables and Constants . . . . . . . . 67
3.4.1 General Remarks Concerning Variables . . . . . . . . . . . . . . . . . . . . 67
3.4.2 General Remarks Regarding Data Types . . . . . . . . . . . . . . . . . . . . . . 68
3.4.3 Elementary Data Types . . . . . . . . . 68
3.4.4 Complex Data Types . . . . . . . . . . . 69
3.4.5 User Data Types . . . . . . . . . . . . . . 70
3.4.6 Parameter Types . . . . . . . . . . . . . . 70

**Basic Functions . . . . . . . . . . . . . . . . . . . 71**

**4 Binary Logic Operation . . . . . . . 72**
4.1 Processing a Binary Logic Operation . . . . . . . . . . . . . . . . . . . 72
4.2 Elementary Binary Logic Operations . . . . . . . . . . . . . . . . . . 74
4.2.1 AND Function . . . . . . . . . . . . . . . 75
4.2.2 OR Function . . . . . . . . . . . . . . . . . 75
4.2.3 Exclusive OR Function . . . . . . . . . 75
4.2.4 Allowing for the Sensor Type . . . . 75
4.3 Negating the Result of the Logic Operation . . . . . . . . . . . . . . . . . . . 77
4.4 Compound Binary Logic Operations . . . . . . . . . . . . . . . . . . 77
4.4.1 Processing Nesting Expressions . . . 78
4.4.2 Combining AND Functions According to OR . . . . . . . . . . . . . 79
4.4.3 Combining OR and Exclusive OR Functions According to AND . . . . 80
4.4.4 Combining AND Functions According to Exclusive OR . . . . . . 80
4.4.5 Combining OR Functions and Exclusive OR Functions . . . . . . . . 80
4.4.6 Negating Nesting Expressions . . . . 81

**5 Memory Functions . . . . . . . . . . . 82**
5.1 Assign . . . . . . . . . . . . . . . . . . . . . 82
5.2 Set and Reset . . . . . . . . . . . . . . . . 82
5.3 RS Flipflop Function . . . . . . . . . . 84
5.3.1 Memory Functions with Reset Priority . . . . . . . . . . . . . . . . . . . . . 84
5.3.2 Memory Function with Set Priority . . . . . . . . . . . . . . . . . . . . . 84
5.3.3 Memory Function in a Binary Logic Operation . . . . . . . . . . . . . . 84
5.4 Edge Evaluation . . . . . . . . . . . . . . 85
5.4.1 Positive Edge . . . . . . . . . . . . . . . . 86
5.4.2 Negative Edge . . . . . . . . . . . . . . . 86
5.4.3 Testing a Pulse Memory Bit . . . . . 86
5.4.4 Edge Evaluation in a Binary Logic Operation . . . . . . . . . . . . . . . . . . . 87
5.4.5 Binary Scaler . . . . . . . . . . . . . . . . 87
5.5 Example of a Conveyor Belt Control System . . . . . . . . . . . . . . . 88

**6 Transfer Functions (Move Functions) . . . . . . . . . . . . 92**
6.1 General Remarks on Loading and Transferring Data . . . . . . . . . . . . . 92
6.2 Load Functions . . . . . . . . . . . . . . . 94
6.2.1 General Representation of a Load Function . . . . . . . . . . . . . . . . . . . . 94
6.2.2 Loading the Contents of Memory Locations . . . . . . . . . . . . . . . . . . . 94
6.2.3 Loading Constants . . . . . . . . . . . . 95
6.3 Transfer Functions . . . . . . . . . . . . 96
6.3.1 General Representation of a Transfer Function . . . . . . . . . . . . . 96
6.3.2 Transferring to Various Memory Areas . . . . . . . . . . . . . . . . . . . . . 96
6.4 Accumulator Functions . . . . . . . . . 97
6.4.1 Direct Transfers Between Accumulators . . . . . . . . . . . . . . . . 97
6.4.2 Exchange bytes in accumulator 1 . . 98
6.5 System Functions for Data Transfer . . . . . . . . . . . . . . . . . . . . 98
6.5.1 Copying a data area . . . . . . . . . . . 98
6.5.2 Filling a data area . . . . . . . . . . . . . 99

**7 Timer Function . . . . . . . . . . . . . . 100**
7.1 Programming a Timer . . . . . . . . . . 100
7.1.1 Starting a Timer . . . . . . . . . . . . . . 100
7.1.2 Specifying the Time . . . . . . . . . . . 101
7.1.3 Resetting and Enabling a Timer . . . 101
7.1.4 Checking a Timer . . . . . . . . . . . . . 102
7.1.5 Sequence of Timer Instructions . . . 102
7.2 Pulse Timers . . . . . . . . . . . . . . . . . 103
7.3 Extended Pulse Timers . . . . . . . . . 104
7.4 On-Delay Timers . . . . . . . . . . . . . 105
7.5 Retentive On-Delay Timers . . . . . . 107
7.6 Off-Delay Timers . . . . . . . . . . . . . 108

**8 Counter Functions . . . . . . . . . . . 110**
8.1 Setting and Resetting Counters . . . 110
8.2 Counting . . . . . . . . . . . . . . . . . . . 111
8.3 Checking a Counter . . . . . . . . . . . 111

8.4 Enabling a Counter . . . . . . . . . . . . 111
8.5 Sequence of Counter Instructions . . 112
8.6 Parts Counter Example . . . . . . . . . 113

**Digital Functions . . . . . . . . . . . . . . . . . .116**

**9 Comparison Functions . . . . . . . . 117**
9.1 General Representation of a Comparison Function . . . . . . . . . . 117
9.2 Description of the Comparison Functions . . . . . . . . . . . . . . . . . . . 117
9.3 Comparison Function in a Logic Operation . . . . . . . . . . . . . . . . . . . 119

**10 Arithmetic Functions . . . . . . . . . 121**
10.1 General Representation of an Arithmetic Function . . . . . . . . . . . 121
10.2 Calculating with Data Type INT . . 122
10.3 Calculating with Data Type DINT . . . . . . . . . . . . . . . . . . . . . . 123
10.4 Calculating with Data Type REAL . . . . . . . . . . . . . . . . . . . . . 124
10.5 Successive Arithmetic Functions . . 125
10.6 Adding Constants to Accumulator 1 . . . . . . . . . . . . . . . 126
10.7 Decrementing and Incrementing . . 126

**11 Math Functions . . . . . . . . . . . . . . 127**
11.1 Processing a Math Function . . . . . 127
11.2 Trigonometric Functions . . . . . . . . 127
11.3 Arc Functions . . . . . . . . . . . . . . . . 128
11.4 Other Math Functions . . . . . . . . . . 128

**12 Conversion Functions . . . . . . . . . 130**
12.1 Processing a Conversion Function 130
12.2 Converting INT and DINT Numbers . . . . . . . . . . . . . . . . . . . 130
12.3 Converting BCD Numbers . . . . . . 131
12.4 Converting REAL Numbers . . . . . 132
12.5 Other Conversion Functions . . . . . 133

**13 Shift Functions . . . . . . . . . . . . . . 135**
13.1 Processing a Shift Function . . . . . . 135
13.2 Shifting . . . . . . . . . . . . . . . . . . . . 136
13.3 Rotating . . . . . . . . . . . . . . . . . . . . 138

**14 Word Logic . . . . . . . . . . . . . . . . . 139**
14.1 Processing a Word Logic Instruction . . . . . . . . . . . . . . . . . . 139
14.2 Description of the Word Logic Instructions . . . . . . . . . . . . . . . . . . 141

**Program Flow Control . . . . . . . . . . . . .143**

**15 Status Bits . . . . . . . . . . . . . . . . . . 144**
15.1 Description of the Status Bits . . . . 144
15.2 Setting the Status Bits and the Binary Flags . . . . . . . . . . . . . . . . . 146
15.3 Evaluating the Status Bit . . . . . . . . 148
15.4 Using the Binary Result . . . . . . . . 150

**16 Jump Functions . . . . . . . . . . . . . 152**
16.1 Programming a Jump Function . . . 152
16.2 Unconditional Jump . . . . . . . . . . . 153
16.3 Jump Functions with RLO and BR . . . . . . . . . . . . . . . . . . . . . . . . 153
16.4 Jump Functions with CC0 and CC1 . . . . . . . . . . . . . . . . . . . . . . . 154
16.5 Jump Functions with OV and OS . . . . . . . . . . . . . . . . . . . . . . . . 156
16.6 Jump Distributor . . . . . . . . . . . . . . 156
16.7 Loop Jump . . . . . . . . . . . . . . . . . . 156

**17 Master Control Relay . . . . . . . . . 158**
17.1 MCR Dependency . . . . . . . . . . . . 158
17.2 MCR Area . . . . . . . . . . . . . . . . . . 159
17.3 MCR Zone . . . . . . . . . . . . . . . . . . 159
17.4 Setting and Resetting I/O Bits . . . . 160

**18 Block Functions . . . . . . . . . . . . . 162**
18.1 Block Functions for Code Blocks . 162
18.1.1 Block Calls: General . . . . . . . . . . . 162
18.1.2 CALL Call Statement . . . . . . . . . . 163
18.1.3 UC and CC Call Statements . . . . . 164
18.1.4 Block End Functions . . . . . . . . . . . 165
18.1.5 Temporary Local Data . . . . . . . . . 165
18.1.6 Static Local Data . . . . . . . . . . . . . 167
18.2 Block Functions for Data Blocks . . 170
18.2.1 Two Data Block Registers . . . . . . . 170
18.2.2 Accessing Data Operands . . . . . . . 171
18.2.3 Open Data Block . . . . . . . . . . . . . 172
18.2.4 Exchanging the Data Block Registers . . . . . . . . . . . . . . . . . . . 173
18.2.5 Data Block Length and Number . . 173

18.2.6 Special Points in Data Addressing . 174
18.3 System Functions for Data Blocks . 176
18.3.1 Creating a Data Block . . . . . . . . . . 176
18.3.2 Deleting a Data Block . . . . . . . . . . 176
18.3.3 Testing a Data Block . . . . . . . . . . . 176
18.4 Null Operations . . . . . . . . . . . . . . 177
18.4.1 NOP Statements . . . . . . . . . . . . . . 177
18.4.2 Program Display Statements . . . . . 177

**19 Block Parameters . . . . . . . . . . . . 178**
19.1 Block Parameters in General . . . . . 178
19.1.1 Defining the Block Parameters . . . 178
19.1.2 Processing the Block Parameters . . 178
19.1.3 Definition of the Block Parameters 180
19.1.4 Declaration of the Function Value . 180
19.1.5 Initializing Block Parameters . . . . . 181
19.2 Formal Parameters . . . . . . . . . . . . 181
19.3 Actual Parameters . . . . . . . . . . . . . 184
19.4 "Passing On" Block Parameters . . . 186
19.5 Examples . . . . . . . . . . . . . . . . . . . 187
19.5.1 Conveyor Belt Example . . . . . . . . 187
19.5.2 Parts Counter Example . . . . . . . . . 188
19.5.3 Feed Example . . . . . . . . . . . . . . . . 188

**Program Processing . . . . . . . . . . . . . . . .192**

**20 Main Program . . . . . . . . . . . . . . 193**
20.1 General Remarks . . . . . . . . . . . . . 193
20.1.1 Program Structure . . . . . . . . . . . . . 193
20.1.2 Program Organization . . . . . . . . . . 194
20.2 Scan Cycle Control . . . . . . . . . . . . 195
20.2.1 Process Image Updating . . . . . . . . 195
20.2.2 Scan Cycle Monitoring Time . . . . . 196
20.2.3 Minimum Scan Cycle Time, Background Scanning . . . . . . . . . . 196
20.2.4 Response Time . . . . . . . . . . . . . . . 197
20.2.5 Scan Cycle Statistics . . . . . . . . . . . 198
20.3 Program Functions . . . . . . . . . . . . 198
20.3.1 Real-Time Clock . . . . . . . . . . . . . 198
20.3.2 Read System Clock . . . . . . . . . . . . 199
20.3.3 Run-Time Meter . . . . . . . . . . . . . . 199
20.3.4 Compressing CPU Memory . . . . . . 200
20.3.5 Waiting and Stopping . . . . . . . . . . 200
20.3.6 Multiprocessing Mode . . . . . . . . . 201
20.4 Data Interchange Using System Functions . . . . . . . . . . . . . . . . . . . 201
20.4.1 System Functions for Distributed I/O . . . . . . . . . . . . . . . . . . . . . . . 201
20.4.2 System Functions for GD Communication . . . . . . . . . . . . . . 203
20.4.3 System Functions for Data Interchange Within a Station . . . . . 203
20.4.4 System Functions for Data Interchange Between Two Stations . . . . . . . . . . . . . . . . . . . . 204
20.4.5 SFB Communication . . . . . . . . . . . 206
20.5 Start Information . . . . . . . . . . . . . 211
20.5.1 Start Information for OB 1 . . . . . . 211
20.5.2 Reading Out Start Information . . . . 212

**21 Interrupt Handling . . . . . . . . . . . 213**
21.1 General Remarks . . . . . . . . . . . . . 213
21.2 Hardware Interrupts . . . . . . . . . . . 214
21.2.1 Generating a Hardware Interrupt . . 215
21.2.2 Servicing Hardware Interrupts . . . . 215
21.2.3 Configuring Hardware Interrupts with STEP 7 . . . . . . . . . . . . . . . . . 215
21.3 Watchdog Interrupts . . . . . . . . . . . 216
21.3.1 Handling Watchdog Interrupts . . . . 216
21.3.2 Configuring Watchdog Interrupts with STEP 7 . . . . . . . . . . . . . . . . . 217
21.4 Time-of-Day Interrupts . . . . . . . . . 217
21.4.1 Handling Time-of-Day Interrupts . . 218
21.4.2 Configuring Time-of-Day Interrupts with STEP 7 . . . . . . . . . 218
21.4.3 System Functions for Time-of-Day Interrupts . . . . . . . . . . . . . . . . . . . 219
21.5 Time-Delay Interrupts . . . . . . . . . . 220
21.5.1 Handling Time-Delay Interrupts . . 220
21.5.2 Configuring Time-Delay Interrupts with STEP 7 . . . . . . . . . . . . . . . . . 221
21.5.3 System Functions for Time-Delay Interrupts . . . . . . . . . . . . . . . . . . . 221
21.6 Multiprocessor Interrupt . . . . . . . . 222
21.7 Handling Interrupts . . . . . . . . . . . . 222

**22 Start-Up Characteristics . . . . . . . 225**
22.1 General Remarks . . . . . . . . . . . . . 225
22.1.1 Operating Modes . . . . . . . . . . . . . 225
22.1.2 HOLD Mode . . . . . . . . . . . . . . . . 226
22.1.3 Disabling the Output Modules . . . . 226
22.1.4 Start-Up Organization Blocks . . . . 226
22.2 Power-Up . . . . . . . . . . . . . . . . . . . 227
22.2.1 STOP Mode . . . . . . . . . . . . . . . . . 227
22.2.2 Memory Reset . . . . . . . . . . . . . . . 227
22.2.3 Retentivity . . . . . . . . . . . . . . . . . . 228
22.3 Types of Start-up . . . . . . . . . . . . . 228

22.3.1 STARTUP Mode . . . . . . . . . . . . . . 228
22.3.2 Complete Restart . . . . . . . . . . . . . 228
22.3.3 Warm Restart . . . . . . . . . . . . . . . . 229
22.4 Ascertaining a Module Address . . . 230
22.5 Parameterizing Modules . . . . . . . . 231

**23 Error Handling . . . . . . . . . . . . . . 234**
23.1 Synchronous Errors . . . . . . . . . . . . 234
23.1.1 Programming Errors . . . . . . . . . . . 235
23.1.2 Access Errors . . . . . . . . . . . . . . . . 235
23.2 Synchronous Error Handling . . . . . 235
23.2.1 Error Filters . . . . . . . . . . . . . . . . . 235
23.2.2 Masking Synchronous Errors . . . . . 236
23.2.3 Unmasking Synchronous Errors . . . 237
23.2.4 Reading the Error Register . . . . . . 238
23.2.5 Entering a Substitute Value . . . . . . 238
23.3 Asynchronous Errors . . . . . . . . . . 238
23.4 System Diagnostics . . . . . . . . . . . . 239
23.4.1 Diagnostic Events and Diagnostic Buffer . . . . . . . . . . . . . . . . . . . . . 239
23.4.2 Writing User Entries in the Diagnostic Buffer . . . . . . . . . . . . . 240
23.4.3 Evaluating Diagnostic Interrupts . . 240
23.4.4 Reading the System Status List . . . 241

**Variable Handling . . . . . . . . . . . . . . . . .242**

**24 Data Types . . . . . . . . . . . . . . . . . 243**
24.1 Elementary Data Types . . . . . . . . . 243
24.1.1 Declaration of Elementary Data Types . . . . . . . . . . . . . . . . . . . . . 243
24.1.2 BOOL, BYTE, WORD, DWORD, CHAR . . . . . . . . . . . . . . . . . . . . . 243
24.1.3 Number Representations . . . . . . . . 244
24.1.4 Time Representations . . . . . . . . . . 246
24.2 Complex Data Types . . . . . . . . . . . 247
24.2.1 DATE_AND_TIME . . . . . . . . . . . 248
24.2.2 STRING . . . . . . . . . . . . . . . . . . . . 248
24.2.3 ARRAY . . . . . . . . . . . . . . . . . . . . 249
24.2.4 STRUCT . . . . . . . . . . . . . . . . . . . 251
24.3 User-Defined Data Types . . . . . . . 253

**25 Indirect Addressing . . . . . . . . . . . 254**
25.1 Pointers . . . . . . . . . . . . . . . . . . . . 254
25.1.1 Area Pointer . . . . . . . . . . . . . . . . . 254
25.1.2 DB Pointer . . . . . . . . . . . . . . . . . . 255
25.1.3 ANY Pointer . . . . . . . . . . . . . . . . 256
25.2 Types of Indirect Addressing . . . . . 257
25.2.1 General . . . . . . . . . . . . . . . . . . . . 257
25.2.2 Indirect Addressable Operands . . . 257
25.2.3 Memory-Indirect Addressing . . . . . 258
25.2.4 Register-Indirect Area-Internal Addressing . . . . . . . . . . . . . . . . . . 259
25.2.5 Register-Indirect Area-Crossing Addressing . . . . . . . . . . . . . . . . . . 260
25.2.6 Summary . . . . . . . . . . . . . . . . . . . 260
25.3 Working the Address Registers . . . 261
25.3.1 Overview . . . . . . . . . . . . . . . . . . . 261
25.3.2 Loading into an Address Register . 261
25.3.3 Transferring from an Address Register . . . . . . . . . . . . . . . . . . . . 262
25.3.4 Swap Address Registers . . . . . . . . 262
25.3.5 Adding to the Address Register . . . 262
25.4 Special Features of Indirect Addressing . . . . . . . . . . . . . . . . . . 264
25.4.1 Using Address Register AR1 . . . . . 264
25.4.2 Using Address Register AR2 . . . . . 264
25.4.3 Restrictions with Static Local Data . . . . . . . . . . . . . . . . . . . . . . 265

**26 Direct Variable Access . . . . . . . . . 266**
26.1 Loading the Variable Address . . . . 266
26.2 Data Storage of Variables . . . . . . . 267
26.2.1 Storage in Global Data Blocks . . . . 267
26.2.2 Storage in Instance Data Blocks . . 269
26.2.3 Storage in the Temporary Local Data . . . . . . . . . . . . . . . . . . . . . . 270
26.3 Data Storage when Transferring Parameters . . . . . . . . . . . . . . . . . . 270
26.3.1 Parameter Storage in Functions . . . 271
26.3.2 Storing Parameters in Function Blocks . . . . . . . . . . . . . . . . . . . . . 272
26.3.3 “Variable” ANY Pointer . . . . . . . . 274
26.4 Brief Description of the Message Frame Example . . . . . . . . . . . . . . 275

**Appendix . . . . . . . . . . . . . . . . . . . . . . .281**

**27 S5/S7 Converter . . . . . . . . . . . . . 282**
27.1 General . . . . . . . . . . . . . . . . . . . . 282
27.2 Preparation . . . . . . . . . . . . . . . . . . 283
27.2.1 Checking Executability on the Destination System . . . . . . . . . . . . 283
27.2.2 Checking Program Execution Characteristics . . . . . . . . . . . . . . . 283
27.2.3 Checking the Modules . . . . . . . . . 285

27.2.4 Checking the Operands . . . . . . . . . 285
27.2.5 Preparing the STEP 5 Program . . . 286
27.3 Converting . . . . . . . . . . . . . . . . . . 286
27.3.1 Creating Macros . . . . . . . . . . . . . . 286
27.3.2 Preparing the Conversion . . . . . . . 287
27.3.3 Starting the Converter . . . . . . . . . . 287
27.3.4 Convertible Functions . . . . . . . . . . 288
27.4 Post-Editing . . . . . . . . . . . . . . . . . 289
27.4.1 Creating the STEP 7 Project . . . . . 289
27.4.2 Unconvertible Functions . . . . . . . . 290
27.4.3 Address Changes . . . . . . . . . . . . . 290
27.4.4 Indirect Addressing . . . . . . . . . . . . 290
27.4.5 Access to "Excessively Long" Data Blocks . . . . . . . . . . . . . . . . . 291
27.4.6 Working with Absolute Addresses . . . . . . . . . . . . . . . . . . . 291
27.4.7 Parameter Initialization . . . . . . . . . 293
27.4.8 Special Function Organization Blocks . . . . . . . . . . . . . . . . . . . . . 293
27.4.9 Error Handling . . . . . . . . . . . . . . . 293

**28 Block Libraries . . . . . . . . . . . . . . 296**

28.1 Organization Blocks . . . . . . . . . . . 296
28.2 System Function Blocks . . . . . . . . 297
28.3 IEC Converting Blocks . . . . . . . . . 299
28.4 S5–S7 Converting Blocks . . . . . . . 300
28.5 TI–S7 Converting Blocks . . . . . . . 301
28.6 PID Control Blocks . . . . . . . . . . . 301
28.7 Communication Blocks . . . . . . . . . 301

**29 Operation Set . . . . . . . . . . . . . . . 302**

29.1 Basic Functions . . . . . . . . . . . . . . 302
29.1.1 Binary Logic Operations . . . . . . . . 302
29.1.2 Memory Functions . . . . . . . . . . . . 303
29.1.3 Transfer Functions . . . . . . . . . . . . 303
29.1.4 Timer Functions . . . . . . . . . . . . . . 303
29.1.5 Counter Functions . . . . . . . . . . . . . 303
29.2 Digital Functions . . . . . . . . . . . . . 303
29.2.1 Comparison Functions . . . . . . . . . 303
29.2.2 Math Functions . . . . . . . . . . . . . . . 304
29.2.3 Arithmetic Functions . . . . . . . . . . 304
29.2.4 Conversion Functions . . . . . . . . . . 304
29.2.5 Shift F unctions . . . . . . . . . . . . . . 304
29.2.6 Word Logic Operations . . . . . . . . . 304
29.3 Program Flow Control . . . . . . . . . 305
29.3.1 Jump Functions . . . . . . . . . . . . . . 305
29.3.2 Master Control Relay . . . . . . . . . . 305
29.3.3 Block functions . . . . . . . . . . . . . . 305
29.4 Indirect Adressing . . . . . . . . . . . . . 305

**Index . . . . . . . . . . . . . . . . . . . . . . . . . . 306**

**Abbreviations . . . . . . . . . . . . . . . . . . . . 310**

The author and publisher are always grateful
to hear your responses to the contents
of the book.

Publicis MCD Verlag
Postfach 3240
D-91050 Erlangen
Federal Republic of Germany
Fax: ++49 9131/727838
E-mail: publishing-books@publicis-mcd.de

# Introduction

This portion of the manual provides an overview of the SIMATIC S7-300/400 programmable controller.

- **The S7-300/400 programmable controller** is of modular design. The modules with which it is configured can be central (in the vicinity of the CPU) or distributed without any special settings or parameter assignments having to be made. In SIMATIC S7 systems, distributed I/O is an integral part of the system. The CPU, with its various memory areas, forms the hardware basis for processing of the user programs. A load memory contains the complete user program; the parts of the program relevant to its execution at any given time are in a work memory whose short access times are the prerequisite for fast program processing.
- **STEP 7** is the programming software for S7-300/400; the automation tool is the SIMATIC Manager. The SIMATIC Manager is a Windows 95/NT application, and contains all functions needed to set up a project. When necessary, the SIMATIC Manager starts additional tools, for example to configure stations, initialize modules, and to write and test programs.
- You formulate your automation solution in the **STL programming language**. The STL program is structured, that is to say it consists of blocks with defined functions which are composed of networks or rungs. Different priority classes allow a graduated interruptibility of the user program currently executing. STEP 7 works with variables of different data types, from binary variables (data type BOOL) to digital variables (for instance of data type INT or REAL for computing tasks) to complex data types such as arrays or structures (combining of variables of different data types to form a single variable).

The first chapter contains an overview of the hardware in an S7-300/400 **programmable controller,** the second chapter an overview of the STEP 7 programming software. The basis for the description is the function scope for STEP 7 Version 4.02.

Chapter 3, "The STL Programming Language", serves as an introduction to the most important STL elements. A more detailed approach follows in the subsequent chapters of the manual. The descriptions of all STL functions are enhanced by brief examples. At the end of some chapters you will find extensive examples, the intention of which is to show the use of the STL language in a larger context.

**1 SIMATIC S7-300/400 Programmable Controller**
Structure of the programmable controller; configuring a station (introduction); addressing modules; CPU memory areas; process images, bit memory

**2 STEP 7 Programming Software**
SIMATIC Manager; processing a project; configuring a station; configuring communication links; writing programs (symbol table, editor); testing programs

**3 STL Programming Language**
Program processing with priority classes; program blocks; networks and statements; programming blocks; variables and constants; data types (overview)

# 1 SIMATIC S7-300/400 Programmable Controller

## 1.1 Structure of the Programmable Controller

### 1.1.1 Components

The SIMATIC S7-300/400 is a modular programmable controller comprising the following components:

- Racks
  Accommodate the modules and connect them to each other
- Power supply (PS);
  Provides the internal supply voltages
- Central processing unit (CPU)
  Stores and processes the user program
- Interface modules (IMs);
  Connect the racks to one another
- Signal modules (SMs);
  Adapt the signals from the system to the internal signal level or control actuators via digital and analog signals
- Function modules (FMs);
  Execute complex or time-critical processes independently of the CPU
- Communications processors (CPs)
  Interface the programmable controllers to each other or to other devices via serial links

A programmable controller (or station) may consist of several racks, which are linked to one another via bus cables. If there is not enough room in the controller rack for the I/O modules or if you want some or all I/O modules to be separate from the controller rack, expansion racks are available which are connected to the controller rack via interface modules (Figure 1.1).

The racks connect the modules with two busses: the I/O bus (or P bus) and the communication bus (or K bus). The I/O bus is designed for the high-speed interchange of input and output signals, the communication bus for the exchange of large amounts of data. The communication bus connects the CPU and the programming device interface (MPI) with function modules and communications processors.

### 1.1.2 S7-300 Station

In an S7-300 controller, as many as 8 I/O modules can be plugged into the controller rack. Should this single-tier configuration prove insufficient, you have two options for controllers equipped with a CPU 314 or a more advanced CPU:

- Either choose a two-tier configuration (with IM 365 up to 1 meter between racks)
- or choose a configuration of up to four tiers (with IM 360 and IM 361 up to 10 meters between racks)

You can operate a maximum of 8 modules in a rack. The number of modules may be limited by the maximum permissible current per rack, which is 1.2 A (0.8 A for CPU 312 IFM).

The modules are linked to one another via a backplane bus which combines the functions of the P and K busses.

A special situation regarding configuration is the use of the FM 356 application module from the M7-300 family of automation computers. An FM 356 is able to "split" a module's backplane bus and to take over control of the remaining modules in the split-off "local bus segment" itself. The limitations mentioned above regarding the number of modules and the power consumption also apply in this case.

### 1.1.3 S7-400 Station

In S7-400s, there are controller racks with 18 or 9 slots (UR1 or UR2); the power supply module and the CPU also reserve slots, sometimes even two or more per modules. The IM 460-1 and IM 461-1 interface modules make it possible to have one expansion rack per interface up to 1.5 meters

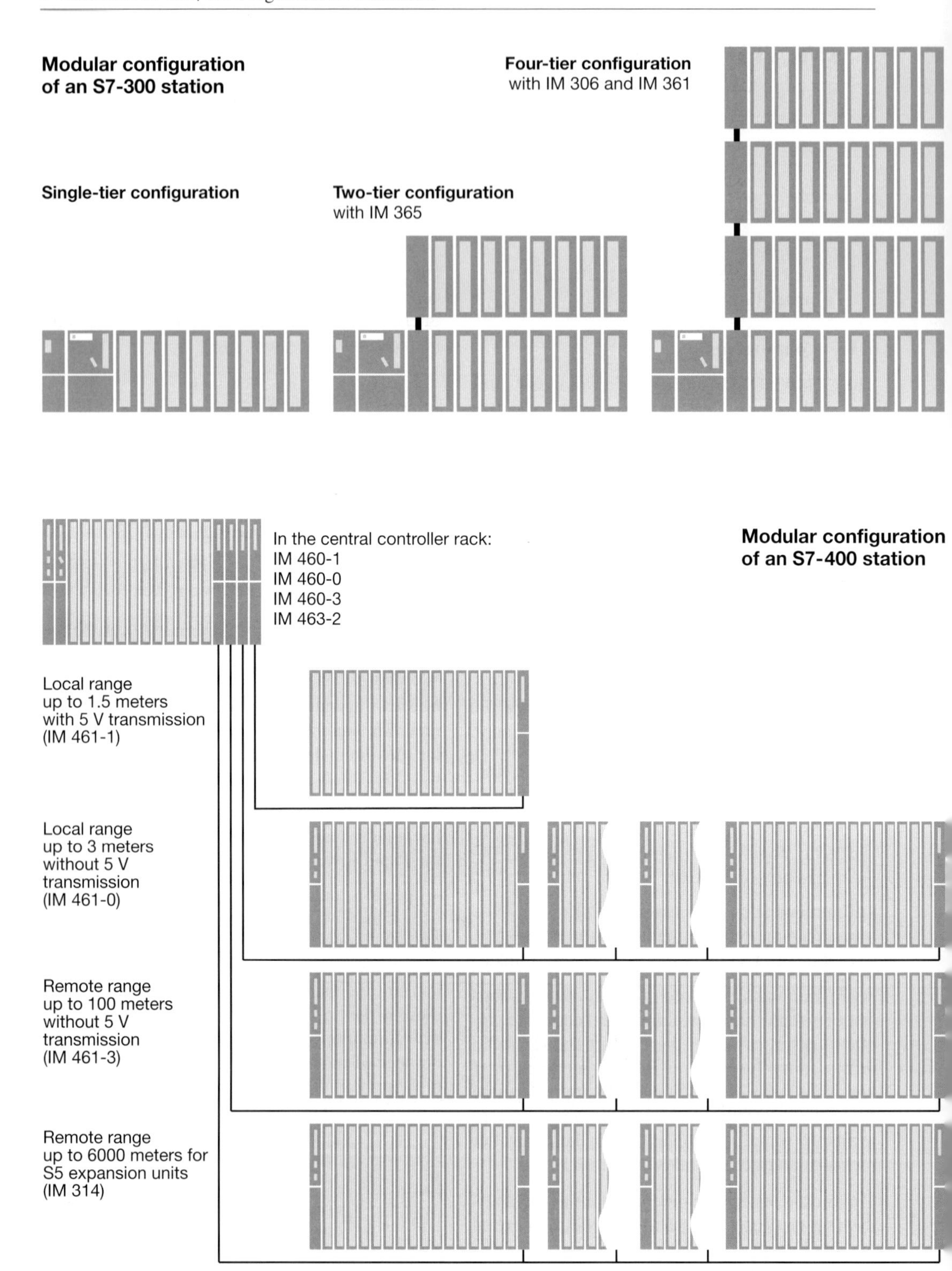


**Figure 1.1** Hardware Configuration for S7-300/400

from the controller rack, including the 5 V supply voltage. In addition, as many as four expansion racks can be operated up to 3 meters away using IM 360-0 and IM 361-0 interface modules. And finally, IM 360-3 and IM 361-3 interface modules can be used to operate as many as four expansion racks at a distance of up to 100 meters away.

A maximum of 21 expansion racks can be connected to a controller rack. To distinguish between racks, you set the number of the rack on the coding switch of the receiving IM.

The backplane bus consists of a parallel P bus and a serial K bus. Expansion racks ER1 and ER2, with 18 or 9 slots, are designed for "simple" signal modules which generate no hardware interrupts, do not have to be supplied with 24 V voltage via the P bus, require no back-up voltage, and have no K bus connection. The K bus is in racks UR1, UR2 and CR2 either when these racks are used as controller racks or expansion racks with the numbers 1 to 6.

A special feature is the segmented rack CR2. The rack can accommodate two CPUs with a shared power supply while keeping them functionally separate. The two CPUs can exchange data with one another via the K bus, but have completely separate P busses for their own signal modules.

In an S7-400, as many as four specially designed CPUs in a suitable rack can take part in multiprocessor operation. Each module in this station is assigned to only one CPU, both with its address and its interrupts.

The IM 463-2 interface module allows you to connect S5 expansion units (EG 183U, EG 185U, EG 186U as well as ER 701-2 and ER 701-3) to an S7-400, and also allows centralized expansion of the expansion units. An IM 314 in the S5 expansion unit handles the link. You can operate all analog and digital modules allowed in these expansion units. An S7-400 can accommodate as many as four IM 463-2 interface modules; as many as four S5 expansion units can be connected in a distributed configuration to each of an IM 463-2's interfaces.

### 1.1.4 Distributed I/O

The term distributed I/O is understood to mean modules that are connected to a DP master module via PROFIBUS-DP. PROFIBUS-DP is a manufacturer-independent standard to EN 50170 Volume 2, Part 3 for the connection of DP standard slaves.

You can set the baud rate for PROFIBUS-DP to between 9.6 kbps and 12 Mbps, depending on the permissible line length in a segment (subnetwork). Segments are connected using RS 485 repeaters which may be up to 1000 meters apart. The physical medium is either a shielded two-wire cable in a network extending over a maximum of 9.6 kilometers or a fiber-optic cable extending over no more than 23.8 kilometers.

A DP master (such as the CPU 414-2 DP or the CP 342-5 DP) controls as many as 31 DP slaves in a segment or up to 63 throughout the entire network. You can also connect programming devices, operator interface systems, ET 200's or SIMATIC S5 DP slaves to the PROFIBUS-DP network.

The S7-300 model can have a CPU 315-2DP or a CP 342-5 as DP master. The CPU 315-2DP can also be used as "intelligent" DP slave, and can – just as the ET 200M DP station with an IM 513 as interface module – control other S7-300 modules.

At the present time, CPUs 413-2DP, 414-2DP and 416-2DP are available as DP master for S7-400. As PROFIBUS interface module, the CP 443-5 is also conversant with the FMS protocol.

### 1.1.5 Communications

#### MPI Network

Every central processing unit has an "interface with multipoint capability" (multipoint interface, or MPI). The MPI gives a CPU network capability without the need for additional modules. An MPI network may comprise as many as 32 nodes; these may be

- programming devices (PGs, PCs with appropriate interface card),
- operator panels (OPs),
- appropriately designed function modules (FMs) or communications processors (CPs) and
- additional CPUs in other programmable controllers.

This allows you to use only a single programming device as operator interface to all connected CPUs, and also allows low-overhead interchanging of data between the CPUs' user programs via global data communication (which is integrated in the CPUs).

The transmission rate on the MPI bus is permanently set to 187.5 kbps. The cable in one segment (subnetwork) may be up to 50 meters long, but can be increased to a total of 1000 meters through the use of RS 485 repeaters. The ends of the MPI bus must be terminated and biased.

Every node in the network is identified by an MPI address. A node's MPI address is preset at the factory (PG = 0, OP = 1, CPU = 2). This allows you to put a programmable controller with only one CPU into operation with one PG and one OP without setting any addresses.

Within a segment, nodes connected to an MPI network must have different MPI addresses. The MPI address is specifiable. The entire network may comprise up to 126 nodes, a subnetwork up to 32 nodes.

## GD communications

The CPUs in an MPI network can exchange data amongst themselves. This capability, called global data communication, is restricted to small amounts of data only. Global data (GD) include

- Inputs and outputs (process images)
- Bit memory
- Data in data blocks
- Timers and counters (as Send data).
- The data are exchanged in the form of data packets (GD packets) between CPUs, which are combined into GD circles. A GD circle may be
- a two-way connection between two CPUs, each of which can both send and receive a GD packet
- a one-way connection in which a single CPU sends a GD packet to several other CPUs, which receive that packet.

A CPU may belong to several GD circles. The possible number of GD circles and the size of the GD packets is CPU-specific.

Please note that a receiving CPU does not acknowledge the receipt of global data. The sender thus receives no information of any kind to indicate whether data was received or which CPU received that data. You do, however, have the option of viewing the status of communications between two CPUs as well as the status of all of the GD circles to which a CPU belongs.

Global data communication requires a considerable portion of the processing time in the CPU operating system. For this reason, you can define a "scan rate" in the global data table which specifies after how many program cycles the data are to be sent or received. Because the data are not updated in every program cycle, you should not use this form of communication to transfer time-critical data.

Global data are sent and received asynchronously between sender and receiver at the scan cycle checkpoint, that is to say after cyclic program scanning and before a new program cycle begins.

The S7-400 CPUs also allow event-driven global data communication using system functions SFC 60 GDSND and SFC 61 GDRCV. These SFCs can be used, for example, to exchange data between CPUs in multiprocessor mode.

## SIMATIC NET

SIMATIC NET is the new product designation for networks and network components (formerly SINEC, Siemens Network and Communication).

You can connect the S7-300/400 programmable controllers via communications processors to the following bus systems:

- *Industrial Ethernet* is the network for the management level and the cell level defined by international standard IEEE 802.3, with emphasis on the industrial area (former product designation: SINEC H1). SIMATIC S7 provides communications processors CP 343-1 and CP 343-1 TCP (S7-300) as well as CP 443-1 and CP 443-1 TCP (S7-400).
- *PROFIBUS* stands for "Process Field Bus", and is a manufacturer-independent standard to EN 50170 for the networking of field devices (former product designation: SINEC L2). PROFIBUS provides two protocols: PROFIBUS-FMS for the transfer of large amounts of

data and PROFIBUS-DP for distributed I/O. Communications processors CP 342-5 DP and CP 342-5 FMS (S7-300) and CP 443-5 (S7-400) are available for interfacing to SIMATIC S7.

- The *AS-Interface* is a networking system for binary sensors and actuators at the lowest field level (former product designation: SINEC S1). The AS Interface is connected to an S7-300 via a CP 342-2 communications module. The DP/AS-Interface Link connects the AS-Interface to PROFIBUS-DP.
- A *point-to-point* link is used to connect devices with a serial interface to the programmable controller, for instance bar code readers, printers, or SIMATIC S5 programmable controllers. For a point-to-point link you need a CP 340 (S7-300) or a CP 441-1 or CP 441-2 (S7-400) communications processor with one or two interfaces.

## 1.2 Configuring Modules

### 1.2.1 Signal Path

When you wire your machine or plant, you determine which signals are connected where on the programmable controller (Figure 1.2). An input signal, for example the signal from momentary-contact switch +HP01-S10, the one for "Switch motor on", is run to an input module, where it is connected to a specific terminal. This terminal has an 'address' called the I/O address (for instance byte 5, bit 2). The CPU copies the signal to the process-image input table, where it is then addressed as input (I 5.2, for example). The term "I 5.2" is the *absolute address*. You can now give this input a name by assigning a symbol to the absolute address in the symbol table which corresponds to this input signal (such as "Switch motor on"). The term "Switch motor on" is the *symbolic address*.

### 1.2.2 Slot Address

Every slot has a fixed address in the programmable controller (an S7 station). This slot address consists of the number of the mounting rack and the number of the slot. A module is uniquely described using the slot address ("geographical address").

If the module contains interface cards, each of these cards is also assigned a submodule address. In this way, each binary and analog signal and each serial connection in the system has its own unique address.

Correspondingly, distributed I/O modules also have a "geographical address". In this case, the number of the DP master system and the station number replace the rack number.

You use STEP 7's *hardware configuration* tool to plan the hardware configuration of an S7 station as per the physical location of the modules. This tool also makes it possible to set the module start addresses and parameterize the modules (see Section 2.3, "Configuring Stations").

### 1.2.3 Logical Module Address

In addition to the slot address, which defines the slot, each module has a start address which defines the location in the logical address space (in the I/O address space). The I/O address space begins at address 0 and ends at a CPU-specific address.

The module start address is determinative for the referencing of the input and output signals by the program. In the case of digital modules, the individual signals (bits) are bundled into groups of eight called "bytes". There are modules with one, two or four bytes. These bytes have the relative addresses 0, 1, 2 and 3; addressing of the bytes begins at the module start address. Example: In the case of a digital module with four bytes and the start address 8, the individual bytes are referenced by addresses 8, 9, 10 and 11. In the case of analog modules, the individual analog signals (voltages, currents) are called "channels", each of which reserves two bytes. Analog modules are available with two, four, eight and 16 channels, corresponding to four, eight, 16 or 32 bytes address area.

On power-up (if there is no setpoint configuration), the CPU defaults to a slot-oriented module start address which depends on the module type, the slot, and the rack. This module start address corresponds to (relative) byte 0. You can view this address in the configuration table.

In S7-315 DP and S7-400 systems, you can change this address. You have the option of assigning the start addresses of the modules in your hardware configuration yourself. In a raster of 16 (S7-300) or four (S7-400), you can assign each slot an arbitrary address. You also have the option of assigning different start addresses for inputs and outputs on a hybrid digital or analog module. FMs and CPs normally reserve the same start address for inputs and outputs.

The advantages of self-assigned addresses are most apparent when it comes to distributed I/O. The counting of module addresses according to slot serves little purpose, since both digital and analog modules can be plugged into a distributed station. Default assignments thus do not guarantee that all digital modules lead to the process image (which always begins at address 0 and ends at a maximum address specific to the relevant CPU), nor is it guaranteed that the analog modules do not lead to the process image (which one usually wishes to avoid).

Like centralized modules, distributed I/O modules (stations) reserve a specific number of bytes in the I/O address space. Appropriately equipped DP slaves can be parameterized so that a specific number of bytes constitute consistent (logically associated) data for data transfers. These slaves display an I/O address of only one byte, via which they are addressed with system functions SFC 14 and SFC 15. Please note that the bytes which are not displayed and which are not accessible using "standard" statements nonetheless still reserve address space! Every DP master and every DP slave reserves an additional byte for the "diagnostic address". You use this byte as an address for reading diagnostic data with SFC 13. Each DP station on the PROFIBUS also has a station address (also called a node address) with which it can be uniquely addressed on the bus.

Modules which are nodes in an MPI network (CPUs, FMs and CPs) also have an MPI address. This address is decisive for global communication and for the link to programming devices and operator interface systems. Each module on the bus also has a node address, which you set when you parameterize the module.

### 1.2.4 User Data Area

In SIMATIC S7, each module has two address areas: a *user data area,* which can be directly addressed with Load and Transfer statements, and a *system data area* for transferring data records.

When modules are addressed, it makes no difference whether they are in racks with centralized configuration or used as distributed I/O. All modules occupy the same (logical) address space.

A module's useful data properties depend on the module type. In the case of signal modules, they are either digital or analog input/output signals, in the case of function modules and communications processors, they might, for example, be control or status information. The amount of useful data is module-specific. There are modules which reserve one, two, four or more bytes in this area. Addressing always begins at relative byte 0. The address of byte 0 is the module start address; it is stipulated in the configuration table.

The useful data represent the I/O address area, subdivided, depending on the direction of transfer, into peripheral inputs (PIs) and peripheral outputs (PQs). If the useful data are in the process-image area, the CPU automatically handles the transfers to update the process images.

#### Peripheral inputs

You use the peripheral input (PI) address area when you read from the useful data area on an input module. Part of the PI address area leads to the process image. This part always begins at I/O address 0; the length of the area is CPU-specific.

With a Direct I/O Read operation, you can address the modules whose interfaces do not lead to the process-image input area (for instance analog input modules). The signal states of modules which lead to the process-image input area also cannot be read with a Direct Read operation. The momentary signal states of the input bits are then scanned Note that this signal state may differ from the relevant inputs in the process image! (the process-image input area is updated at the beginning of the program scan.)

Peripheral inputs may reserve the same absolute addresses as peripheral outputs.

#### Peripheral outputs

You use the peripheral output (PQ) address area when you write values to the useful data area on an output module. Part of the PQ address area leads to the process image. This part always be-

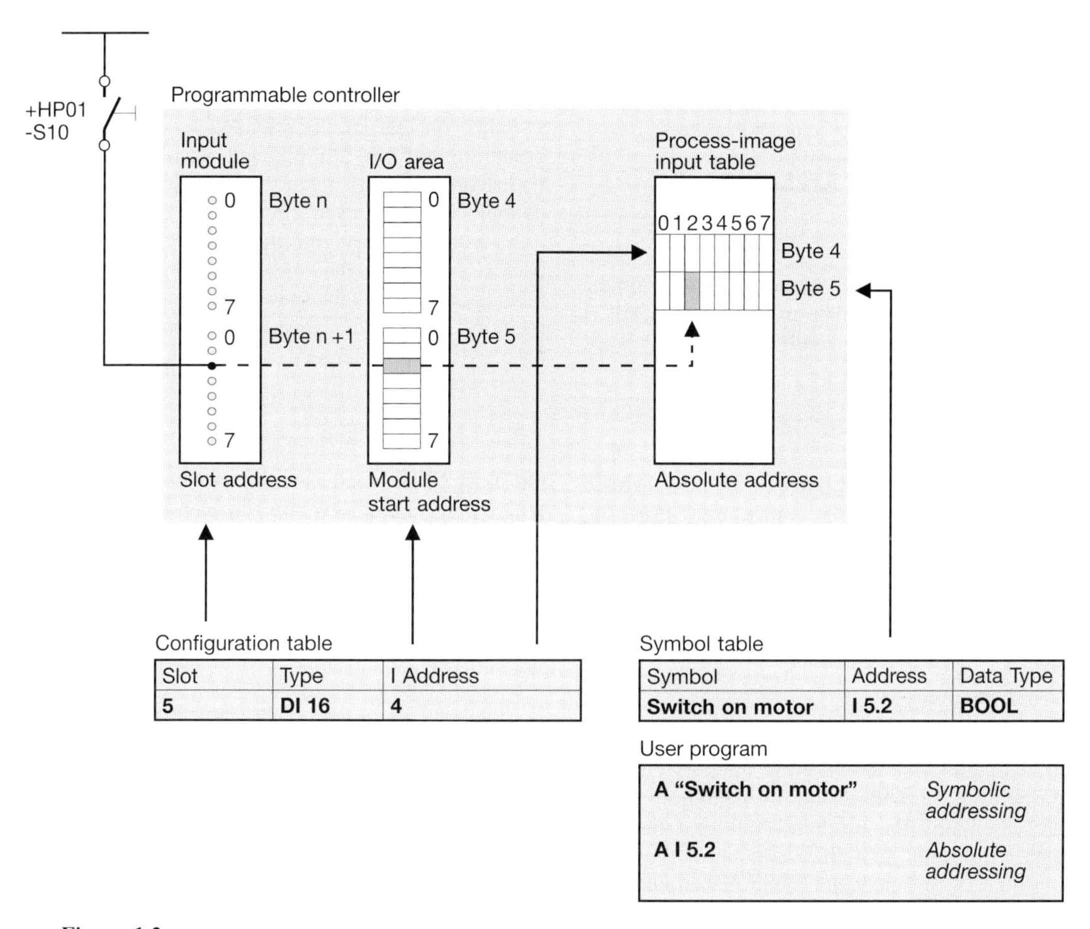


| Slot | Type | I Address |
|---|---|---|
| **5** | **DI 16** | **4** |

| Symbol | Address | Data Type |
|---|---|---|
| **Switch on motor** | **I 5.2** | **BOOL** |

**Figure 1.2**
Correlation Between Module Address, Absolute Address and Symbolic Address
(Path of a Signal from Sensor to Scanning in the Program)

gins at I/O address 0; the length of the area is CPU-specific.

With a Direct I/O Write operation, you can address modules whose interfaces do not lead to the process-image output area (such as analog output modules). The signal states of modules controlled by the process-image output area can also be directly affected. The signal states of the output bits then change immediately. Note that a Direct I/O Write operation also changes the signal states of the relevant modules in the process-image output area! In this way, there is no difference between process-image output area and the signal states on the output modules.

Peripheral outputs can reserve the same absolute addresses as peripheral inputs.

### Substitute value strategy

In STOP mode or during startup, the CPU disables the output modules. In this case, zero or, if the module is designed accordingly, a specified substitute value, is output.

When no further values can be read from an input module or when an attempt is made to read from a non-existent module, the CPU invokes OB 122, the organization block for synchronization errors. In this OB, you can return a substitute value in accumulator 1 using system function SFC 44 REPLVAL instead of returning the value (not) read.

## 1.3 CPU Memory Areas

### 1.3.1 User Memory

Figure 1.3 shows the CPU memory areas important to your program. The user program itself is in two areas, namely load memory and work memory.

#### Load memory

Load memory can be a plug-in memory card or integrated memory. The entire user program, including module configuration and module parameters, are in load memory. In future, load memory should also be capable of holding all symbols.

#### Work memory

Work memory is designed in the form of integrated high-speed RAM. Work memory contains the relevant portions of the user program; these are essentially the program code and the user data. "Relevant" is a characteristic of the existing objects, and does not mean that a particular code block will necessarily be called and processed.

The operating system copies the relevant parts of a user program from a plug-in memory card to work memory when the supply voltage is switched on. All blocks reloaded from the programming device are updated accordingly in both load and work memory. Changes made to data blocks by the user program, however, affect only work memory; on the next non-battery-backed power-up, the (old) data block in load memory overwrites the (new) data block in work memory. Data blocks generated at program runtime are in work memory only.

### 1.3.2 Memory Card

On most CPUs, load memory can be expanded with a plug-in memory card. There are two types of memory card: RAM cards and Flash EPROM cards.

If you want to expand load memory only, use a RAM card. A RAM card allows you to modify individual blocks on-line.

If you want to protect your user program, including configuration data and module parameters, against power failure, use a Flash EPROM card. In this case, load the entire program off-line onto the Flash EPROM card with the card plugged into the programming device. On S7-400 and 315 CPUs (with 5 V Flash EPROM card), you can also load the program on-line with the memory card plugged into the CPU.

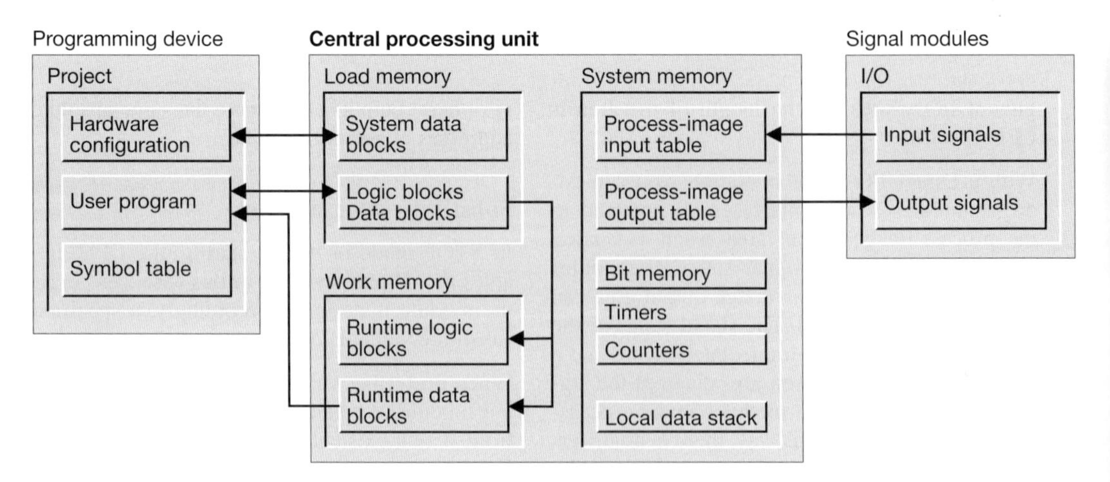


**Figure 1.3** CPU Memory Areas

### 1.3.3 System Memory

System memory contains the variables which you address in your program. The variables (addresses) are combined into areas (address areas) containing a CPU-specific number of addresses. Addresses may be, for example, inputs used to scan the signal states of momentary-contact switches and limit switches, and outputs, which you can use to control contactors and lamps. The system memory on a CPU contains the following address areas:

- Inputs (I)
  Inputs are an image ("process image") of the digital input modules.
- Outputs (Q)
  Outputs are an image ("process image") of the digital output modules.
- Bit memory (M)
  Stores of information addressable throughout the whole program.
- Timers (T)
  Timers are locations used to implement waiting and monitoring times.
- Counters (C)
  Counters are software-level locations which can be used for up and down counting.
- Temporary local data (L)
  Locations which serve as dynamic intermediate buffers during block processing. The temporary local data are located in the L stack, which the CPU reserves dynamically during program scanning.

The letters enclosed in parentheses represent the abbreviations to be used for the different addresses when writing programs. You may also assign a symbol to each variable and then use the symbol in place of the address identifier.

### 1.3.4 Address Areas in System Memory

The address areas in the CPU's system memory are the process image with the inputs and outputs, the bit memory, the temporary local data, and the timers and counters. The timers and counters are described in detail in the corresponding chapters in the next part of the book and the temporary local data are described in Section 18.1.5, "Temporary Local Data".

#### Process Image

The process image contains the image of the digital input and digital output modules, and is thus subdivided into process-image input area and process-image output area. The process-image input area is addressed via the address area for inputs (I), the process-image output area via the address area for outputs (Q). As a rule, the machine or process is controlled via the inputs and outputs.

When you address the inputs and outputs in your program, an S7-400 system must be configured with the appropriate input/output modules. In S7-300 systems, the areas of the process image to which no modules are assigned may be used as additional memory in a manner similar to the use of bit memory.

*Inputs*

An input is an image of the corresponding bit on a digital input module. Scanning an input is the same as scanning the bit on the module itself. Prior to program scanning in every program cycle, the CPU's operating system copies the signal state from the module to the process-image input area. The use of a process-image input area has many advantages:

- Inputs can be scanned and linked bit by bit (I/O bits cannot be directly addressed).
- Scanning an input is much faster than addressing an input module (for example, there is no transient recovery time on the I/O bus, and the system memory response times are shorter than the module's response times). The program is therefore scanned that much more quickly.
- The signal state of an input is the same throughout the entire program cycle (there is data consistency throughout a program cycle). When a bit on an input module changes, the change in the signal state is transferred to the input at the start of the next program cycle.
- Inputs can also be set and reset because they are located in random access memory. Digital input modules can only be read. Inputs can be set during debugging or start-up to simulate sensor states, thus simplifying program testing.

In opposition to these advantages stands an increased program response time (also refer to Section 20.2.4, "Response Time").

*Outputs*

An output is an image of the corresponding bit on a digital output module. Setting an output is the same as setting the bit on the output module itself. The CPU's operating system copies the signal state from the process-image output area to the module. The use of a process-image output area has many advantages:

- Outputs can be set and reset bit by bit (direct addressing of I/O bits is not possible).
- Setting an output is much faster than addressing an output module (for example, there is no transient recovery time on the I/O but, and the system memory response times are shorter than the module response times). The program is therefore scanned that much more quickly.
- A multiple signal state change at an output during a program cycle does not affect the bit on the output module. It is the signal state of the output at the end of the program cycle which is transferred to the module.
- Outputs can also be scanned because they are located in ransom access memory. While it is possible to write to digital output modules, it is not possible to read them. The scanning and linking of the outputs makes additional storage of the output bit to be scanned unnecessary.

These advantages are offset by an increased program response time. A detailed description of how a programmable controller's response time is put together can be found in Section 20.2.4, "Response Time".

## Bit Memory

The area called bit memory hold what could be regarded as the controller's "auxiliary contactors". Bit memory is used primarily for storing binary signal states. The bits in this area can be treated as outputs, but are not "externalized". Bit memory is located in the CPU's system memory area, and is therefore available at all times. The number of bits in bit memory is CPU-specific.

Bit memory is used to store intermediate results which are valid beyond block boundaries and are processed in more than one block. Besides the data in global data blocks, the following are also available for storing intermediate results:

- *Temporary local data,* which is available in all blocks but valid for the current block call only, and
- *Static local data,* which is available only in function blocks but valid over multiple block calls.

*Retentive bit memory*

Part of bit memory may be designated "retentive", which means that the bits in that part of bit memory retain their signal states even under off-circuit conditions. Retentivity always begins with memory byte 0 and ends at the designated location. Retentivity is set when the CPU is parameterized. You will find additional information in Section 22.2.3, "Retentivity".

*Clock memory*

Many procedures in the controller require a periodic signal. Such a signal can be implemented using timers (clock pulse generator), watchdog interrupts (time-controlled program scanning), or simply by using clock memory.

Clock memory consists of bits whose signal states change periodically with a mark-to-space ratio of 1:1. The bits are combined into a byte, and correspond to fixed frequencies (Figure 1.4). You specify the number of clock memory bits when you parameterize the CPU. Please note that the updating of clock memory is asynchronous to the scanning of the main program.

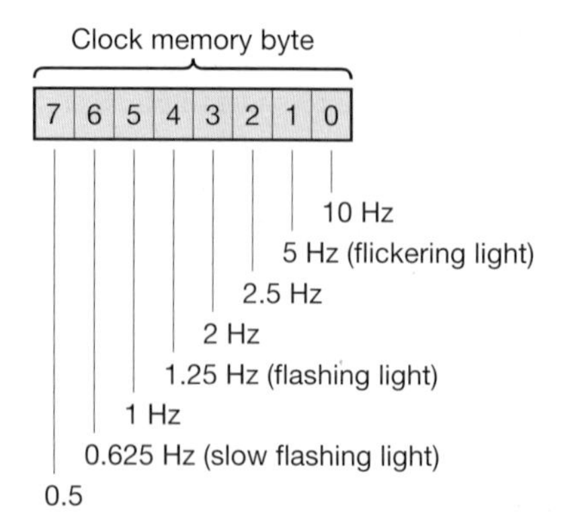


**Figure 1.4** Contents of the Clock Memory Byte

# 2 STEP 7 Programming Software

## 2.1 STEP 7 Basic Package

This chapter describes the STEP 7 basic package, Version 4.0. While the first chapter presented an overview of the characteristics of the programmable controller, this chapter tells you how to set these characteristics.

The basic package contains the statement list (STL), ladder logic (LAD) and function block diagram (FBD) programming languages. In addition to the basic package, option packages such as S7-SCL (Structured Control Language), S7-GRAPH (sequence planning) and S7-HiGraph (state-transition diagram) are also available.

### 2.1.1 Installation

STEP 7 V4.0 is a 32-bit application requiring Microsoft Windows 95 or Microsoft Windows NT as the operating system. To work with the STEP 7 software under Windows 95, you required a programming device (PG) or a PC with an 80486 processor or higher and at least 16 MB or RAM, with 32 MB recommended. For Windows NT, you require a Pentium processor and at least 32 MB of RAM; you must have administration authorization to install STEP 7 under Windows NT.

STEP 7 V4.0 occupies approximately 105 Mbytes per language (for example, English) on the hard disk. A swap-out file is also needed. This file is approximately 128 MB minus main memory; for example, if the main memory configuration is 16 MB, the swap-out file would comprise about 112 MB. You should reserve around 50 MB for your user data, and between 10 and 20 MB for each additional option package. The memory requirements may increase for certain operations, such as copying a project. If there is insufficient space for the swap-out file, errors such as program crashes may occur.

Windows 95/NT's SETUP program, which is on the first diskette, is used for installation. On the PG STEP 7 is already factory-installed.

An MPI interface is needed for the on-line connection. The programming devices have the multipoint interface already built in, but PCs must be retrofitted with an MPI module. If you want to use PC memory cards, you will need a prommer.

STEP 7 V4 has multi-user capability, that is, a project that is stored, say, on a central server can be edited simultaneously from several workstations. You make the necessary settings in the Windows Control Panel with the "SIMATIC Workstation" program. In the dialog box that appears, you can parameterize the workstation as a single-user system or a mult-user system with the protocols used.

### 2.1.2 Authorization

An authorization (right of use) is required to operate STEP 7. The authorization is provided on diskette. Before using STEP 7, you must copy this authorization from the diskette to the hard disk. You may also transfer the authorization to another device by copying it back to the (original) authorization diskette, then transferring it to the new device.

Should you lose your authorization for some reason, such as a hard disk defect, you can use the emergency license, which is also on the authorization diskette and is good for a limited time only, until you are able to obtain a replacement authorization.

### 2.1.3 SIMATIC Manager

The SIMATIC Manager is the main tool in STEP 7; you will find its icon in Windows.

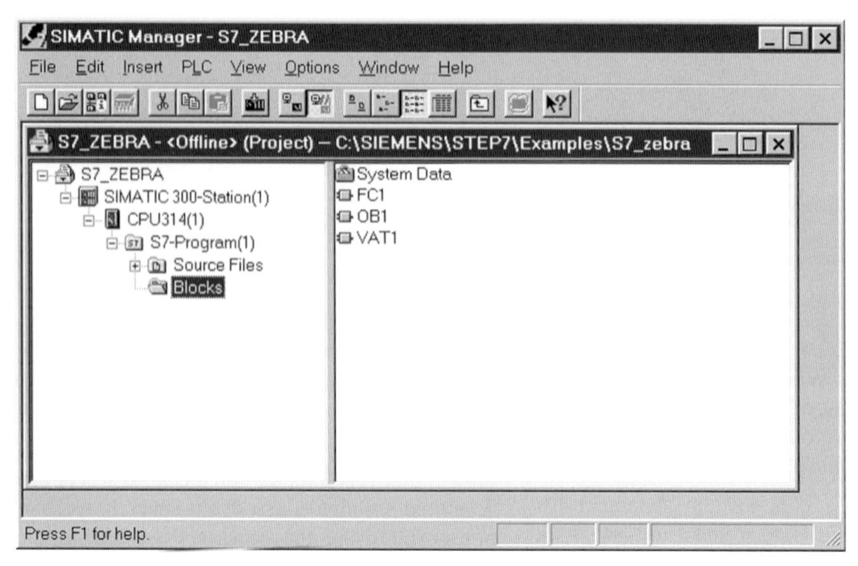


**Figure 2.1** SIMATIC Manager Example

The SIMATIC Manager is started by double-clicking on its icon. Programming begins with opening or creating a 'project'. The projects COM_ SFB, S7_MIX and S7_ZEBRA are provided as learning material. When you open project S7_ZEBRA with FILE → OPEN, you will see the split project window: at the left the structure of the open object (the object hierarchy), at the right the selected object (Figure 2.1). Clicking on the box containing a plus sign in the left window displays additional levels of the structure; selecting an object in the left half of the window displays its contents in the right half of the window.

Under the SIMATIC Manager, you work with the objects in the STEP 7 world. These "logical" objects correspond to 'real' objects in your plant. A project contains the entire plant, a station corresponds to a programmable controller. A project may contain several stations connected to one another, for example, via an MPI subnet. A station contains a CPU, and the CPU contains a program, in our case an S7 program. This program, in turn, is a 'container' for other objects, such as the object Blocks, which contains, among other things, the compiled blocks.

The STEP 7 objects are connected to one another via a tree structure. Figure 2.2 shows the most important parts of the tree structure (the "main branch", as it were) when you are working with the STEP 7 basic package for S7 applications in off-line view. The objects shown in bold type are containers for other objects. All objects in the Figure are available to you in the off-line view. These are the objects that are on the programming device's hard disk. If your programming device is on-line on a CPU (normally a PLC target system), you can switch to the on-line view by selecting VIEW → ONLINE. This option displays yet another project window containing the objects on the destination device; the objects shown in italics in the Figure are then no longer included. The title bar of the active project window shows you whether you are currently working on-line or off-line.

Select OPTIONS → CUSTOMIZE to change the SIMATIC Manager's basic settings, such as the session language, the archive program and the storage location for projects and libraries, and configuring the archive program.

The following applies for the general editing of objects:

To *select an object* means to click on it once with the mouse so that it is highlighted (this is possible in both halves of the project window).

To *name an object* means to click on the name of the selected object (a frame will appear around the name and you can change the name in the window) or select the menu item EDIT → OBJECT PROPERTIES and change the name in the dialog box.

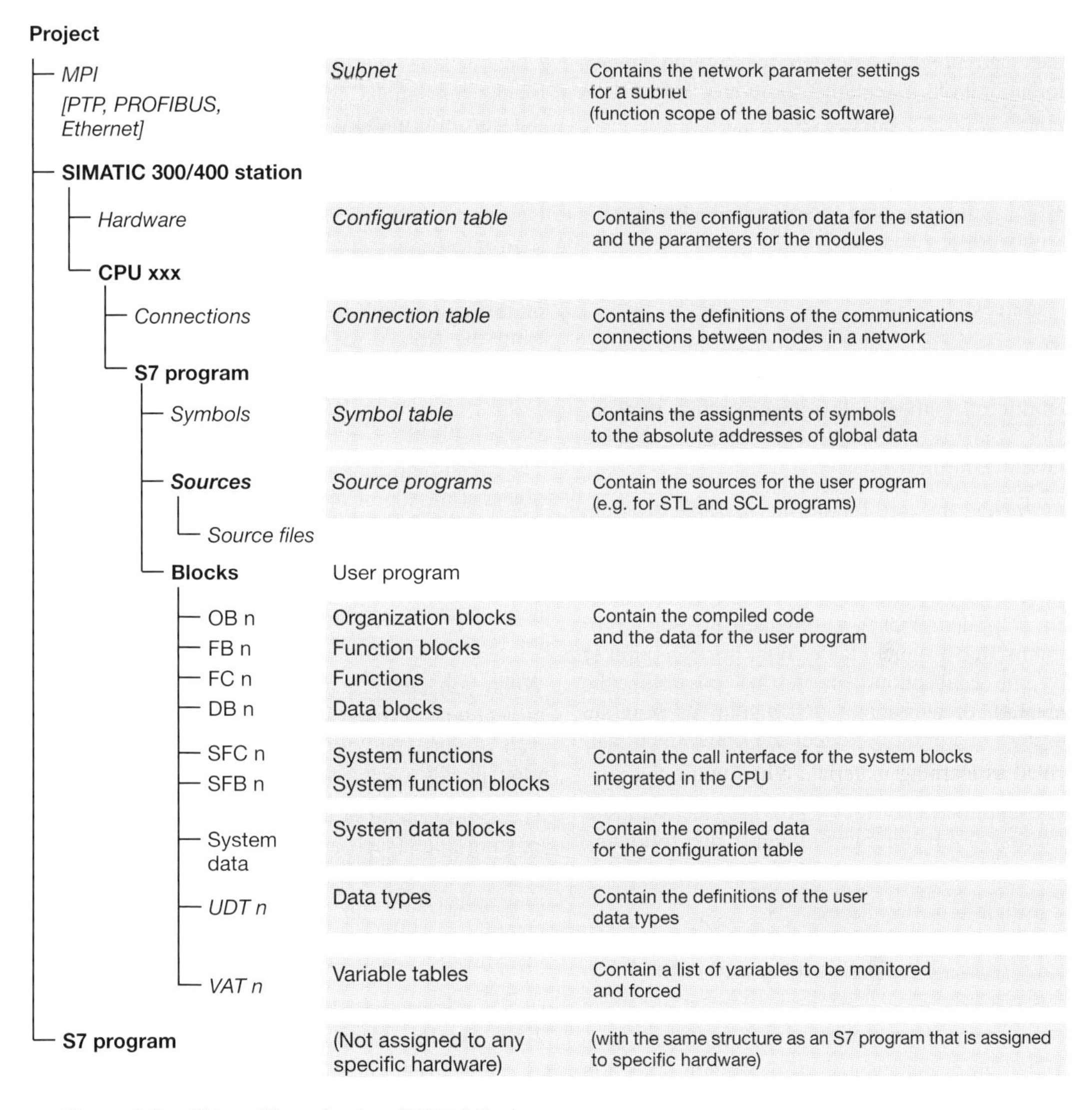


**Figure 2.2** Object Hierarchy in a STEP 7 Project

To *open an object,* double-click on that object. If the object is a container for other objects, the SIMATIC Manager displays the contents of the object in the right half of the window. If the object is on the lowest hierarchical level, the SIMATIC Manager starts the appropriate tool for editing the object (for instance, double-clicking on a block starts the editor, allowing the block to be edited).

In this book, the menu items in the standard menu bar at the top of the window are described as *operator sequences.* Programmers experienced in the use of the operator interface use the icons from the toolbar. The use of the *right mouse button* is very effective. Clicking on an object once with the right mouse button screens a menu showing the current editing options.

### 2.1.4 Projects and Libraries

In STEP 7, the 'main objects' at the top of the object hierarchy are projects and libraries.

*Projects* are used for the systematic storing of data and programs needed for solving an automation task. Essentially, these are

- the hardware configuration data,
- the parameterization data for the modules,
- the configuring data for communication via networks,
- the programs (code and data, symbols, sources).

The objects in a project are arranged hierarchically. The opening of a project is the first step in editing all (subordinate) objects which that object contains. The following sections discuss how to edit these objects.

*Libraries* are used for storing reusable program components. Libraries are organized hierarchically. They may contain STEP 7 programs which in turn may contain a user program (a container for compiled blocks), a container for source programs, and a symbol table. With the exception of on-line connections (no testing possible), the creation of a program or program or program section in a library provides the same functionality as in an object.

As supplied, STEP 7 V4 provides the *Standard Library V3.x* library containing the following programs:

- System Function Blocks
Contains the call interfaces of the system blocks for off-line programming integrated in the CPU

- S5–S7 Converting Blocks
Contains loadable functions for the S5/S7 converter (replacement of S5 standard function blocks in conjunction with program conversion)

- TI–S7 Converting Blocks
Contains loadable functions and function blocks for the TI/S7-Converter

- IEC Converting Blocks
Contains loadable functions for editing variables with complex data types (STRING, DATE_AND_TIME)

- Communication Blocks
Contains loadable functions for controlling CP modules

- PID Control Blocks
Contains loadable function blocks for closed-loop control

- Organization Blocks
Contains the templates for the organization blocks (essentially the variable declaration for the start information)

You will find an overview of the contents of these libraries in Chapter 25, 'Block Libraries'. Should you, for example, purchase an S7 module with standard blocks, the associated installation program installs the standard blocks as a library on the hard disk. You can then copy these blocks from the library to your project. A library is opened with FILE → OPEN, and can then be edited in the same way as a project. You can also create your own libraries.

The menu item FILE → NEW... generates a new object at the top of the object hierarchy (project, library). The location in the directory structure where the SIMATIC Manager is to create a project or library must be specified under the menu item OPTIONS → CUSTOMIZE or in the New dialog box.

When you create a project, you can select the project type "Project 2.x". You can also edit projects of this type with STEP 7 V2 (see also Section 2.2.3 "Project Versions").

The INSERT menu is used to add new objects to existing ones (such as adding a new block to a program). Before doing so, however, you must first select the object container in which you want to insert the new object from the left half of the SIMATIC Manager window.

### 2.1.5 On-line Help

The SIMATIC Manager's on-line help provides information you need during your programming session without the need to refer to hardcopy manuals. You can select the topics you need information on by selecting the HELP menu. The on-line help option GETTING STARTED, for instance, provides a brief summary on how to use the SIMATIC Manager.

The on-line help also provides context-dependent help, that is to say, you can call up information about an object selected with your mouse or the current error message when you press F1. In the toolbar you will see a button with an arrow and a question mark. When you click on this button, the mouse pointer is also given a question mark. With this 'help' mouse pointer, you can now click on an object on the screen, such as an icon or menu item, and you will receive the on-line help associated with that object.

## 2.2 Editing Projects

When you set up a project, you create "containers" for the resulting data, then you generate the data and fill these containers. Normally, you create a project with the relevant hardware, configure the hardware, or at least the CPU, and receive in return containers for the user program. However, you can also put an S7 program directly into the project container without involving any hardware at all. Note that initializing of the modules (address modifications, CPU settings, configuring connections) is possible with the Hardware Configuration only.

We strongly recommend that the entire project editing process be carried out using the SIMATIC Manager. The creating, copying or deleting of directories or files as well as the changing of names (!) with the Windows Explorer within the structure of a project can cause problems with the SIMATIC Manager.

### 2.2.1 Creating Projects

From STEP 7 V3.2, the *STEP 7 Assistant* helps you in creating a new project. You specify the CPU used and the assistant creates for you a project with an S7 station and the selected CPU as well as an S7 program container, a source container and a block container with the selected organization blocks. You will find general information on operator entries for object editing in Section 2.1.3, "SIMATIC Manager".

• Creating a new project

Select FILE → NEW, enter a name in the dialog box, change the project type and storage location if necessary, and confirm with "OK" or RETURN.

• Inserting a new station in the project

Select the project and insert a station with INSERT → STATION → SIMATIC 300 STATION (in this case an S7-300).

• Configuring a station

Click on the plus box next to the project in the left half of the project window and select the station; the SIMATIC Manager displays the Hardware object in the right half of the window. Double-clicking on *Hardware* starts the Hardware Configuration, with which you edit the configuration tables. If the hardware catalog is not on the screen, screen it with VIEW → CATALOG.

You begin configuring by selecting the rail with the mouse, for instance under "SIMATIC 300" and "RACK-300", "holding" it, dragging it to the free portion in the upper half of the station window, and "letting it go" (drag & drop). You then see a table representing the slots on the rail.

Next, select the required modules from the hardware catalog and, using the procedure described above, drag and drop them in the appropriate slots. To enable further editing of the project structure, a station requires at least one CPU, for instance the CPU 314 in slot 2. You can add all other modules later. Editing of the hardware configuration is discussed in detail in Section 2.3, "Configuring Stations".

Store and compile the station, then close and return to the SIMATIC Manager. In addition to the hardware configuration, the open station now also shows the CPU.

When it configures the CPU, the SIMATIC Manager also creates an S7 program with all objects. The project structure is now complete.

• Viewing the contents of the S7 program

Open the CPU; in the right half of the project window you will see the symbols for the S7 program and for the connection table.

Open the S7 program; the SIMATIC Manager displays the symbols for the compiled user program (*Blocks),* the container for the source programs, and the symbol table in the right half of the window.

Open the user program (*Blocks*); the SIMATIC Manager displays the symbols for the compiled configuration data *(System data)* and an empty organization block for the main program (OB 1) in the right half of the window.

• Editing user program objects

We have now arrived at the lowest level of the object hierarchy. The first time OB 1 is opened, the window with the object properties is displayed and the editor needed to edit the program in the organization block is opened. You add another empty block for incremental editing by opening INSERT → S7 BLOCK → ... (*Blocks* must be highlighted) and selecting the required block type from the list provided.

When opened, the *System data* object shows a list of available system data blocks. You receive the compiled configuration data. These system data blocks are edited via the *Hardware* object in container *Station*. You can transfer *System data* to the CPU with PLC → DOWNLOAD and initialize the CPU.

The object container *Sources* is empty. If the *Sources* object is marked, you can insert an empty source text file using INSERT → S7 SOFTWARE → STL SOURCE or using INSERT → EXTERNAL SOURCE you can transfer a source text file generated, for example, with another editor in ASCII format, to the *Sources* container.

• Module-independent S7 programs

If you wish, you can create a program without having first configured a station. To do so, you must generate the container for your program yourself. Mark the project and generate an S7 program with INSERT → PROGRAM → S7 PROGRAM. Under this S7 program, the SIMATIC Manager creates the object containers *Sources* and *Blocks*. *Blocks* contains an empty OB 1. You can now enter your program.

A module-independent program lies directly under the project. If you want to test this program on-line, generate a new window to display the on-line view of the project with VIEW → ONLINE while the project (window) is open. If your programming device has a connection to only one CPU, all blocks available in the CPU are displayed in the on-line object *Blocks*. If more than one node is possible, select the on-line S7 program, then menu item EDIT → OBJECT PROPERTIES, and set the "geographical address" (rack and slot) of the CPU in the *Address* tab.

• S7 program in a library

You can also create a program under a library, for instance if you want to use it more than once. In this way, the standard program is always available and you can copy it entire or in part into your current program. Please note that you cannot establish on-line connections in a library, which means that you can test a program only within a project.

You cannot copy complete libraries. It is, however, possible to transfer the S7 programs and blocks in a library to other libraries or projects, even from version 2 libraries to version 3 libraries and projects (see also Section 2.2.3, "Project Versions").

### 2.2.2 Rearranging, Managing and Archiving

With FILE → REARRANGE... you can reduce the memory requirements for a project or library. When it executes this menu item, the SIMATIC Manager eliminates the gaps created by deletions.

The SIMATIC Manager manages projects and libraries in project lists and library lists. When you execute FILE → MANAGE, the SIMATIC Manager shows you a list of all known projects, with name and path. You can then delete projects you no longer want to display from the list ("Hide") or include new projects in the project list ("Display"). You manage libraries in the same way.

You can also archive a project or library (FILE → ARCHIVE...). In this case, the SIMATIC Manager stores the selected object in an archive file in compressed form. In order to archive a project or library, you need an archive program. STEP 7 contains the archive program ARJ, but you may also use other archive programs (*winzip* from version 6.0, *pkzip* from version 2.04g, *arj* from version 2.4.1a, *jar* from version 1.02 or *lharc* from version 2.13). You may create the archive file on hard disk or diskette(s). You can choose between archiving all files or only those which have been modified (reset the 'archive bit' and select 'incremental archiving'). You make the appropriate settings for archiving with OPTIONS → CUSTOMIZE on the "Archive" tab.

You can decompress an archived object for editing with FILE → RETRIEVE.

### 2.2.3 Project Versions

Since STEP 7 V4 has become available, there are three different versions of SIMATIC projects. STEP 7 V1 creates version 1 projects, STEP 7 V2 version 2 projects, and STEP 7 V3/V4 can be used to create and edit both version 2 and version 3 projects.

If you have a version 1 project, you can convert it into a version 2 project with FILE → OPEN VERSION 1 PROJECT... The project structure with the programs, the compiled version 1 blocks, the STL source programs, the symbol table and the hardware configuration remain unchanged.

Version 2 projects can be created and edited with STEP 7 V3 (Figure 2.3, Editing Projects with Different Versions). However, version 2 products still have the function scope of STEP 7 V2; option packages available only in STEP 7 V3 or STEP 7 V4 cannot be used on these projects. The creation of a version 2 project serves a practical purpose when you want to edit it and run it as a V2 project. If you want to convert a version 2 project into a version 3 project, select the menu item FILE → SAVE AS ... and then, in the window that appears, select the option "With Rearranging (slow)". You then define the folder (directory), the project name and the project type ("Project" stands for Version 3 project). Version 3 projects can no longer be saved as Version 2 projects.

Program development with STEP 7 is upwards-compatible, so that you can transfer blocks from version 2 projects and libraries to version 3 projects and libraries. However, the reverse, that is, the transfer of blocks f rom version 3 objects to version 2 objects, is not possible. For this reason, STEP 7 V4 includes the version 2 library *stdlibs* (*V2*) in order to make it possible to copy standard blocks from this library to version 2 projects.

## 2.3 Configuring Stations

You plan and define the hardware complement of your programmable controller using the Hardware Configuration tool. Configuring is done

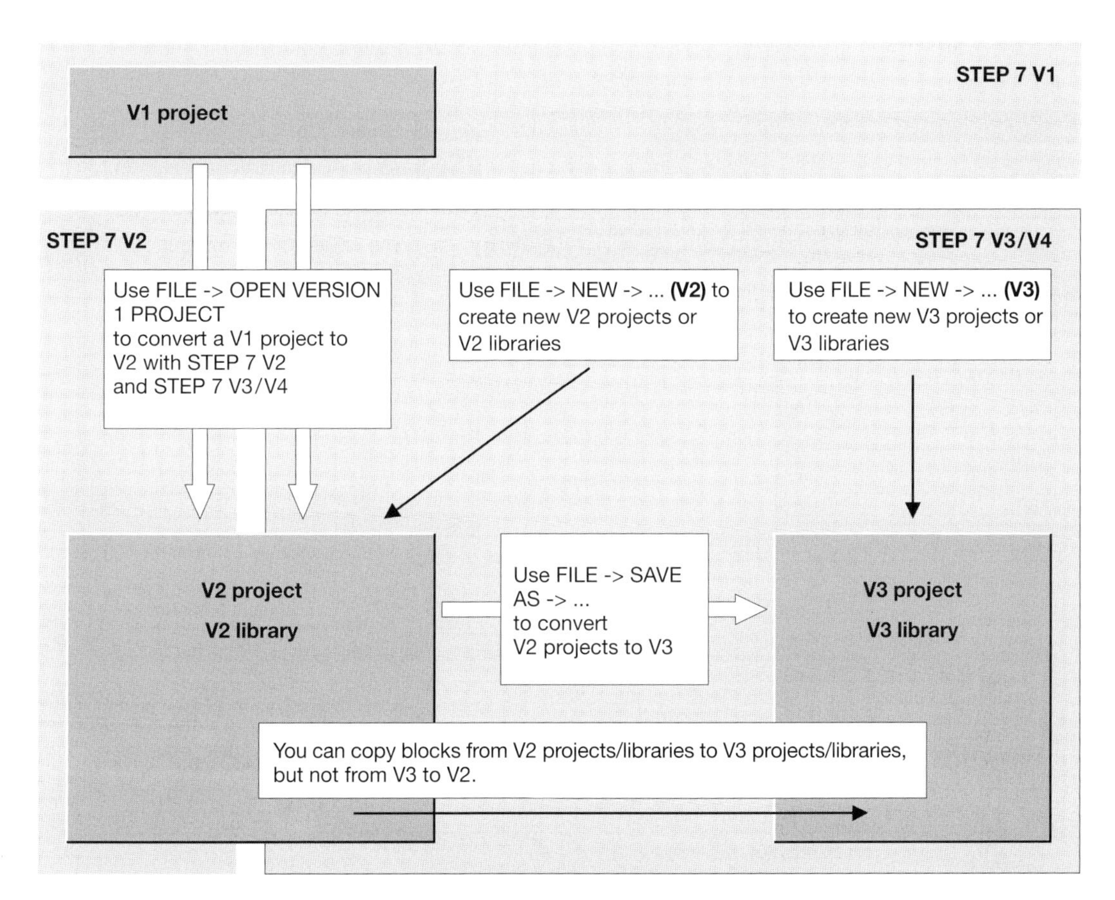


**Figure 2.3** Editing Projects with Different Versions

off-line without any connection to the CPU. This same tool allows you to address and initialize the modules.

You begin configuring your hardware by double-clicking on the *Hardware* object in container *SIMATIC 300/400-Station*. You must make the basic hardware configuration settings with OPTIONS → CUSTOMIZE.

When configuring has been completed, STATION → CONSISTENCY CHECK will show you whether your entries were free of errors. STATION → SAVE stores the configuration tables with all parameter assignment data in your project on the hard disk.

STATION → SAVE AND COMPILE not only saves but also compiles the configuration tables and stores the compiled data in the *System data* object in the off-line container *Blocks*. After compiling, you can transfer the configuration data to a CPU with PLC → DOWNLOAD. The object *System data* in on-line container *Blocks* represents the current configuration data on the CPU. You can 'return' these data to the hard disk with PLC → UPLOAD.

### Configuration table

The Hardware Configuration tool works with tables, each of which represents a rack, a module or a DP station (Figure 2.4). The configuration tables show the slots and the modules plugged into them, or the properties of the module, such as the addresses and the order number. You can generate the configuration tables in the planning phase or wait until the hardware has actually been installed.

## 2.3.1 Arranging Modules

After it has been opened, the Hardware Configuration displays the station window and hardware catalog. You can screen and blank out the hardware catalog with VIEW → CATALOG. You can enlarge or maximize the station window for

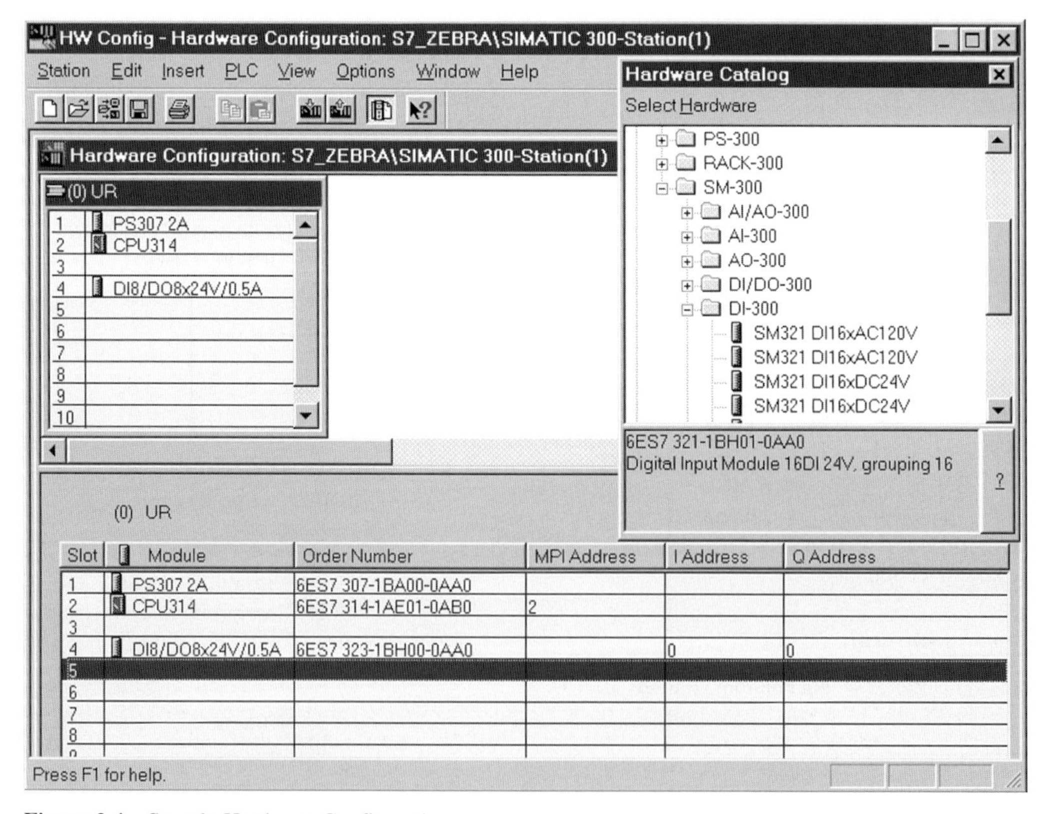


**Figure 2.4** Sample Hardware Configuration

easier editing. In the upper half, it shows the module racks in the form of tables and the DP stations in the form symbols. If there are several racks, you will see here the connection between the interface modules and if PROFIBUS is used, you will see the structure of the DP master system. The lower half of the station window shows the detailed view of the rack or DP slave marked in the upper half.

The hardware catalog contains all available modules known to STEP 7. Whith OPTIONS → EDIT CATALOG PROFILES, you can create your own hardware catalog that shows, in a structure you choose yourself, only those modules with which you want to work. By double-clicking on the Title Bar, you can insert the hardware catalog on the right edge of the station window or delete it from there again.

You begin configuring by selecting and "holding" the rail from the hardware catalog, for instance under "SIMATIC 300" or "RACK 300", with the mouse, dragging it to the upper half of the station window, and dropping it anywhere in that window. An empty configuration table is screened for the controller rack. Next, select the required modules from the hardware catalog and, in the manner described above, drag and drop them in the appropriate slots. A "No Parking" symbol tells you cannot drop the selected module at the intended slot.

You can generate the configuration table for another rack by dragging the selected rack from the catalog and dropping it in the station window. In S7-400 systems, a non-interconnected rack (or more precisely: the relevant receive interface module) is assigned an interface via the "Link" tab in the Properties window of a Send IM (select module and EDIT → OBJECT PROPERTIES).

The arrangement of distributed I/O stations is described in Section 2.4.2 "Configuring DP Communication".

### 2.3.2 Addressing Modules

When arranging modules, the Hardware Configuration tool automatically assigns a module start address. You can view this address in the lower half of the station window in the object properties for the relevant modules. In the case of the S7-400 and the CPU 315-2DP, you can change the module addresses. When doing so, please observe the addressing rules for S7-300 and S7-400 systems as well as the addressing capacity of the individual modules.

There are modules which have both inputs and outputs for which you can (theoretically) reserve different start addresses. However, please note carefully the special information provided in the product manuals; the large majority of function and communications modules require the same start address for inputs and outputs.

Modules on the MPI bus or communications bus have an MPI address. You may also change this address. Note, however, that the new MPI address goes into force as soon as the configuration data are transferred to the CPU.

### 2.3.3 Parameterizing Modules

When you parameterize a module, you define its properties. It is necessary to parameterize a module only when you want to change the default parameters. A requirement for parameterization is that the module is located in a configuration table.

Double-click on the module in the configuration table or select the module and select EDIT → OBJECT PROPERTIES. Several tabs with the specifiable parameters for this module are displayed in the dialog box. When you use this method to parameterize a CPU, you are specifying the run characteristics of your user program.

Some modules allow you to set their parameters at runtime via the user program with system functions SFC 55 WR_PARM, SFC 56 WR_DPARM and SFC 57 PARM_MOD.

## 2.4 Configuring Communication Connections

A subnet is a homogeneous network of modules with communications capabilities which has no gateway. With STEP 7, you can insert the objects for MPI, Industrial Ethernet, PROFIBUS and PTP subnets in one project.

The SIMATIC Manager generates the object for an MPI subnet automatically when the project is created. You insert the objects for the other subnets by marking the project and selecting INSERT → SUBNET.

### 2.4.1 Networking Modules with MPI

You specify the nodes for the MPI subnet when you set the module properties. Select a CPU in the configuration table and open it with EDIT → OBJECT PROPERTIES. A dialog box is then displayed. The "General" tab shown in this dialog box contains the "MPI ..." button; at the right of the button is the MPI address. When you click on this button, another dialog box is displayed. You must then select the suitable subnet from the "Network connection" tab.

Take this opportunity to also set the MPI address you have chosen for this CPU. Note that FM or CP modules with MPI interface that are located in an S7-300 station are automatically assigned an address derived from that CPU's MPI address. The FM or CP module whose slot is closest to that of the CPU is automatically assigned the MPI address which is higher by one than that of the CPU, the next module is assigned the next higher address, and so on. For example, if you set the CPU's MPI address to 5, the nearest FM or CP module is assigned MPI address 6, the next the MPI address 7, and so on. You must take this automatic address allocation into account when you choose the MPI addresses for the CPUs.

The highest MPI address must be higher than or equal to the highest assigned MPI address in the subnet (remember to take automatic address allocation for FM and CP modules into account!). It must have the same value for all nodes in the subnet.

A suggestion: Give the CPUs in the various stations different names ("IDs"). The default name for all stations is "CPUxxx(1)", so that they can be distinguished from one another in the subnet by their MPI address only. If you do not want to assign a name of your own choosing, you can change the default from "CPUxxx(1)" to "CPUxxx(n)", where "n" is the MPI address.

When you assign the MPI address, you should also consider the possibility that, at some later date, you might want to connect a PG or OP to the MPI network for the purpose of service or maintenance. Permanently installed PGs or OPs should be connected directly to the MPI network, an MPI connector with PG socket is available for devices which can be plugged in via a spur. A tip: Reserve address 0 for a service PG, 1 for a service OP, and 2 for a replacement CPU (this corresponds to the preset addresses).

### 2.4.2 Configuring DP Communication

Distributed I/O consists of stations interfaced to one another via the PROFIBUS-DP bus system. A "DP master" connects the PROFIBUS subnet with the CPU's P bus. The stations ("nodes") on the bus are distinguished from one another by the PROFIBUS address ("node number"). A station can have the same characteristics as a module (a "compact" DP slave, such as an ET 200B) or consist of several modules (a "modular" DP slave, such as an ET200M with up to eight S7-300 modules).

Configuring must begin with the DP master module, which you select from the Hardware catalog and place in the configuration table. It is possible that you have already selected a CPU with DP interface. The next line down shows the DP master with one connection to a DP master system in the station window (broken black-and-white line). Using your mouse, drag the stations you select from the Hardware catalog under "PROFIBUS-DP" to this master system symbol. As you drag and drop a station, that station's properties are displayed, and you can set the PROFIBUS address for that station. The DP slave then appears as a symbol in the upper half of the station window and a configuration table for this station appears in the lower half of the window. If you selected a modular station, you can now locate additional modules in the station. You can set the addresses via the object properties of the module.

You can "reinstall" DP slaves that are not in the Hardware catalog. To do so, you need the type file (DDB file, device database file) for that slave. Use OPTIONS → INSTALL NEW DDB FILES to acquaint STEP 7 with the new type file. The new DP slave will be displayed in the Hardware catalog under *PROFIBUS-DP* in container *Additional field devices* following a complete restart of the Hardware Configuration. OPTIONS → IMPORT STATION DDB FILE imports a DDB file from another project.

Essentially, the distributed I/O modules can be addressed in the same way as the centralized I/O (modules in the racks). All modules reserve addresses in the CPU's logical address area (I/O area P). The address assignments (user data assignments) for "centralized" and "distributed" modules must not overlap. The addressing capacity depends on the DP master. You may change the STEP 7 default PROFIBUS address for the DP slave as well as the default input and output addresses.

The "distributed" modules are initialized in the same manner as the "centralized" modules. Double-clicking on the symbol in the upper half of the station window opens a dialog box with one or more tabs in which you set the required module properties. STEP 7 assigns a *diagnostic address*, which you need in order to check a station's diagnostic data. Distributed modules with the appropriate capability can also be initialized at runtime via system functions. Some DP masters enable event-driven synchronization of DP slaves with the commands SYNC and FREEZE.

### 2.4.3 Configuring GD Communication

Global data communication (GD communication) is used for the exchange of small volumes of data between CPUs in an MPI subnet. The prerequisite for the configuring of GD communication is an established subnet (see Section 2.4.1, "Networking Modules with MPI"). In GD communication, the data are normally exchanged cyclically, S7-400 systems also enable additional or exclusive event-driven updating of GD data.

GD communication is configured by making entries in a table. After selecting the MPI subnet in the SIMATIC Manager, select the menu command OPTIONS → DEFINE GLOBAL DATA and GD TABLE → OPEN → GLOBAL DATA FOR SUBNET. An empty GD table is displayed. Select a column, then select EDIT → ASSIGN CPU. A project window is opened. Select the station in the left half and the CPU in the right half of this window. 'OK' places the CPU in the GD table.

Repeat this procedure for each CPU to be configured for GD communication. A GD table may comprise as many as 15 CPU columns.

To configure data interchange between CPUs, select the first location under the CPU that is to be the sender and enter the address whose value is to be transferred (confirm with RETURN). With EDIT → SENDER, define this value as send value, discernible because of ">" and highlighting. In the same line, enter the operand that is to receive the send value (this is preceded by "receiver"). Timers and counters may be used as sender only, the receiver must be a word operand for each timer or counter. A line may contain several receivers but only one sender (Table 2.1).

After completing your entries, compile the GD table with GD TABLE → COMPILE. Following error-free compilation, STEP 7 fills in the "GD Identifier" (for example, GD 2.1.3 would mean GD circle 2, GD packet 1, GD element 3).

After compiling, you can enter the addresses for the communication status in the table with VIEW → GD STATUS. The group status (GST) shows the status of all communication links in the table. The status (GDS) shows the status of one communication link (the status of a GD send packet). In both cases, status information is contained in a doubleword.

The menu command VIEW → SCAN RATES allows you to specify the scan rates for each GD packet and each CPU yourself. The scan rate (SR) specifies after how many CPU cycles the global data are to be updated. The default scan rate is set so that, in the case of a "clean" CPU (a CPU that has no user program), the GD packets will be sent and received approximately every ten milliseconds.

**Table 2.1** Example of a GD Table

| Global Data Identifier | Station1 \ CPU314(1) | Station2 \ CPU315(2) | Station3 \ CPU413(3) | Station4 \ CPU416(4) | Station5 \ CPU414(5) |
|---|---|---|---|---|---|
| GD | >MD16 | MD116 | MD126 | MD126 | |
| GD | | >DB20.DBW14:8 | | | MW124:8 |
| GD | | DB21.DBW28 | | | >T28 |
| GD | DB18.DBW20:4 | | >C10:4 | | |
| GD | | >ID24 | | MD24 | MD24 |

Note that the lower the scan rates, the greater a CPU's "communication load". To keep the communication load within a manageable range, set the scan rate in the sending CPU so that the product of scan rate and cycle time is greater than 60 ms for the S7-300 and greater than 10 ms for the S7-400. The product of these two values must be lower in the receiving CPU than in the sending CPU to ensure that no GD packets will be lost. On S7-400 CPUs, you must disable cyclic data interchange by setting the scan factor to zero when you want to use SFCs and event-controlled GD communication only.

Following a recompilation, use menu command PLC → DOWNLOAD to transfer the GD table to the CPUs, where it immediately comes into force.

### 2.4.4 Configuring SFB Communication

You use special system blocks for communications between nodes in a network. You can choose between the following:

- *Non-Configured connection* between SIMATIC S7 devices where the connection to the partner is built up at runtime if required.
- *Configured connection* where the connection properties between any two partners is fixed in a connection table.

The system blocks for these communications connections are introduced in Chapter 20 "Main Program".

A configured connection requires special system function blocks SFBs that you call in the CPU program in each case at the locations where you want to send or receive data. You can also control the partner device with these blocks, for example, set it to STOP or initiate a complete restart. The requirement for SFB communication is that a subnet is configured (MPI, PROFIBUS or Ethernet).

### 2.4.5 Network Configuration

NETPRO can be used to configure subnets graphically in STEP 7 V4. You reach NETPRO by opening a subnet in the project window of the SIMATIC Manager or by opening *Connections* in the container *CPU*.

The window for network configuration displays all previously created subnets and stations (nodes) in the project with the configured connections. A second window shows the network object catalog with a selection of the available SIMATIC stations, subnets and DP slaves. You can toggle the catalog on and off with VIEW → CATALOG. You can adapt and improve the readability of the representation using VIEW → ZOOM IN, VIEW → ZOOM OUT and VIEW → ZOOM FACTOR. If you have already configured DP slaves, these are displayed with VIEW → DP SLAVES.

You start the network configuration by selecting a subnet that you mark in the catalog and then hold and drag to the network window using the mouse. You proceed in the same way with the desired stations, initially without a connection to the subnet. Double-click on a station to open the Hardware Configuration so that you can configure the station or at least the module with network connection. Save the station and return to the network configuration. The interface of this module is represented in the network configuration as a small box under the module view. Click on this box, hold it and drag it to the relevant subnet. Proceed as follows with all other nodes.

After creating the graphical view, parameterize the subnets: Mark a subnet and select EDIT → OBJECT PROPERTIES. In the subsequent dialog box, you can specify the configuration, the lines and the bus parameters in the 'Network Settings' tab. When the network connection of a node is marked, you can define the network properties

**Table 2.2** Sample Connection Table

| Local ID | Partner ID | Partner | Type | Active Connection Buildup | Send Status Messages |
|---|---|---|---|---|---|
| 1 | 1 | Station4 / CPU416(4) | S7 Connection | Yes | No |
| 2 | | Station2 / CPU315(2) | S7 Connection | Yes | No |
| 3 | 1 | Station5 / CPU414(5) | S7 Connection | Yes | No |

of the node with EDIT → OBJECT PROPERTIES. You define the module properties of the nodes in a similar way (the same operator input as for the Hardware Configuration).

You can highlight the node assignments of a master system graphically marking a DP slave and using VIEW → HIGHLIGHT → MASTER SYSTEM. If you mark the PROFIBUS connection and select EDIT → MASTER SYSTEM, you can configure SYNC/FREEZE groups, provided you have a suitable DP master.

You can also configure global data communications out of NETPRO: Mark the MPI subnet and select OPTIONS → DEFINE GLOBAL DATA.

You can test a configuration for contradictions with NETWORK → CONSISTENCY CHECK. You conclude the network configuration with NETWORK → SAVE AND COMPILE.

### Connection table

For communication via configured connections, you require the connection table (Table 2.2). To configure SFB communication, mark a CPU in the network configuration. A table is then screened in the lower section of the window, and you can make your entries with INSERT → CONNECTION. You must create a connection table for each "active" CPU.

In dialog boxes "Station" and "Module", select the CPU for which you want to configure the connection. You must first have set up the stations and modules with the Hardware Configuration.

The number of possible connections is CPU-specific. STEP 7 establishes a connection ID for each connection and for each partner. You need this ID whenever you use the communication blocks in your program.

You specify the connection setup in the dialog box screened when you select EDIT → OBJECT PROPERTIES with the connection line marked. You can change the connection partner with EDIT → CONNECTION PARTNER. To activate the connections, you must load the connection table, following compilation and saving, into the PLC (all connection tables into all "active" CPUs).

## 2.5 Developing Programs

You can assign an S7 program in the project hierarchy of a module or develop it independent of a module. It contains the *Symbols* object and the containers *Sources* and *Blocks*.

In the STL programming language, you can enter the program incrementally (directly) or generate a source program and compile it later. If you want to program with symbols, which is strongly recommended, you must first generate a symbol table, which you can update while you are programming (Figure 2.5).

The memory requirement for a compiled block is in the block properties (select the block in the SIMATIC Manager, then select tab "General – Part 2" in EDIT → OBJECT PROPERTIES). In the length specification under 'MC7 Code', you must add 36 bytes for the block header. You can find out how much room your whole program takes up in memory by selecting the program (*Blocks* object) in the SIMATIC Manager, then selecting menu item EDIT → OBJECT PROPERTIES. Take the size of the program in load and work memory and the number of blocks of each type from the "Special" tab.

### 2.5.1 Symbol Table

In the control program, you work with addresses, such as those for inputs, outputs, timers and blocks. You can use absolute addressing (for instance I 1.0) or symbolic addressing (for instance Startsignal). Symbolic addressing uses names instead of absolute addresses. You can make your program more readable by using appropriately descriptive names. You will find additional information on the absolute and symbolic addressing of variables in Section 3.4, "Variables and Constants".

In symbolic addressing, a distinction is made between *local* symbols and *global* symbols. A local symbol is known only in the block in which it was defined. You can use the same local symbols in different blocks for different purposes. A global symbol is known throughout the entire user program, and has the same meaning in all blocks. You define global symbols in the symbol table *Symbols* in the container *S7 Program*.

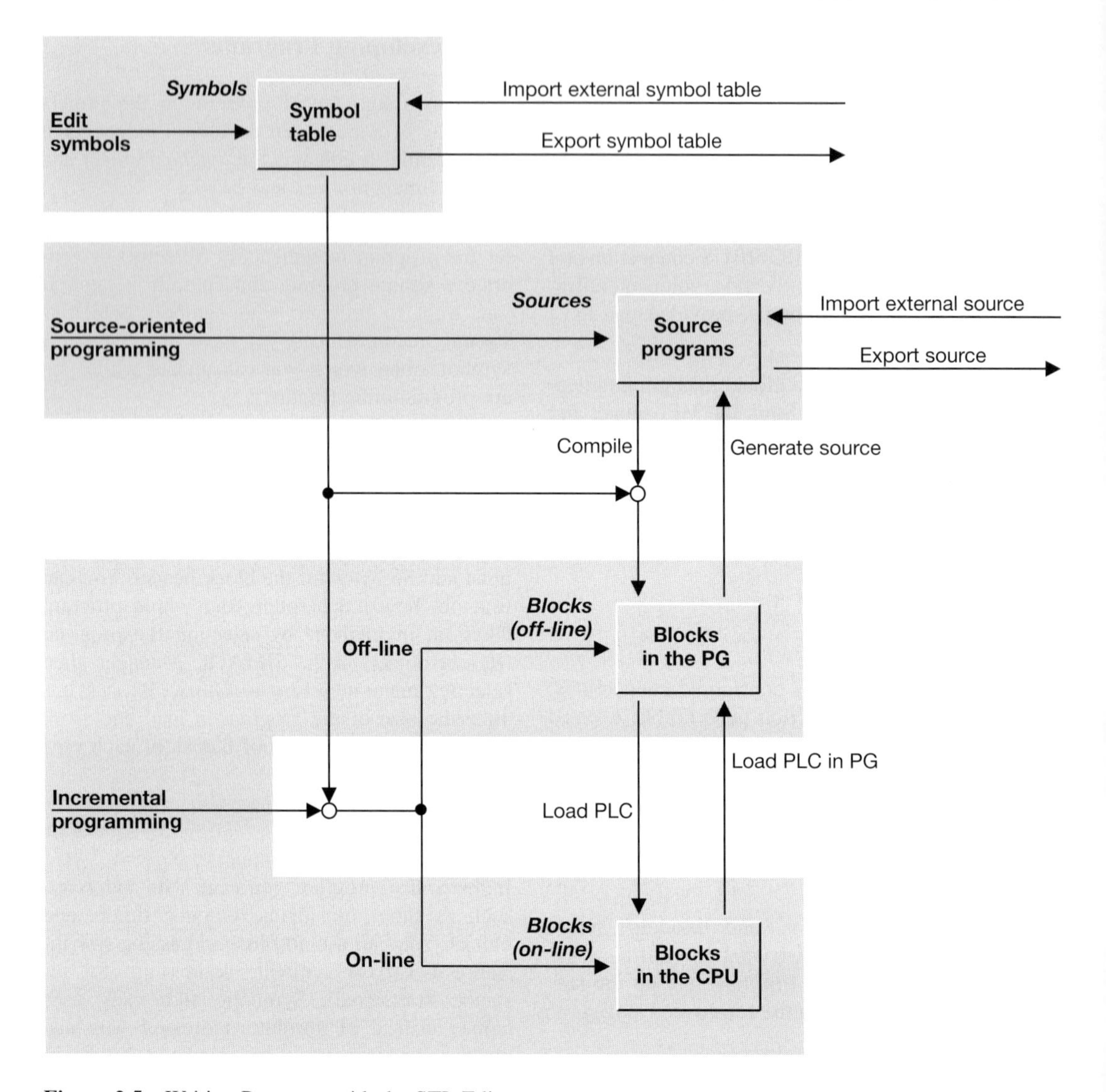


**Figure 2.5** Writing Programs with the STL Editor

A global symbol begins with a letter and may comprise up to 24 characters. A global symbol may also contain spaces, special characters and country-specific characters such as the umlaut. Exceptions are the characters $00_{hex}$, $FF_{hex}$ and quotation marks. When you use symbols containing special characters, you must enclose these symbols in quotation marks when you are programming. In a compiled block, the editor always shows global symbols in quotation marks. The symbol comment may comprise up to 80 characters.

You may assign names to the following addresses and objects in the symbol table:

- Inputs I, outputs Q, peripheral inputs PI and peripheral outputs PQ
- Memory bits M, timers T, and counters C
- Code blocks OB, FB, FC, SFC, SFB and data blocks DB
- User data types UDT
- Variable tables VAT

**Table 2.3** Sample Symbol Table

| Symbol | Address | Data Type | Comment |
|---|---|---|---|
| Main program | OB 1 | OB 1 | Program for cyclic scanning |
| Receive mailbox | DB 10 | DB 10 | Data block for receiving frames |
| Start signal | I 1.0 | BOOL | Start conveyor belt |
| +HS13-S104 | I 1.1 | BOOL | Limit switch upper bay 13 |
| Quantity | MW 20 | INT | Production per batch |
| Start-up time B3 | T 1 | TIMER | Start monitor for belt 3 |
| Temp_Mi7 | MD 24 | REAL | Temperature in mixer 7 |

Data addresses in data blocks are considered to be local addresses; the associated symbols are defined in the declaration section of the data block in the case of global data blocks and in the declaration section of the function block in the case of instance data blocks.

When an S7 program is created, the SIMATIC Manager also generates an empty symbol table called *Symbols*. Simply open this table and you can define the global symbols and allocate them to absolute addresses (Table 2.3). Definition of a symbol also includes the data type, which defines specific properties of the data behind the symbol, most importantly the form in which the data are represented. The data type BOOL, for instance, identifies a binary variable, and the data type INT a digital variable whose contents represent a 16-bit integer number. The data types used for the for the various addresses are described in the respective chapters of this book.

Symbols used for incremental program input must already be allocated to absolute addresses. You can add symbols to or correct entries in the symbol table during input with the incremental editor (with OPTIONS → SYMBOL TABLE or INSERT → SYMBOL). Then simply resume input with the new or modified symbol.

If you use a source file to input the program, you must provide the allocation data at compilation time. Source files with symbolic address are eminently suitable for standardized, reusable program sections: while the function (the algorithm) stays the same, you can have another symbol table ready to compile and thus adapt your program to different address definitions quickly and easily.

Symbol tables can be imported and exported. "Exported" means that a file will be created with the contents of your symbol table. You may choose between a pure ASCII file (file ends in *.asc), a sequential allocation list (*.seq), System Data Format (*.sdf for Microsoft Access), and Data Interchange Format (*.dif for Microsoft Excel). You can edit the exported file with a suitable editor. You can also import symbol tables in these formats.

### 2.5.2 Editor

STEP 7 provides you with an editor in the FBD, LAD and STL programming languages. You enter the editor's environment when you open a block in the SIMATIC Manager, for instance by double-clicking on the automatically generated symbol for organization block OB 1 or via the Windows taskbar with START → SIMATIC → STEP 7 → PROGRAM S7 BLOCKS LAD, STL, FBD.

You can adapt the editor's features to suit your requirements with OPTIONS → CUSTOMIZE. Select the font for the text and the programming language (STL, LAD, FBD) from the 'Editor' tab. When you generate a new block with the editor, the language in which you generate it will be taken over as the programming language. With the selection of symbols and comments you stipulate the defaults with which a block will be opened.

You can use the editor both for writing ASCII source programs and for direct (incremental) programming of blocks. You will use incremental programming for the off-line user program in PG data management and for the on-line user program in the CPU.

### Source-file-oriented programming

Source-file-oriented programming is used to edit an STL source file in the *Source Files* object container. An STL source file is a pure ASCII text file. It may contain the source program for one or more code or data blocks or an entire program as well as the definitions of the user data types.

In the SIMATIC Manager, select the souce program container *Source Files* and create a new source file with INSERT → S7 SOFTWARE → STL SOURCE FILE. You can open and edit this file. You can make the creation of new blocks much simpler for yourself by using INSERT → BLOCK TEMPLATE → ... (in the editor). You also have the editor option of generating a new STL source file from one or more compiled blocks with FILE → GENERATE SOURCE FILE.

If you generated a source file with another text editor, you can use the SIMATIC Manager's INSERT → EXTERNAL SOURCE FILE menu item to place that file in the *Source Files* container. You can copy the selected source file to a directory of your choice with EDIT → EXPORT SOURCE FILE.

You can save source files during editing whenever necessary. Not until a source file is compiled does the editor generate blocks, which it stores in the off-line user program *Blocks*. If you used global symbols in an STL source file, the completed symbol table must also be available when you compile the file.

In source-file-oriented programming, you must observe certain rules and use keywords intended for the compiler. Section 3.3.4, "Source-File-Oriented Block Programming", discusses in detail how to create an STL source file.

The optimum programming method for extensive programs is to create a source file. This method is particularly suited for the creation of (your own) reusable standard blocks with symbolic names of your choice. It is actually the only method which will give your program block protection (KNOW_HOW_PROTECT).

### Incremental programming

You use incremental programming to edit blocks in the *Blocks* user program. The editor checks your entries in incremental mode as soon as you have terminated a program line. When the block is closed it is immediately compiled, so that only error-free blocks can be saved.

The blocks can be edited both off-line in the programming device's database and on-line in the CPU, generally referred to as the "programmable controller", or "PLC". For this purpose, the SIMATIC Manager provides an off-line and an on-line window; the one is distinguished from the other by the labelling in the title bar.

In the off-line window, you edit the blocks right in the PG database. If you are in the editor, you can store a modified block in the off-line database with FILE → SAVE and transfer it to the CPU with PLC → DOWNLOAD.

To edit a block in the CPU, open that block in the on-line window. This transfers the block from the CPU to the programming device so that it can be edited. You can write the edited block back to the CPU with PLC → DOWNLOAD. If the CPU is in RUN mode, the CPU will process the edited block in the next program scan cycle. If you want to save a block that you edited on-line in the off-line database as well, you can do so with File → Save.

Incremental program input is discussed in detail in Section 3.3.3, "Incremental Block Programming".

## 2.5.3 Reference Data

As a supplement to the program itself, the SIMATIC Manager shows you the reference data, which you can use as the basis for corrections or tests. These reference data include the following:

- Cross-reference list
- Assignment list
- Program structure
- List of unused symbols
- List of addresses without symbols

To generate reference data, select the *Blocks* object and the menu command OPTIONS → REFERENCE DATA → DISPLAY. The representation of the reference data can be changed specifically for each work window with VIEW → FILTER...; you can save the settings for later editing by selecting the "Save as Standard" option. You can display and view several lists at the same time.

### Cross references

The cross-reference list shows the use of the operands and blocks in the user program. It includes the absolute address, the symbol (if any), the block in which the address was used, how it was used (read or write), and language-related information. For STL, the language-related information column shows the network in which the address was used. Click on a column header to sort the table by column contents.

The cross-reference list shows the addresses you selected with VIEW → FILTER... (for instance bit memory). When you double-click on an address, the editor opens the block displayed on that line at the location at which the address appears. STEP 7 then uses the filter saved as "Standard" every time it opens the cross-reference list.

*A tip:* The cross references show you whether the referenced addresses were also scanned or reset. They also show you in which blocks addresses are used (possibly more than once).

### Assignments

The I/Q/M reference list shows which bits in address areas I, Q and M are assigned in the program. One byte, broken down into bits, appears on each line. Also shown is whether access is by byte, word, or doubleword. The T/C reference list shows the timers and counters used in the program. Ten timers or counters are displayed on a line.

*A tip:* The list shows you whether certain address areas were (improperly) assigned or where there are still addresses available.

### Program structure

The program structure shows the call hierarchy of the blocks in a user program. The "starting block" for the call hierarchy is specified via filter settings. You have a choice between two different views:

The *tree structure* shows all nesting levels of the block calls. You control the display of nesting levels with the "+" and "–" boxes. The requirements for temporary local data are shown for the entire path following the starting block and/or per call path. Click the right mouse button to fade in a menu field in which you can open the block, switch to the call location, or screen additional block information.

The *parent/child structure* shows two call levels with one block call. Language-related information is also included.

*A tip:* Which blocks were used? Were all programmed blocks called? What are the blocks' temporary local data requirements? Is the specified local data requirement per priority class (per organization block) sufficient?

### Unused symbols

This list shows all operands which have symbol table allocations but were not used in the program. The list shows the symbol, the operand, the data type, and the comment from the symbol table.

*A tip:* Were the operands in the list inadvertently forgotten when the program was being written? Or are they perhaps superfluous, and not really needed?

### Addresses without symbols

This list shows all the operands used in the program to which no symbols were allocated. The list shows these operands and how often they were used.

*A tip:* Were operands used inadvertently (by accident, or because of a typing error)?

## 2.5.4 Rewiring

The Rewiring function allows you to replace addresses in individually compiled blocks or in the entire user program. For example, you can replace input bits I 0.0 to I 0.7 with input bits I 16.0 to I 16.7. You can replace addresses in the address areas for inputs, outputs, bit memory, timers, counters, FC functions and FB function blocks.

In the SIMATIC Manager, choose the objects in which you want to do rewiring select a single block, select a group of blocks by holding Ctrl/Strg and clicking with the mouse, or select the entire user program *(Blocks)*. Call the menu command OPTIONS → REWIRE... to screen a table in which you can enter the old and the new addresses. When you confirm with 'OK', the SIMATIC Manager exchanges the addresses. A subsequently displayed info file shows you in which block changes were made, and how many.

## 2.6 Debugging Programs

After setting up a connection to a CPU and loading the user program, you can debug the whole program or only specific program sections, for instance individual blocks. To debug the program, you supply the variables with signals and values, using, for example, simulator modules, and evaluate the information returned by your program. If the CPU goes to stop as the result of an error, functions such as CPU Information help you locate the problem.

With the PLCSIM option package, you can simulate a CPU in the programming device, thus making it possible for you to test your program without additional hardware.

### 2.6.1 Connecting a PLC

The connection between the PG's and CPU's MPI interface is the mechanical requirement for an on-line connection. The connection is unique when a CPU is the only programmable module connected. If there are several CPUs in the MPI subnet, each CPU must be assigned a unique node number (MPI address). You set the MPI address when you initialize the CPU. Before linking all the CPUs to one network, connect the PG on one CPU at a time and transfer the *System Data* object from the off-line user program *Blocks* or direct with the Hardware Configuration editor using the menu command PLC → DOWNLOAD. This assigns a CPU its own special MPI address ("naming") along with the other properties. If all you want to do is change the CPU's MPI address, you can also use the PLC → DISPLAY ACCESSIBLE NODES button.

The MPI address of a CPU in the MPI network can be changed at any time by transferring a new parameter data record containing the new MPI address to the CPU. Note carefully: The new MPI address takes effect immediately. The MPI parameters are retained in the CPU even after a memory reset. The CPU can thus be addressed even after a memory reset.

With suitably designed CPUs, access can be protected with a password (8 characters). You configure the protection level with the Hardware Configuration in the object properties of the CPU. Before accessing a protected CPU, you can enter the password with PLC → ACCESS RIGHTS or you can wait to be prompted when making an access.

A PG can always be operated on-line on a CPU, even with a module-independent program and even though no project has been set up.

- If no project has been set up, you must establish the connection to the CPU with PLC → DISPLAY ACCESSIBLE NODES. Using the menu command, you screen a project window with the structure "*Accessible Nodes*" – "Module (MPI=n)" – "Online User Program *(Blocks)*". When you select the *Module* object, you may utilize the on-line functions, such as changing the operational status and checking the module status. Selecting the *Blocks* object displays the blocks in the CPU's user memory. You can then edit (modify, delete, insert) individual blocks.

- If the project window shows a module-independent program, generate the associated on-line project window. This window contains all on-line objects for this project. Selecting the S7 program in the on-line window makes all on-line functions for the connected CPU available to you. *Blocks* displays the blocks in the CPU's user memory. If the blocks in the off-line program and those in the on-line program are identical, you can edit the blocks in user memory with the information from the PG database (symbolic address, comments).

- When you switch a module-dependent program into on-line mode, you can carry out program modifications just as you would in a module-independent program. In addition, it is now possible for you to configure the SIMATIC station, that is, to set CPU parameters and address and parameterize modules.

### 2.6.2 CPU Information

In on-line mode, the CPU information listed below is available to you. The menu commands are screened when you have selected a module (in on-line mode and without a project) or S7 program (in the on-line project window).

- PLC → DIAGNOSE HARDWARE
Information on the status and operating mode of the accessible on-line modules in the form of the Hardware Configuration.

- PLC → MODULE INFORMATION
General information (such as version), diagnostic buffer, memory (current map of work memory and load memory, compression), cycle time (length of the last, longest, and shortest program cycle), timing system (properties of the CPU clock, clock synchronization, run-time meter), performance data (memory configuration, sizes of the address areas, number of available blocks, SFCs, and SFBs), communication (baud rate and communication links), stacks in STOP state (B stack, I stack, and L stack).

- PLC → OPERATING MODE
Display of the current operating mode (for instance RUN or STOP), modification of the operating mode

- PLC → CLEAR/RESET
Resetting of the CPU in STOP mode

- PLC → SET TIME AND DATE
Setting of the internal CPU clock

- PLC → CPU MESSAGES
The detection of problems in the connection to the CPU, asynchronous errors which set the CPU to STOP, and user-defined messages generated with SFC 52 WR_USMSG, SFC 18 ALARM_S and SFC 19 ALARM_SG. By activating the "Archive" button, a window appears in which you can set the number of stored messages or delete all of them.

- PLC → MONITOR/MODIFY VARIABLES
(Refer to Section 2.6.6, "Monitoring, Modifying and Forcing Variables")

### 2.6.3 Loading the User Program into the CPU

When you transfer your user program (compiled blocks and configuration data) to the CPU, it is loaded into the CPU's load memory. Physically, load memory can be RAM or Flash EPROM and either integrated in the CPU or on a memory card.

If the memory card is a Flash EPROM, you can write to it in the programming device and use it as data medium. With the card off circuit, insert it into the CPU; when it is powered up, the relevant data are transferred from the memory card to the CPU's work memory. As regards the CPU 315 in conjunction with a 5V Flash EPROM card and as regards all S7-400 CPUs, you can also write to a Flash EPROM card while it is inserted in the CPU, but only the full program.

When load memory is RAM storage, you can not only transfer the full program on-line, but you an also modify, delete or reload individual blocks. A power-down without battery backup or a memory reset erases the program in RAM.

You transfer a complete user program by switching the CPU to STOP, executing a memory reset, and transferring the complete user program. The configuration data are also transferred. If a Flash EPROM card is in the CPU when the memory reset is executed, the CPU copies its contents to work memory following the reset.

You can modify or reload individual blocks at runtime when the mode selector on the front of the CPU is in the RUN-P position. When the mode selector is at RUN, you can only read the user program. Now, however, you can remove the key, thus preventing any changes to the program.

If you only want to change the configuration data (CPU properties, the configured connections, GD communication, module parameters, and so on), you need only load the *System Data* object into the CPU (select the object and transfer it with menu command PLC → DOWNLOAD. The parameters for the CPU go into effect immediately; the CPU transfers the parameters for the remaining modules to those modules during startup. Note that the *complete* configuration is always loaded into the PLC.

### 2.6.4 Block Handling

Extensive programs are debugged section by section. If, for instance, you only want to debug one block, load that block into the CPU, then call it in OB 1. If OB 1 is arranged so that the program can be debugged section by section "from front to back", you can select the blocks or program sections you want to test by jumping over the calls or program sections that are not to be executed, for example with the jump function JMP.

Special caution is advised when transferring individual blocks while the CPU is in RUN-P mode. If one block calls other blocks (which are not in the CPU's memory), you must first load the "'subordinate" blocks. This also applies to data blocks. You must load the "uppermost" block last. That block is then processed immediately in the next program cycle.

Just as in the off-line user program, you can execute blocks incrementally in the on-line user program (in the CPU). But when the on-line and the off-line data management diverge, it is possible that the editor can no longer display the additional information from the off-line data manager, and this information might be lost.

The compiled blocks in the user program *Blocks* are transferred with the SIMATIC Manager. From the off-line project window, write the selected blocks to the CPU with PLC → DOWNLOAD; in the on-line project window, read out the selected blocks from the CPU with PLC → UPLOAD.

With the editor, you can edit blocks both on-line and off-line. Save the edited block in the PG off-line with FILE → SAVE; write it to the CPU with PLC → DOWNLOAD. If you want to save the open block under another number or transfer it to another project, a library, or another CPU, use the menu command FILE → SAVE AS.

### Compressing

When you load a new or modified block into the CPU, the CPU places the block in load memory and transfers the relevant data to work memory. If there is already a block with the same number, this "old block" is declared invalid (following a prompt for confirmation) and the new block "added on at the end" in memory. Even a deleted block is "only" declared invalid, not actually removed from memory. This results in gaps in user memory which further and further reduce the amount of memory still available. These gaps can be filled only by the *Compress* function. When you compress in RUN mode, the blocks currently being executed are not relocated; only in STOP mode can you truly achieve compression without gaps. The current memory allocation can be displayed in percent with the menu command PLC → MODULE INFORMATION, *Memory* tab. The dialog box which then appears also has a button for preventive compression.

### Data blocks off-line/on-line

The data operands in a data block can be assigned an *initial value* and an *actual value*. When the data block is first created, these two values are normally the same. When a data block is loaded into the CPU, the initial value goes to load memory and the actual value to work memory. Every value change in a data operand during the program scan is equivalent to a change in the actual value.

When you load a data block from the CPU, its values are taken from work memory, for work memory is where all actual values are to be found. You can view the actual values at the time they are read out with VIEW → DATA VIEW. If you modify an actual value in the data block and write the block back to the CPU, the modified value is placed in work memory.

When a Flash EPROM memory card is used as load memory, the blocks on the memory card are transferred to work memory following a CPU memory reset. The data block retains the initial values originally programmed for them. The same applies on power-up without battery backup. On the S7-300, you can prevent this from happening to a data area by declaring that area retentive.

## 2.6.5 Determining the Cause of a STOP

If the CPU goes to STOP because of an error, the first measure to take in order to determine the reason for the STOP is to output the diagnostic buffer. The CPU enters all messages in the diagnostic buffer, including the reason for a STOP and the errors which led to it. To output the diagnostic buffer, switch the PG to on-line, select an S7 program, and choose register card *Diagnostic Buffer* with the menu command PLC → MODULE INFORMATION. The last event (the one with the number 1) is the cause of the STOP, for instance "STOP because programming error OB not loaded". The error which led to the STOP is described in the preceding message, for example "FC not loaded". By clicking on the message number, you can screen an additional comment in the next lower display field. If the message relates to a programming error in a block, you can open and edit that block with the "Open Block" button.

If the cause of the STOP is, for example, a programming error, you can ascertain the surrounding circumstances with the *Stacks* tab. When you open *Stacks*, you will see the B stack (block stack), which shows you the call path of all non-terminated blocks up to the block containing the interrupt point. Use the "I stack" button to screen the interrupt stack, which shows you the contents of the CPU registers (accumulators, address register, data block register, status word) at the interrupt point at the instant the error occurred. The L stack (local data stack) shows the block's temporary local data, which you select in the B stack by clicking with the mouse.

### 2.6.6 Monitoring, Modifying and Forcing Variables

One excellent resource for debugging user programs is the monitoring, modifying and forcing of variables with VAT variable tables. These features allow you to display the signal states or values of variables of elementary data types. In RUN-P mode, you can even modify the variables, that is, change the signal state or assign new values. On appropriately designed CPUs, you can specify fixed values for certain variables which the user program cannot change ("force"). *But you must exercise extreme caution; make sure that no dangerous states can result from modifying or forcing variables!*

To monitor, modify or force variables, you must create a VAT variable table containing the variables and the associated data formats. You can generate up to 255 variable tables (VAT 1 to VAT 255) and assign them names in the symbol table (Figure 2.6). You can generate a VAT off-line by selecting the user program *Blocks* with INSERT → S7 BLOCK → VARIABLE TABLE, and you can generate an unnamed VAT on-line by selecting *S7 Program* with PLC → MONITOR/MODIFY VARIABLES.

- VAT entries

You can specify the variables with either absolute or symbolic addresses and choose the data type (display format) with which a variable is to be displayed and modified (with VIEW → MONITOR FORMAT or by clicking the right mouse button directly on the monitor format). Use comment lines to give specific sections of the table a header. You may also stipulate which columns are to be displayed. You can change variable or display format or add or delete lines at any time.

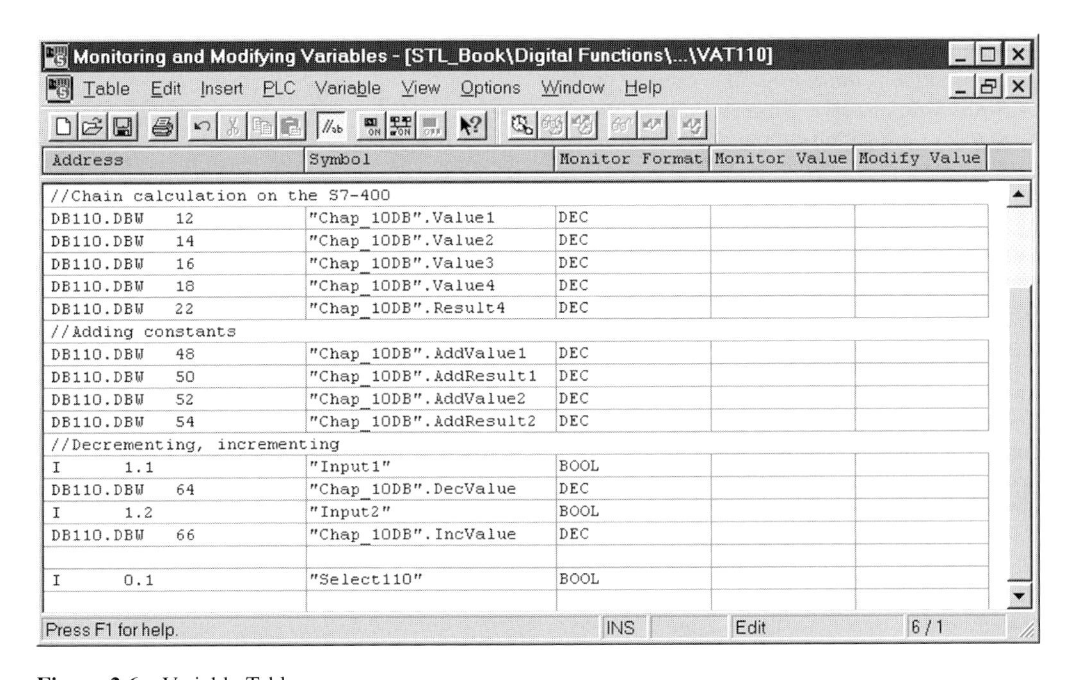


| Address | Symbol | Monitor Format | Monitor Value | Modify Value |
|---|---|---|---|---|
| //Chain calculation on the S7-400 | | | | |
| DB110.DBW 12 | "Chap_10DB".Value1 | DEC | | |
| DB110.DBW 14 | "Chap_10DB".Value2 | DEC | | |
| DB110.DBW 16 | "Chap_10DB".Value3 | DEC | | |
| DB110.DBW 18 | "Chap_10DB".Value4 | DEC | | |
| DB110.DBW 22 | "Chap_10DB".Result4 | DEC | | |
| //Adding constants | | | | |
| DB110.DBW 48 | "Chap_10DB".AddValue1 | DEC | | |
| DB110.DBW 50 | "Chap_10DB".AddResult1 | DEC | | |
| DB110.DBW 52 | "Chap_10DB".AddValue2 | DEC | | |
| DB110.DBW 54 | "Chap_10DB".AddResult2 | DEC | | |
| //Decrementing, incrementing | | | | |
| I 1.1 | "Input1" | BOOL | | |
| DB110.DBW 64 | "Chap_10DB".DecValue | DEC | | |
| I 1.2 | "Input2" | BOOL | | |
| DB110.DBW 66 | "Chap_10DB".IncValue | DEC | | |
| | | | | |
| I 0.1 | "Select110" | BOOL | | |
| | | | | |

**Figure 2.6** Variable Table

• Switch to on-line

To operate a variable table that was generated off-line, switch it to on-line (PLC → CONNECT TO). You must switch each VAT on-line, and you can also carry out a disconnect.

• Trigger conditions

Use VARIABLE → TRIGGER to set the trigger conditions in the variable table. Triggers are the instants at which the CPU reads the values out of or into system memory. Specify whether the read or write is to take place only once or periodically.

• Monitoring variables

Select the Monitor function with the menu command VARIABLE → MONITOR. The variables in the VAT are updated in accordance with the specified trigger conditions. Permanent monitoring allows you to follow changes in the values on the screen. The values are displayed in the data format which you set in the Monitor Format column. VARIABLE → UPDATE MONITOR VALUES updates the monitor values once only and immediately without regard to the specified trigger conditions.

• Modifying variables

Use VARIABLE → MODIFY to transfer the specified values to the CPU in dependence on the trigger conditions. Enter values only in the lines containing the variables you want to modify. You can expand the commentary for a value with '//' or with VARIABLE → MODIFY VALUE VALID; these values are not taken into account for modification. You must define the values in the data format which you set in the Monitor Format column. VARIABLE → ACTIVATE MODIFY VALUES transfers the modify values only once and immediately, without regard to the specified trigger conditions.

• Forcing variables

The menu command VARIABLE → FORCE starts a force job in the CPU. In preparation, set up a connection to a CPU with a variable table and open a window with the force values using the menu command VARIABLE → DISPLAY FORCE VALUES. Enter the respective addresses with the required force values and start the force job. The CPU fetches the force values and allows no changes to be made to them. *Note carefully: closing the variable table or interrupting the connection to the CPU does not stop forcing!* You can terminate forcing only with the menu command VARIABLE → STOP FORCING.

• Enabling peripheral outputs

VARIABLE → ENABLE PERIPHERAL OUTPUTS allows you to modify peripheral outputs while the CPU is in STOP mode. Enter the outputs to be modified (PQB, PQW, PQD) in a variable table and set up a connection to a CPU. The CPU must be in STOP mode. With VARIABLE → ACTIVATE MODIFY VALUES, transfer the modification values to the outputs. To cancel the Enable, select menu command VARIABLE → ENABLE PERIPHERAL OUTPUTS a second time or press the ESC key. Practical example for using this function: Wiring test in STOP mode and without a user program.

### 2.6.7 Program Status

The editor provides an additional debugging option for your program with the "Program status" function. When this function is used, the editor shows you line by line the contents of the registers that you have selected under EXTRAS → SETTINGS in the "STL" register card.

The block whose program you want to debug is in the CPU's user memory and is called and edited there. Open this block, for example by double-clicking on it in the SIMATIC Manager's on-line window. The editor is started and shows the program in the block.

Select the network you want to debug. Activate the program status with DEBUG → MONITOR. You will now see the address statuses, the result of logic operation and the register contents.

The trigger conditions are set with DEBUG → CALL ENVIRONMENT. You require this parameter when the block to be debugged is called more than once in your program. You can initiate recording of the status information either by specifying the call sequence or in dependence on the open data block. If a block is called only once, select "no condition".

The recording of the program status information requires additional execution time in the program cycle. For this reason, do not select a cycle time that is barely sufficient. You determine the permissible increase in cycle time when you set the option 'Process Operation' in the CPU properties with the Hardware Configuration ('Protection' tab on CPUs of the relevant design). The operating mode set is displayed with DEBUG → OPERATION.

If the block was written in the STL language, some CPUs allow you to debug the program statement by statement in *single-step mode*. The CPU is in HOLD mode; for reasons of safety, the peripheral outputs are disabled. Using breakpoints, you can stop the program at any location and test it step by step.

# 3 The STL Programming Language

This chapter discusses the elements of the STL programming language, encompassing everything from the different priority classes (program processing methods) through the components of a STL program (blocks and program elements) to variables (data types).

You structure your user program in the planning phase by adapting it to technological and functional conditions and circumstances; the program's structure is the determining factor in program development, debugging and startup. To ensure effective programming, it is therefore necessary to pay particular attention to the program structure.

## 3.1 Program Processing

The overall program for a CPU consists of the operating system and the user program.

The *operating system* is the totality of all instructions and declarations which control the system resources and the processes using these resources, and includes such things as data backup in the event of a power failure, the activation of priority classes, and so on. The operating system is a component of the CPU to which you, as user, have no write access. However, you can reload the operating system from a memory card, for instance in the event of a program update.

The *user program* is the totality of all instructions and declarations (in this case program elements) for signal processing, through which a plant (process) is affected in accordance with the defined control task.

### 3.1.1 Program Processing Methods

The user program may be composed of program sections which the CPU processes in dependence on certain events. Such an event might be the start of the automation system, an interrupt, or detection of a program error (Figure 3.1). The programs allocated to the events are divided into *priority classes,* which determine the program processing order (mutual interruptibility) when several events occur.

The lowest-priority program is the *main program,* which is processed cyclically by the CPU. All other events can interrupt the main program at any location, the CPU then executes the associated interrupt service routine or error handling routine and returns to the main program.

A specific *organization block (OB)* is allocated to each event. The organization blocks represent the priority classes in the user program. When an event occurs, the CPU invokes the assigned organization block. An organization block is a part of a user program which you yourself may write.

Before the CPU begins processing the main program, it executes a start-up routine. This routine can be triggered by switching on the mains power, by actuating the mode switch on the CPU's front panel, or via the programming device. Program processing following execution of the start-up routine always starts at the beginning of the main program in S7-300 systems (complete restart); in S7-400 systems, it is also possible to resume the program scan at the point at which it was interrupted (warm restart).

The main program is in organization block OB 1, which the CPU always processes. The start of the user program is identical to the first network in OB 1. After OB 1 has been processed (end of program), the CPU returns to the operating system and, after calling for the execution of various operating system functions, such as the updating of the process images, it once again calls OB 1.

Events which can intervene in the program are interrupts and errors. Interrupts can come from the process (hardware interrupts) or from the

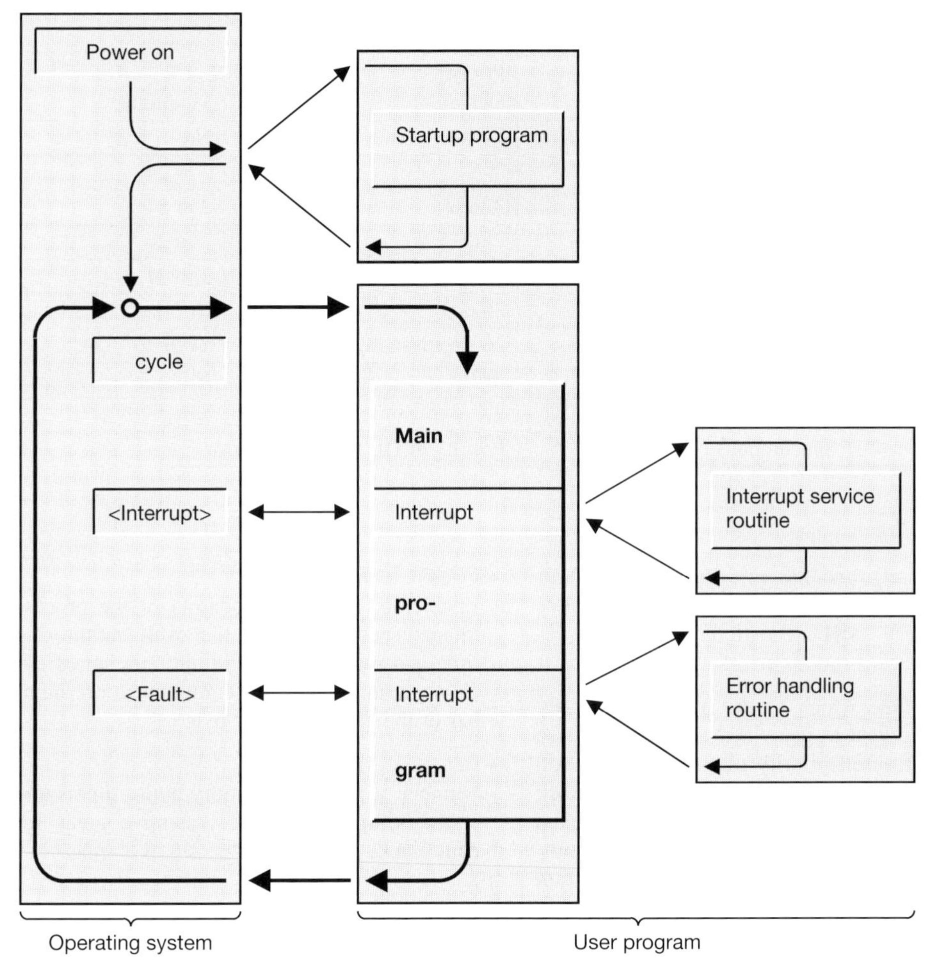


**Figure 3.1** Methods of Processing the User Program

CPU (watchdog interrupts, time-of-day interrupts, etc.). As far as errors are concerned, a distinction is made between synchronous and asynchronous errors. An asynchronous error is an error which is independent of the program scan, for example failure of the power to an expansion unit or an interrupt that was generated because a module was being replaced. A synchronous error is an error caused by program processing, such as referencing a non-existent address or a data type conversion error. The type and number of recorded events and the associated organization blocks are CPU-specific; not every CPU can handle all possible STEP 7 events.

### 3.1.2 Priority Classes

Table 3.1 lists the available SIMATIC S7 organization blocks, each with its priority. In some priority classes, you can change the assigned priority when you parameterize the CPU. The Table shows the lowest and highest possible priority classes; each CPU has a different low/high range.

Organization block OB 90 (background processing) executes alternately with organization block OB 1, and can, like OB 1, be interrupted by all other interrupts and errors.

The start-up routine may be in organization block OB 100 (complete restart) and OB 101

**Table 3.1** SIMATIC S7 Organization Blocks

| Organization block OB | | Priority Default | Modifiable |
|---|---|---|---|
| 1 | Free cycle | 1 | No |
| 10 to 17 | TOD interrupts | 2 | 2 to 24 |
| 20 to 23 | Time-delay interrupts | 3 to 6 | 2 to 24 |
| 30 to 38 | Watchdog interrupts | 7 to 15 | 2 to 24 |
| 40 to 47 | Hardware interrupts | 16 to 23 | 2 to 24 |
| 60 | Multiprocessor interrupt | 25 | No |
| 80 to 87 | Asynchronous errors | 26 (in start: 28) | 24 to 26 |
| 90 | Background processing | 29[1)] | No |
| 100, 101 | Start-up routine | 27 | No |
| 121, 122 | Synchronous errors | Priority of OB that caused error | |

[1)] See text

(warm restart), and has priority 27. Asynchronous errors occurring in the start-up routine have priority class 28. Diagnostic interrupts are regarded as asynchronous errors.

You determine which of the available priority classes you want to use when you parameterize the CPU. Unused priority classes (organization blocks) must be assigned priority 0 or L stack length 0. The relevant organization blocks must be programmed for all priority classes used; otherwise the CPU will invoke OB 85 ("Program Processing Error") or go to STOP.

### 3.1.3 Specifications for Program Processing

The CPU's operating system normally uses default parameters. You can change these defaults when you parameterize the CPU (in the *Hardware* object) to customize the system to suit your particular requirements. You can change the parameters at any time.

Every CPU has its own specific number of parameter settings. The following list provides an overview of all STEP 7 parameters and their most important settings.

- Startup
  Specifies the type of start-up (complete restart, warm restart); monitoring of Ready signals or module parameterization; maximum amount of time which may elapse before a warm restart
- Cycle/Clock Memory
  Enable/disable cyclic updating of the process image; specification of the cycle monitoring time and minimum cycle time; amount of cycle time, in percent, for communication; number of the clock memory byte
- Retentive Memory
  Number of retentive memory bytes, timers and counters; specification of retentive areas for data blocks
- Protection
  Setting the protection level; defining a password
- Local Data
  Local data specifications for priority classes (organization blocks)
- Interrupts
  Specification of the priority for hardware interrupts, time-delay interrupts, asynchronous errors and (available soon) communication interrupts
- Time-of-Day Interrupts
  Specification of the priority, specification of the start time and periodicity
- Cyclic Interrupts
  Specification of the priority, the time cycle and the phase offset
- Diagnostics/Clock
  Indicate the cause of a STOP; type and interval for clock synchronization, correction factor

- Multicomputing
  Specification of the CPU number
- Integrated I/O
  Activation and parameterization of the integrated I/O

On start-up, the CPU puts the user parameters into effect in place of the defaults, and they remain in force until changed.

## 3.2 Blocks

You can subdivide your program into as many sections as you want to in order to make it easier to read and understand. The STL programming language supports this by providing the necessary functions. Each program section should be self-contained, and should have a technological or functional basis. These program sections are referred to as "Blocks". A block is a section of a user program which is defined by its function, structure or intended purpose.

### 3.2.1 Block Types

STL provides different types of blocks for different tasks:

- User blocks
  Blocks containing user program and user data
- System blocks
  Blocks containing system program and system data
- Standard blocks
  Turnkey, off-the-shelf blocks, such as drivers for FM and CP modules

#### User blocks

In extensive and complex programs, "structuring" (subdividing) of the program into blocks is recommended, and in part necessary. You may choose among different types of blocks, depending on your application:

- Organization blocks (OBs)
  These blocks serve as the interface between operating system and user program. The CPU's operating system calls the organization blocks when specific events occur, for example in the event of a hardware or time-of-day interrupt. The main program is in organization block OB 1. The other organization blocks have permanently assigned numbers based on the events they are called to handle.
- Function blocks (FBs)
  These blocks are parts of the program whose calls can be programmed via block parameters. They have a variable memory which is located in a data block. This data block is permanently allocated to the function block, or, to be more precise to the function block *call*. It is even possible to assign a different data block (with the same data structure but containing different values) to each function block call. Such a permanently assigned data block is called an instance data block, and the combination of function block call and instance data block is referred to as a call instance, or "instance" for short. Function blocks can also save their variables in the instance data block of the calling function block; this is referred to as a 'local instance'.
- Functions (FCs)
  Functions are used to program frequently recurring or complex automation functions. They can be parameterized, and return a value (called the function value) to the calling block. The function value is optional, in addition to the function value, functions may also have other output parameters. Functions do not store information, and have no assigned data block.
- Data blocks (DBs)
  These blocks contain your program's data. By programming the data blocks, you determine in which form the data will be saved (in which block, in what order, and in what data type). There are two ways of using data blocks: as global data blocks and as instance data blocks. A global data block is, so to speak, a "free" data block in the user program, and is not allocated to a code block. An instance data block, however, is assigned to a function block, and stores part of that function block's local data.

The number of blocks per block type and the length of the blocks is CPU-dependent. The number of organization blocks, and their block numbers, are fixed; they are assigned by the CPU's operating system. Within the specified range, you can assign the block numbers of the other block types yourself. You also have the option of assigning every block a name (a symbol) via the symbol table, then referencing each block by the name assigned to it.

### System blocks

System blocks are components of the operating system. They can contain programs (system functions (SFCs) or system function blocks (SFBs)) or data (system data blocks (SDBs)). System blocks make a number of important system functions accessible to you, such as manipulating the internal CPU clock, or various communications functions

You can call SFCs and SFBs, but you cannot modify them, nor can you program them yourself. The blocks themselves do not reserve space in user memory; only the block calls and the instance data blocks of the SFBs are in user memory.

SDBs contain information on such things as the configuration of the automation system or the parameterization of the modules. STEP 7 itself generates and manages these blocks. You, however, determine their contents, for instance when you configure the stations. As a rule, SDBs are located in load memory. You cannot access them from your user program.

### Standard blocks

In addition to the functions and function blocks you create yourself, turnkey blocks (called "standard blocks") are also available. They can either be obtained on a storage medium or are on libraries delivered as part of the STEP 7 package (for example IEC functions, or functions for the S5/S7 converter).

Chapter 28, "Standard Libraries", includes an overview of the standard blocks supplied with STEP 7 V4.

## 3.2.2 Block Structure

Essentially, code blocks consist of three parts (Figure 3.2):

- The block header, which contains the block properties, such as the block name
- The declaration section, in which the block-local variables are declared, that is, defined
- The program section, which contains the program and program commentary

A data block is similarly structured:

- The block header contains the block properties
- The declaration section contains the definitions of the block-local variables, in this case the data addresses with data type specification
- The initialization section, in which initial values can be specified for individual data addresses

In incremental programming, the declaration section and the initialization section are combined. You define the data addresses and their data types in the "declaration view", and you can initialize each data address individually in the "data view" (see below).

## 3.2.3 Block Properties

The block properties, or attributes, are contained in the block header. You can view and modify the block attributes from the Editor with the menu command FILE → PROPERTIES (Figure 3.3). Input of the block properties in a source text file is described in detail in Section 3.3.4, "Source-File-Oriented Block Programming".

The *Name* of a block is used to identify that block; it is not the same as the symbolic address. Different blocks may have the same name. A block *Family* allows you to assign a group of block a common characteristic. A block's name and family are displayed in the comment field when you select that block in the dialog box of the program elements catalog. Under *Author* you enter the originator of the block. Name, family and author may each comprise up to 8 characters, the first of which must be a letter. Letters, digits, and the underline character are permitted. The *Version* is entered twice as a two-digit number from 0 to 15.

The Editor keeps track of a block's creation date or modification date in two time stamps, one for the program code and one for the interface, that is to say the block parameters and the static local data. Note that the modification date for the interface must be lower than (that is, must precede) the modification date of the program code in the calling block. If this is not the case, the Editor reports a "time stamp conflict" upon output of the calling block. To rectify the error, you must re-enter the block call.

Blocks can be created or compiled as version 1 or version 2 blocks. This serves a practical purpose only for the function blocks. If the "multiple in-

**Code block, incremental programming**

*Block header*

*Declaration*

| Address | Declaration | Name | Type |
|---|---|---|---|
| | | | |
| | | | |
| | | | |

Program

```
A Input1      //Limit switch responded
A Input2      //Manual operation
= Output1     //Message to operator panel
```

**Code block, source-oriented programming**

Block type Address
*Block header*

VAR_xxx

*name : Data type := Initialization;*
*name : Data type := Initialization;*
...
END_VAR

BEGIN

*Program*

END_Block Type

**Data block, incremental programming**

*Block header*

*Declaration*

| Address | Name | Type | Initial Value |
|---|---|---|---|
| | | | |
| | | | |
| | | | |
| | | | |
| | | | |
| | | | |
| | | | |
| | | | |
| | | | |

**Data block, source-oriented programming**

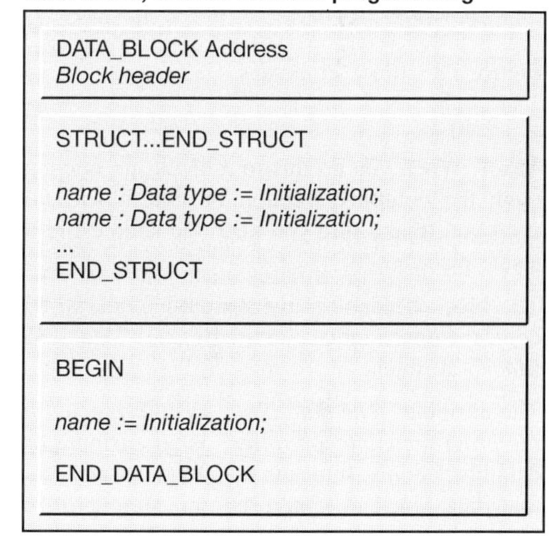


**Figure 3.2** Structure of a Block

stance capability" is enabled, which is normally the case, then the block is a version 2 block. If the "multiple instance capability" is disabled, you cannot call the block as local instance, nor can you call any other function block as local instance from within this block. The advantage of a version 1 block is the unrestricted use of instance data in conjunction with indirect addressing.

The upper portion of the tab "General – Part 2" shows the memory allocation for the block:

- Block: Allocation in load memory
- MC7 code: Allocation in work memory
- Local data: Allocation in the local data stack (temporary local data)

The *Know-How Protection* attribute is used for block protection. If a block is KNOW HOW-protected, the program in that block can not be viewed, printed out or modified. The Editor shows only the block header and the declaration table with the block parameters. When you input to a source file, you can protect every block yourself with the keyword KNOW_HOW_PRO-

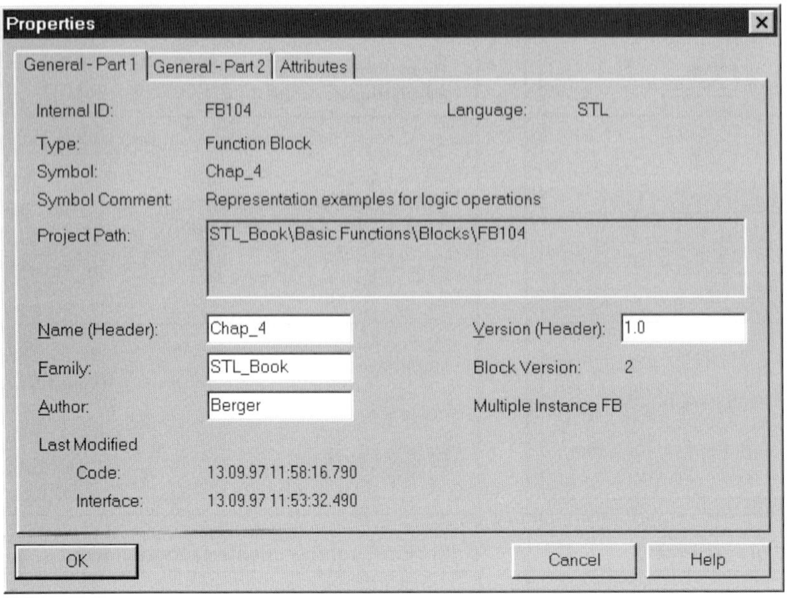

Properties

General - Part 1 | General - Part 2 | Attributes

Internal ID: FB104 Language: STL
Type: Function Block
Symbol: Chap_4
Symbol Comment: Representation examples for logic operations
Project Path: STL_Book\Basic Functions\Blocks\FB104

Name (Header): Chap_4 Version (Header): 1.0
Family: STL_Book Block Version: 2
Author: Berger Multiple Instance FB
Last Modified
Code: 13.09.97 11:58:16.790
Interface: 13.09.97 11:53:32.490

OK Cancel Help

**Figure 3.3** Block Properties

TECT. When you do this to a block, no one can view the compiled version of that block, not even you (make sure you keep the source file in a safe place!).

The block header of any standard block which comes from Siemens contains the *"Standard Block"* attribute.

The attribute *"The data block write-protected in the programmable controller"* is an attribute for data blocks only. It means that you can only read that data block in your program. Output of an error message prevents the overwriting of the data in that data block. This write protection feature must not be confused with block protection. A data block with block protection can be read out and written to in the user program, but its data can no longer be viewed with a programming or operator monitoring device.

A data block that has the *Unlinked* attribute is only in load memory; it is not "execution-relevant". You cannot write to data blocks in load memory, and you can read them only with system function SFC 20 BLKMOV.

Blocks may have system attributes. System attributes control and coordinate functions between applications, for example in the SIMATIC PCS7 control system.

## 3.3 Programming Blocks

You program blocks using the Editor, and either incrementally or on a source-file-oriented basis. It is recommended that you program in source files, for this is "true" symbolic programming. Only when you program in source files is block protection possible. Incremental programming is good for "quick" changes that allow you to try out corrections. Blocks can be modified both off-line in the programming device database and on-line, even when the CPU is in RUN mode.

### 3.3.1 Structure of an STL Statement

A program consists of a sequence of individual statements. A statement is the smallest autonomous unit of a user program. It represents a work specification for the CPU. Figure 3.4 shows the structure of an STL statement.

An STL statement consists of the following:

- A label (optional) comprising up to 4 characters and ending with a colon (see the section entitled "Jump Functions")
- An OP code describing what the CPU is to do (such as load, scan and AND, compare, etc.)

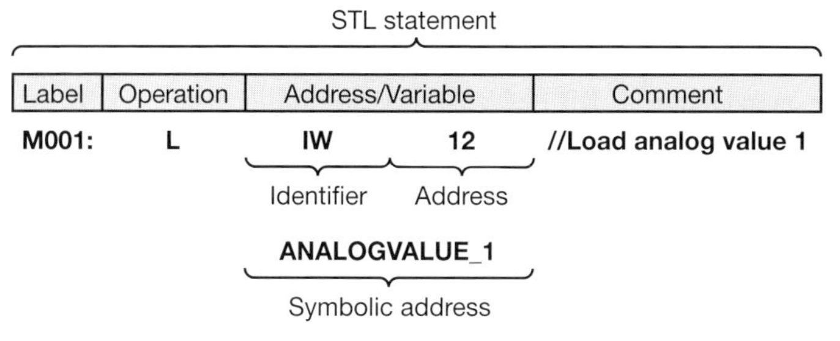


**Figure 3.4** Structure of an STL Statement

- An address providing the information needed to execute the operation (for instance an absolute address such as IW 12, the symbolic address of a variable such as ANALOGVALUE_1 or of a constant such as W#16#F001, and so on). Some operations require no address specification.
- A comment (optional), which must begin with two slashes and may extend up to the end of the line.

When inputting to a source file, you must terminate each statement (before beginning a comment, if any) with a semicolon. An STL line may contain no more than 200 characters, a comment no more than 160 characters.

### 3.3.2 Addressing Variables

When addressing variables, you may choose between absolute addressing and symbolic addressing. Absolute addressing uses numerical addresses beginning with zero for each address area. Symbolic addressing uses alphanumeric names which you yourself define in the symbol table for global addresses or in the declaration section for block-local addresses. An extension of absolute addressing is indirect addressing, in which the addresses of the memory locations are not computed until runtime.

#### Absolute addressing of variables

Variables of elementary data type can be referenced by absolute addresses.

The absolute address of an input or output is computed from the module start address which you set or had set in the configuration table and the type of signal connection on the module. A distinction is made between binary signals and analog signals.

- Binary signals
  A binary signal contains one bit of information. Examples of binary signals are the input signals from limit switches, momentary-contact switches and the like which lead to digital input modules, and output signals which control lamps, contactors, and the like via digital output modules.
- Analog signals
  An analog signal contains 16 bits of information. An analog signal corresponds to a "channel", which is mapped in the controller as word (2 bytes) (see below). Analog input signals (such as voltages from resistance thermometers) are carried to analog input modules, digitized, and made available to the controller as 16 information bits. Conversely, 16 bits of information can control an indicator via an analog output module, where the information is converted into an analog value (such as a current).

The information width of a signal also corresponds to the information width of the variable in which the signal is stored and processed. The information width and the interpretation of the information (for instance the positional weight), taken together, produce the *data type* of the variable. Binary signals are stored in variables of data type BOOL, analog signals in variables of data type INT.

The only determining factor for the addressing of variables is the information width. In STL, there are four widths which can be referenced with absolute addressing:

- 1 bit  Data type BOOL
- 8 bits  Data type BYTE or another data type with 8 bits

- 16 bits Data type WORD or another data type with 16 bits
- 32 bits Data type DWORD or another data type with 32 bits
- Variables of data type BOOL are referenced via an address identifier, a byte number, and – separated by a decimal point – a bit number. Numbering of the bytes begins at zero for each address area. The upper limit is CPU-specific. The bits are numbered from 0 to 7. Examples:

  I 1.0 Input bit no. 0 in byte no. 1
  Q 16.4 Output bit no. 4 in byte no. 16

- Variables of data type BYTE have as absolute address the address identifier and the number of the byte containing the variable. The address identifier is supplemented by a B. Examples:

  IB 2 Input byte no. 2
  QB 18 Output byte no. 18

- Variables of data type WORD consist of two bytes (a word). They have as absolute address the address identifier and the number of the low-order byte of the word containing the variable (Figure 3.5). The address identifier is supplemented by a W. Examples:

  IW 4 Input word no. 4; contains bytes 4 and 5
  QW 20 Output word no. 20; contains bytes 20 and 21

- Variables of data type DWORD consist of four bytes (a doubleword). They have as absolute address the address identifier and the number of the low-order byte of the word containing the variable. The address identifier is supplemented by a D. Examples:

  ID 8 Input doubleword no. 8; contains bytes 8, 9, 10 and 11
  QD 24 Output doubleword no. 24; contains bytes no. 24, 25, 26 and 27

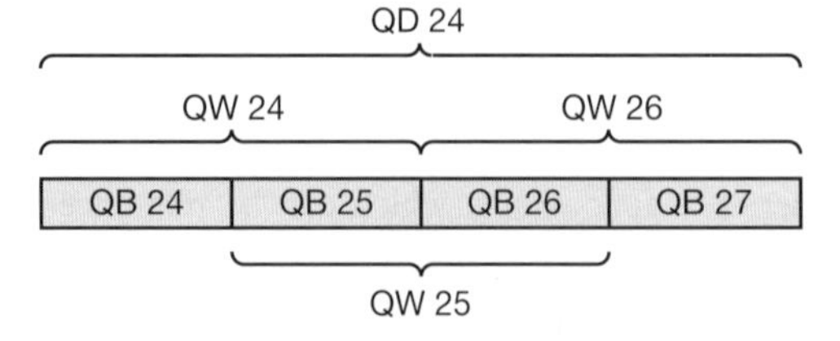


**Figure 3.5** Byte Contents in Words and Doublewords

Addresses for the data area include the data block. Examples:

DB 10.DBX 2.0 Data bit 2.0 in data block DB 10

DB 11.DBB 14 Data byte 14 in data block DB 11

DB 20.DBW 20 Data word 20 in data block DB 20

DB 22.DBD 10 Data doubleword 10 in data block DB 22

Additional information on addressing the data area can be found in section 18.2.2, "Accessing the Data Area".

### Indirect addressing

Indirect addressing allows you to wait until run-time to compute an address in the data area. A distinction is made between memory-indirect and register-indirect addressing:

- Memory-indirect addressing
  IW [MD 200]
  The address is in the memory doubleword
- Register-indirect area-internal addressing
  IW [AR1, P#2.0]
  The address is in address register AR1, and is incremented by the offset P.0 when the statement is executed
- Register-indirect area-crossing addressing
  W [AR1, P#0.0]
  The address area and the address itself are in address register AR 1

Doublewords from the address areas for data (DBD and DID), bit memory (MD) and temporary local data (LD) are available for saving addresses when using memory-indirect addressing. You can implement register-indirect addressing with two address registers (AR 1 and AR 2).

Indirect addressing is described in detail in the chapter of the same name.

### Symbolic addressing of variables

Symbolic addressing uses a name (called a symbol) in place of an absolute address. You yourself choose this name. Such a name must begin with a letter and may comprise up to 24 characters. The use of keywords as symbols is not permitted.

The name, or symbol, must be allocated to an absolute address. A distinction is made between global and symbols that are local to a block.

### Global symbols

You may assign names in the symbol table to the following objects:

- Inputs, outputs, peripheral inputs and peripheral outputs
- Memory bits, timers and counters
- Data blocks and code blocks
- User data types
- Variable tables

A global symbol may also include spaces, special characters and country-specific characters such as umlauts. Exceptions to this rule are the characters $00_{hex}$, $FF_{hex}$ and the quotation mark ("). When using symbols containing special characters, you must put the symbols in quotation marks in the program. In compiled blocks, the STL Editor always shows global symbols in quotation marks.

You can use global symbols throughout the program; each such symbol must be unique within a program.

### Block-local symbols

The names for the local data are specified in the declaration section of the relevant block. These names may contain only letters, digits and the underline character.

Local symbols are valid only within a block. The same symbol (the same variable name) may be used in a different context in another block. The Editor shows local symbols with a leading "#". When the Editor cannot distinguish a ocal symbol from an address, you must precede the symbol with a "#" character during input.

Local symbols are available only in the PG database (in the off-line container *Blocks).* If this information is missing on decompilation, the Editor inserts a substitute symbol.

### Using symbol names

If you use symbolic names while programming with the incremental Editor, they must have already been allocated to absolute addresses. You also have the option of entering new symbolic names in the symbol table while programming with the incremental Editor. Once the new symbolic names have been entered, you can use them immediately while writing the rest of your program. If you are using a text file to input your program, you must make the absolute addresses available at compilation time.

The data width (data type) is no longer immediately apparent from the symbol name. For example, a variable named IDENT might be a binary or a digital variable.

In the case of arrays, the individual components are referenced via the array name and a subscript, for example MSERIES[1] for the first component. In structures, each subidentifier is separated from the preceding subidentifier by a decimal point, for instance FRAME.HEADER. CNUM. Components of user data types are addressed exactly like structures. For further details see Chapter 24, "Data Types".

### Data addresses

Symbolic addresses for data include the data block. Example: The data block with the symbolic address MVALUES contains the variables MVALUE1, MVALUE2 and MTIME. These variables can be addressed as follows:

```
"MVALUES".MVALUE1
"MVALUES".MVALUE2
"MVALUES".MTIME
```

You will find additional information on addressing data in Section 18.2.2, "Accessing Data Addresses".

## 3.3.3 Programming Blocks with the Incremental Editor

### Generating blocks

You begin block programming by opening a block either by double-clicking on the block in the SIMATIC Manager's project window or with FILE → OPEN in the Editor. If the block does not yet exist, you can generate it in the following ways:

- In the SIMATIC Manager by selecting the *Blocks* object in the left half of the project window and generating a new block with INERT → S7 BLOCK →" ... You will see a dialog box with the block header (number of the block,

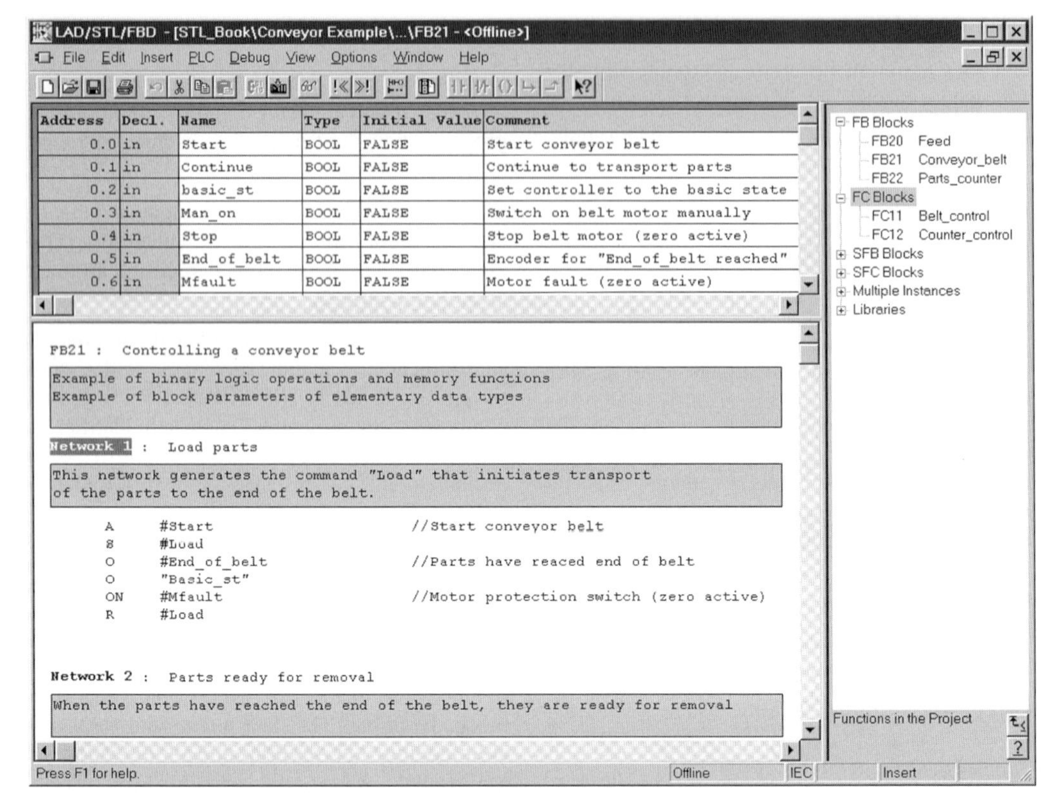


**Figure 3.6** Example of an Opened Block

language, block attributes). Choose the "STL" language. You can enter the remaining attributes later.

- In the Editor with menu command FILE → NEW, which displays a dialog box in which you enter the desired block under "Objekt Name". After closing the dialog box you can program the contents of the block.

You can enter the information for the block header when you generate the block or you can enter the block attributes later in the Editor by opening the block and selecting the menu command FILE → PROPERTIES.

### Block window

When a code block is opened, three windows are displayed Figure 3.6:

- The variable declaration table at the top
  It is here that you define the block-local variables
- The program window
  This is where you enter the program
- The program elements catalog
  In STL, this catalog contains the available blocks

### Variable declaration table

The variable declaration table is in the window above the program window. If it is not visible, position the mouse pointer to the upper line of demarcation for the program window, click on the left mouse button when the mouse pointer changes its form, and pull down. You will see the variable declaration table, which is where you define the block-local variables (see Table 3.2). Not every type of variable can be programmed in every kind of code block. If you do not use a variable type, the corresponding line remains empty.

The declaration for a variable consists of the name, the data type, a default value, if any, and a

**Table 3.2** Variable Types in the Declaration Section

| Variable Type | Declaration | Possible in Block Type | | |
|---|---|---|---|---|
| Input parameters | in | – | FC | FB |
| Output parameters | out | – | FC | FB |
| In-out parameters | in_out | – | FC | FB |
| Static local data | stat | – | – | FB |
| Temporary local data | temp | OB | FC | FB |

variable comment (optional). Not all variables can be assigned a default value (for instance, it is not possible for temporary local data). The default values for functions and function blocks are described in detail in Chapter 19, "Block Parameters".

The order of the declarations in code blocks is fixed (as shown in the table above), while the order within a variable type is arbitrary. You can save room in memory by bundling binary variables into blocks of 8 or 16 and BYTE variables into pairs. The Editor stores a (new) BOOL or BYTE variable at a byte boundary and a variable of another data type at a word boundary (beginning at a byte with an even address).

### Program window

In the program window, you will see – depending on the Editor's default settings – the fields for the block title and the block comment and, if it is the first network, the fields for the network title, the network comment, and the field for the program entry. In the program section of a code block, you control the display of comments and symbols with the menu commands VIEW → COMMENT, VIEW → SYMBOLIC REPRESENTATION and VIEW → SYMBOL INFORMATION. You can change the size of the display with VIEW → ZOOM IN, VIEW → ZOOM OUT and VIEW → ZOOM FACTOR.

You can subdivide an STL program into networks. The Editor numbers the networks automatically, beginning with 1. You may give each network a *network title* and a *network comment*. During editing, you can select each network directly with the menu command EDIT →GO TO → ... A subdivision into networks is optional.

To enter the program code, click once below the window for the network comment. You will see a framed empty window. You can begin entering your program anywhere within this window. Refer to section 3.3.1, "Structure of an STL Statement", to review the structure of an STL statement. Separate the OP code and the operand from one another by one or more spaces or tabs. Following the operand, you can enter two slashes and a statement comment. Terminate a statement by pressing RETURN. You can also enter a line comment by beginning a line with two slashes.

Program a new network with INSERT → NETWORK. The Editor then inserts an empty network behind the currently selected network.

You need not terminate a block with a special statement, simply terminate block input. However, you can program a last (empty) network with the title "Block End", providing an easily seen visual end of the block (an advantage, particularly in the case of exceptionally long blocks).

When the Editor opens a compiled block, it "decompiles" it back into STL. To do this, the Editor uses the program sections in the programming device database which are not relevant to the program's execution in order, for example, to represent symbols, comments and jump labels. If information needed from the programming device database is missing when the Editor decompiles the program, it uses substitute symbols.

You can create new blocks or open and edit existing ones in the Editor without having to return to the SIMATIC Manager.

### Program elements catalog

The program elements catalog is at the right of the program window, and is visible as a narrow strip only. You can enlarge the catalog by positioning the mouse pointer to the right border of the program window, pressing the left mouse button when the mouse pointer changes its form, and pulling toward the left. If the program elements catalog is not visible, screen it with VIEW → CATALOG.

The program elements catalog supports programming in LAD and FBD by providing the available graphic elements. In the STL language, it shows only the blocks that are already in the off-line container *Blocks*, as well as the *multi-instances* already programmed and the available *libraries*.

**Programming Data Blocks**

You have three different options for programming a data block:

- As global data block, in which case you would declare the data addresses when you program the data block
- As instance data block, in which case the data structure which you declared when you programmed the corresponding function block will be used
- As data block with user data type, in which case you must declare the data structure as user data type UDT

When programming a global data block, you can assign an *initial* value for each data address. Depending on their data type, variables normally default to either zero, the lowest possible value, or blank. An instance data block generated from a function block uses as initial value the one defined in the declaration part of the function block. If you generate a data block from a user data type (UDT), the initialization values (default values) from the UDT are taken as the initial values in the data block.

The Editor displays a data block in two views. In the *declaration* view, (VIEW → DECLARATION VIEW) you define the data addresses and you see the variables as you defined them, for example a field or user data type as a single variable. In the *data* view (VIEW → DATA VIEW), the Editor displays each variable and each component of a field or structure separately. You will now see an addition column titled *Current Value*. The current value is the value which a data address has or will have in the CPU's work memory. The standard is for the Editor to take the initial value as current value.

You can modify the current value individually for each data address. For example: You generate several instance data blocks from one function block, but want to have slightly different initial values for some of the instance data items for each function block call (for each FB/DB pair). You could them edit each data block with VIEW → DATA VIEW and enter the values you want for this data block in the Actual Value column. You would then select menu command EDIT → INITIALIZE DATA BLOCK so that the Editor will again replace all current values with the original initial values.

**Programming user data type UDT**

You create a user data type in the SIMATIC Manager by selecting an S7 program (*Blocks*) with INSERT → S7 BLOCK → DATA TYPE or in the Editor by choosing the menu command FILE → NEW and entering "*UDTn*" in the "Object Name" line.

Double-clicking on the UDT object in the program window opens a declaration table that looks just like the declaration table for a data block. A UDT is programmed exactly like a data block, each line with name, type, initial value and comment. The only difference is that it is not possible to switch to the data view (with UDT, you do not create variables, but only a collection of data types; for this reason, you cannot have any actual values).

The initial values which you program in a UDT are transferred to the variables upon declaration. Programming source file-oriented blocks

### 3.3.4 Programming Source-File-Oriented Blocks

An STL source file is a pure ASCII text file. It may contain the source program for one or more code or data blocks as well as the definitions of the user data types. You can save the source files at any time during editing. Not until a source file is compiled does the Editor generate blocks which it places in the off-line user program *Blocks*. If you used global symbols in the STL source file, the completed symbol table must be available at compilation time.

When using source-file-oriented programming, you must observe certain rules and use keywords intended for the compiler. The keywords in this

**Table 3.3** Keywords for Programming Code Blocks

| Block Type | Organization Block | Function Block | Function |
|---|---|---|---|
| Block type | ORGANIZATION_BLOCK | FUNCTION_BLOCK | FUNCTION : *Function value* |
| Header | TITLE = *Block title*<br>*//Block comment*<br><br>KNOW_HOW_PROTECT<br>NAME : *Block name*<br>FAMILY : *Block family*<br>AUTHOR : *Originator*<br>VERSION : *Version* | TITLE = *Block title*<br>*//Block comment*<br>CODE_VERSION1<br>KNOW_HOW_PROTECT<br>NAME : *Block name*<br>FAMILY : *Block family*<br>AUTHOR : *Originator*<br>VERSION : *Version* | TITLE = *Block title*<br>*//Block comment*<br><br>KNOW_HOW_PROTECT<br>NAME : *Block name*<br>FAMILY : *Block family*<br>AUTHOR : *Originator*<br>VERSION : *Version* |
| Declaration | | VAR_INPUT<br>*Input parameters*<br>END_VAR | VAR_INPUT<br>*Input parameters*<br>END_VAR |
| | | VAR_OUTPUT<br>*Output parameters*<br>END_VAR | VAR_OUTPUT<br>*Output parameters*<br>END_VAR |
| | | VAR_IN_OUT<br>*In-out parameters*<br>END_VAR | VAR_IN_OUT<br>*In-out parameters*<br>END_VAR |
| | | VAR<br>*Static local data*<br>END_VAR | |
| | VAR_TEMP<br>*Temporary local data*<br>END_VAR | VAR_TEMP<br>*Temporary local data*<br>END_VAR | VAR_TEMP<br>*Temporary local data*<br>END_VAR |
| Program | BEGIN<br>NETWORK<br>TITLE = *Network title*<br>*//Network comment*<br>... *STL statements*<br>*//Line comment* | BEGIN<br>NETWORK<br>TITLE = *Network title*<br>*//Network comment*<br>... *STL statements*<br>*//Line comment* | BEGIN<br>NETWORK<br>TITLE = *Network title*<br>*//Network comment*<br>... *STL statements*<br>*//Line comment* |
| | NETWORK<br>... etc. | NETWORK<br>... etc. | NETWORK<br>... etc. |
| Block end | END_ORGANIZATION_BLOCK | END_FUNCTION_BLOCK | END_FUNCTION |

chapter are all written in upper case for emphasis, but the Editor also understands them when they are written in lower case, that is to say, no distinction is made between upper and lower case.

### Block structure for source-file-oriented programming

When you create a source file for an STL program, you must conform to the structure or order shown in Table 3.3 when programming blocks.

When a block is called, the Editor needs the information in the block header so it knows which block parameters need to be initialized and the declaration type and data type of each block parameter. This means that you must program the called functions and function blocks beforehand or begin programming with the blocks on the "lowest level" (putting them at the beginning in the source text file).

However, it would also suffice for you to program only the block header with the parameter declaration (as "interface description", as it were). The block would then contain only the Block End statement BE as program. Later, you can actually program the block (you must, however, be careful not to change the interface to blocks already called! If you do, the Editor will report a time stamp conflict when it outputs the block call).

In the case of extensive user programs, you will surely subdivide the program source file into several "handy-sized" files, for example into "program standards", which you use throughout the entire program, individual process-related or function-related subprograms, and a "main program" containing, for instance, the organization blocks. When you create the various source files, you must – for the reasons relating to block calls discussed above – keep an eye on the order in which they are compiled. The following order is recommended for the creation of a source file:

- User data types UDT
- Global data blocks
- Functions and function blocks, beginning with the blocks on the "lowest" call level
- Instance data blocks (can also be placed immediately following the associated function block)
- Organization blocks

### Block header

You program the attributes of a block in its block header, after the block type and before the variable declaration. All information for the block header is optional, and you can leave out some or all of it. A detailed description of the block attributes, including contents, can be found in Section 3.2.3, "Block Properties".

The keyword "TITLE =" immediately following the line for the block type gives you the option of entering a block title comprising up to 64 characters. One or more comment lines, each beginning with a double-slash, may follow as block comment. The block comment may be up to 18 Kbytes in length.

### Variable declaration

The declaration section contains the definition of the block-local variables, that is, of the variables which you use only in that block. You cannot program every variable type in every block (see Table 3.3). If you do not use a variable type, omit the declaration, including keywords. The declaration for a variable consists of the name, the data type, a default value, if any, and a variable comment (optional). Example:

```
Quantity : INT := +500; //counted parts
```

Not all variables can be assigned a default value (for instance temporary local data cannot). The defaults for functions and function blocks are described in detail in Chapter 19, "Block Parameters".

The order of the declarations for code blocks is fixed (as shown in the table 3.3). The order within a variable type is arbitrary, and also determines, in conjunction with the data type, the amount of room required in memory; Chapter 24, "Data Types", shows you how you can optimize memory requirements by skillfully planning the order.

### Program section

The program section of a code block begins with the keyword BEGIN and ends with END_xxx, with block type ORGANIZATION_BLOCK, FUNCTION_BLOCK or FUNCTION taking the place of xxx. The keyword END_xxx replaces Block End BE.

In both keywords and program code, the Editor accepts upper and lower case. Details on statement syntax can be found in Section 3.3.1, "Structure of an STL Statement". The OP code must be separated from the address (operand) by one or more spaces or tabs. To improve the readability of the source text, you can leave one or more spaces and/or tabs between words. You must terminate each statement with a semicolon. After the semicolon you can write a statement comment, but it must begin with two slashes; it may extend up to the end of the line. You may also program several statements on one line, separating each from its predecessor by a semicolon.

You begin a line comment with two slashes at the beginning of the line. A line comment may comprise no more than 160 characters; it may contain no tabs and no non-printable characters.

For better readability and logic, you can divide the program in a block into *networks*. In the

**Table 3.4** Keywords for Programming Data Blocks

| Block Type | Global Data Block | Global Data Block from UDT | Instance Data Block |
|---|---|---|---|
| Block type | DATA_BLOCK | DATA_BLOCK | DATA_BLOCK |
| Header | TITLE = *Block title*<br>//*Block comment*<br>KNOW-HOW-PROTECT<br>NAME : *Block name*<br>FAMILY : *Block family*<br>AUTHOR : *Originator*<br>VERSION : *Version*<br>READ_ONLY<br>UNLINKED | TITLE = *Block title*<br>//*Block comment*<br>KNOW-HOW-PROTECT<br>NAME : *Block name*<br>FAMILY : *Block family*<br>AUTHOR : *Originator*<br>VERSION : *Version*<br>READ_ONLY<br>UNLINKED | TITLE = *Block title*<br>//*Block comment*<br>KNOW-HOW-PROTECT<br>NAME : *Block name*<br>FAMILY : *Block family*<br>AUTHOR : *Originator*<br>VERSION : *Version* |
| Declaration | STRUCT<br>*name* : *type* := *Default*<br>END_STRUCT | *UDTname* | *FBname* |
| Initialization | BEGIN<br>*name* := *Default;*<br>... etc. | BEGIN<br>*COMPname* := *Default;*<br>... etc. | BEGIN<br>*COMPname* := *Default;*<br>... etc. |
| Block end | END_DATA_BLOCK | END_DATA_BLOCK | END_DATA_BLOCK |

graphic languages, a subdivision into networks is necessary; in STL, it is not. Networks have no functional purpose; they are simply used in STL to divide the program into more logically related sections and to improve its readability, and to make it easier and more efficient to write comments. In very extensive programs, it is an advantage to be able to directly address the networks in the compiled block, thus reaching a particular program location quickly (with EDIT → GO TO → ... you can specify the network number or the line number relative to the beginning of the network). Networks begin with the keyword NETWORK; the keyword "TITLE =" in the next line allows you to give each network a header of up to 64 characters. The line comments immediately following the network title forms the *network comment,* which may be up to 18Kbytes long. STL numbers the networks automatically, beginning with 1. A complete network, including all comments, may not exceed 64 Kbytes.

#### Declaration in a data block

The declaration section (see table 3.4) contains the definition of the block-local variables, that is, of those variables which you use only in that block. You can declare a data block as global data block with "individual" variables, as global data block with UDT, and as instance data block.

The declaration of a variable in a global data block consists of name, data type, in some cases a default value, and a variable comment (optional). Example:

```
Quantity : INT := +500; //counted parts
```

All variables may be initialized. The variables may be listed in any order, but the order, in conjunction with the data type, determines the amount of memory required; the information in Chapter 24, "Data Types", shows you how clever selection of the order in which the variables are listed can help you optimize the memory requirements. If you choose not to initialize some variables, the Editor will pad them, depending on the data type, with zeroes or with the smallest possible value, or fills them with blanks. The declaration section of a data block derived from a UDT consists only of the UDT. You can use the absolute address (for example UDT 51) or the symbolic address (for instance FrameHeader).

The declaration section of an instance data block consists only of the specification of the associated function block in either absolute or symbolic form.

```
FUNCTION_BLOCK V_Memory
TITLE = Intermediate buffer for 4 values
//Example of a function block with block parameters and static local data
AUTHOR  : Berger
FAMILY  : STL_Book
NAME    : Memory
VERSION : 01.00
VAR_INPUT
  Transfer    : BOOL := FALSE;      //Transfer on positive edge
  Input value : REAL := 0.0;        //in data format REAL (fractional number)
END_VAR
VAR_OUTPUT
  Output value : REAL := 0.0;       //in data format REAL (fractional number)
END_VAR
VAR
  Value1 : REAL := 0.0;             //First stored value in data format REAL
  Value2 : REAL := 0.0;             //Second value
  Value3 : REAL := 0.0;             //Third value
  Value4 : REAL := 0.0;             //Fourth value
  Edge memory bit : BOOL := FALSE; //Edge memory bit for the transfer
END_VAR
BEGIN
NETWORK
TITLE = Program for transfer and output
//Transfer and output take place if there is a positive edge at Transfer
      A    Transfer;                //if transfer changes to "1"
      FP   Edge memory bit;         //the RLO is "1" following FP
      JCN  End;                     //Jump if no positive edge present
//Transfer of the values, starting with the last value
      L    Value4;
      T    OutputValue;             //Output of the last value
      L    Value3;
      T    Value4;
      L    Value2;
      T    Value3;
      L    Value1;
      T    Value2;
      L    InputValue;              //Transfer input value
      T    Value1;
End: BE;
END_FUNCTION_BLOCK

DATA_BLOCK Values1
TITLE = Instance data block for "V_Memory"
//Example of an instance data block
AUTHOR  : Berger
FAMILY  : STL_Book
NAME    : V_MEM_DB1
VERSION : 01.00
  V_Memory                          //Instance for the FB "V_Memory"
BEGIN
  Value1 := 1.0;                    //Individual pre-assignment
  Value2 := 1.3;                    //of selected values
END_DATA_BLOCK
```

**Figure 3.7** Example for the Programming of a Function Block and the Associated Instance Data Block

### Initialization in a data block

The initialization section starts with BEGIN and ends with END_DATA_BLOCK. Even if you do not initialize the variables in the initialization section, you must still specify these keywords.

The values which you enter in the initialization section of a data block correspond to the actual values specified in incremental programming. During compilation, the default values from the declaration section become the initial values and the initialization values become the actual values. When a data block is loaded into the CPU, the initial values are transferred to load memory and the actual values to work memory (also see "Data blocks on-line/off-line" in Section 2.6.4, "Block Handling"). If you do not specify an initialization value for a data operand, the Editor takes the initial value as actual value. If you use UDTs in the declaration which have been assigned default values, you can overwrite the defaults in the initialization section. The same applies for instance data blocks which have as data structure the associated function block (with its default). In the initialization section, you can define the actual values specifically for this instance (for the function block call with this data block).

### User data types (UDTs)

The source-file-oriented entry of UDTs is discussed in detail in Section 24.3, "User Data Types".

## 3.3.5 Example of a Function Block with Instance Data Block

Figure 3.7 shows an example for a function block with static local data, followed by the programmed instance data block associated with that function block. Chapter 18, "Block Functions", contains additional examples for programming blocks, as does the diskette which accompanies this book; in the latter case, the examples are in the source file "Block Programming" in the "Program Flow Control" program.

# 3.4 Variables and Constants

## 3.4.1 General Remarks Concerning Variables

A variable is a value with a specific format (Figure 3.8). Simple variables consist of an address (such as input 5.2) and a data type (such as BOOL for a binary value). The address, in turn, comprises an address identifier (such as I for input) and an absolute storage location (such as 5.2 for byte 5, bit 2). You can also reference an address or a variable symbolically by assigning the address a name (a symbol) in the symbol table.

A bit or the data type BOOL is referred to as a *binary address* or *binary operand*. Addresses comprising one, two or four bytes or variables with the relevant data types are called *digital operands*.

Variables which you declare within a block are referred to as (block-) local variables. These include the block parameters, the static and temporary local data, even the data addresses in global data block. When these variables are of an elementary data type, they can also be referenced as operands (for instance static local data as DI operands, temporary local data as L operands, and data in global data blocks as DB operands).

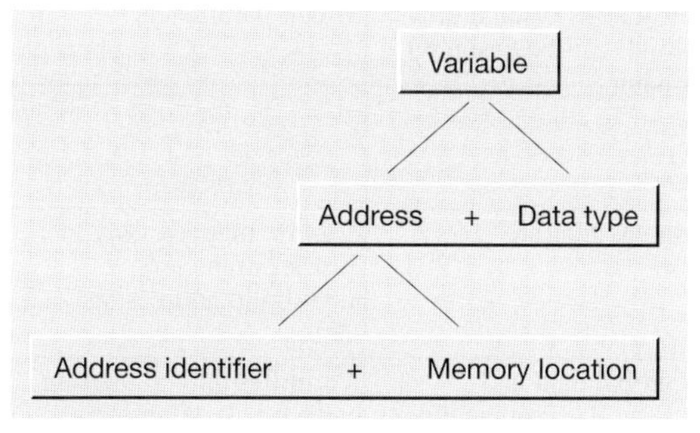


*A variable consists of the address and the data type. It is addressed symbolically.*

*An address can be absolute, e.g. I 5.2, or symbolic, e.g. Manual_on.*

**Figure 3.8** Structure of a Variable

Local variables, however, can also be of complex data type (such as structures or arrays). Variables with these data types require more than 32 bits, so that they can no longer, for example, be loaded into the accumulator. And for the same reason, they cannot be addressed with "normal" STL statements. There are special functions for handling these variables, such as the IEC functions, which are provided as a standard library with STEP 7 (you can generate variables of complex data type in block parameters of the same data type).

If variables of complex data type contain components of elementary data type, these components can be treated as though they were separate variables (for example, you can load a component of an array consisting of 30 INT values into the accumulator and further process it).

*Constants* are used to preset variables to a fixed value. The constant is given a specific prefix depending on the data type.

### 3.4.2 General Remarks Regarding Data Types

Data types stipulate the characteristics of data, essentially the representation of the contents of a variable, and the permissible ranges. STEP 7 provides predefined data types which you can combine into user data types. The data types are available on a global basis, and can be used in every block.

This section provides an overview of all data types and a brief introduction, particularly to the elementary data types. This knowledge will allow you to program a programmable logic controller. More depth of detail, such as the structure and format of variables of complex data type, are presented in Chapter 24, "Data Types", and information on data types in conjunction with block parameters is provided in Chapter 19, "Block Parameters".

Table 3.5 shows a rough overview of data types in STEP 7.

### 3.4.3 Elementary Data Types

Table 3.6 shows the elementary data types. Variables with one of these data types can be handled with the binary logic operations and Set/Reset functions (data type BOOL) or they can be directly processed with load and transfer statements. In STL, there is no restriction of operations (operators) to specific data types (the only exception being the distinction between binary operand and digital operand). Comparison functions, for example, compare the accumulator contents without regard to the data types of the variables in those accumulators.

**Table 3.5** Subdivision of the Data Types

| Elementary Data Types | Complex Data Types | User Data Types | Parameter Data Types |
|---|---|---|---|
| BOOL, BYTE, CHAR, WORD, INT, DATE, DWORD, DINT, REAL, S5TIME, TIME, TOD | DT, STRING ARRAY, STRUCT | UDT Global data blocks Instances | TIMER, COUNTER, BLOCK_DB, BLOCK_SDB, BLOCK_FC, BLOCK_FB, POINTER, ANY |
| Data types comprising no more than one doubleword (32 bits) | Data types that can comprise more than one doubleword (DT, STRING) or which consist of several components | Structures or data areas which can be assigned a name | Block parameters |
| Can be mapped to operands referenced with absolute and symbolic addressing | Can be mapped only to variables that are addressed symbolically | | Can be mapped only to block parameters (symbolic addressing only) |
| Permitted in all address areas | Permitted in data blocks (as global data and instance data), as temporary local data and as block parameters | | Permitted in conjunction with block parameters |

**Table 3.6** Overview of Elementary Data Types

| Data Type | Description | | Example for Constant Notation | |
|---|---|---|---|---|
| BOOL | Bit | 1 bit | TRUE, FALSE | |
| BYTE | Byte<br>8-bit hexadecimal number | 8 bits | <br>B#16#00<br>B#16#FF | <br>(Min. value)<br>(Max. value) |
| CHAR | One character (ASCII) | 8 bits | 'A' | |
| WORD | Word<br>16-bit hexadecimal number<br>16-bit binary number<br>Count value, 3 decades BCD<br>Two 8-bit unsigned decimal numbers | 16 bits | <br>W#16#0000<br>W#16#FFFF<br>2#0000_0000_0000_0000<br>2#1111_1111_1111_1111<br>C#000<br>C#999<br>B(0,0)<br>B(255,255) | <br>(Min. value)<br>(Max. value)<br>(Min. value)<br>(Max. value)<br>(Min. value)<br>(Max. value)<br>(Min. value)<br>(Max. value) |
| DWORD | Doubleword<br>32-bit hexadecimal number<br>32-bit binary number<br>Four 8-bit unsigned decimal numbers | 32 bits | <br>DW#16#0000_0000<br>DW#16#FFFF_FFFF<br>2#0000_0000_0000_0000_0000_0000_0000_0000<br>2#1111_1111_1111_1111_1111_1111_1111_1111<br>B(0,0,0,0)<br>B(255,255,255,255) | <br>(Min. value)<br>(Max. value)<br>(Min. value)<br>(Max. value)<br>(Min. value)<br>(Max. value) |
| INT | Fixed-point number | 16 bits | –32768<br>+32767 | (Min. value)<br>(Max. value) |
| DINT | Fixed-point number | 32 bits | –2 147 483 648<br>+2 147 483 647 | (Min. value)<br>(Max. value) |
| REAL | Floating-point number | 32 bits | +123.4567 as decimal number with decimal point or 1.234567E+02 in exponential representation (see text for value range) | |
| S5TIME | Time value in S5 format | 16 bits | S5T#0ms<br>S5TIME#2h46m30s | (Min. value)<br>(Max. value) |
| TIME | Time value in IEC format | 32 bits | T#-24d20h31m23s647ms<br>TIME#24d20h31m23s647ms | (Min. value)<br>(Max. value) |
| DATE | Date | 16 bits | D#1990-01-01<br>DATE#2168-12-31 | (Min. value)<br>(Max. value) |
| TIME_OF_DAY | Time of day | 32 bits | TOD#00:00:00<br>TIME_OF_DAY#23:59:59.999 | (Min. value)<br>(Max. value) |

### 3.4.4 Complex Data Types

You can use complex data types (Table 3.7) in conjunction with variables in data blocks or in the L stack or in conjunction with variables which are block parameters. Constant representation is defined for data type DT only.

A string can comprise up to 254 characters and reserves two bytes more in memory than the number of characters in the string. An array can have as many as 65 536 elements per dimension (from –32 768 to 32 767).

**Table 3.7** Overview of Complex Data Types

| Data Type | Description | | Example: |
|---|---|---|---|
| DATE_AND_TIME | Date and time | 64 bits | DT#1990-01-01-00:00:00.000<br>DATE_AND_TIME168-12-31:23:59:59.999 |
| STRING | String | Variable | Collection of ASCII characters, for instance 'String 1' |
| ARRAY | Array | Variable | Collection of components with the same data type, as many as 6 dimensions possible |
| STRUCT | Structure | Variable | Collection of components of arbitrary data type, up to 6 nesting levels possible |

### 3.4.5 User Data Types

A user data type (UDT) is equivalent to a structure (a collection of components of arbitrary data type), which is globally valid. You program a user data type almost like you would program a data block, then use this data type like you would a structure. You can also stipulate entire data blocks as user data types.

### 3.4.6 Parameter Types

The parameter types are data types for block parameters (Table 3.8). The length specifications in the Table refer to the memory requirement for block parameters in the case of function blocks. Also use TIMER and COUNTER in the symbol table as data types for timers and counters.

**Table 3.8** Overview of Parameter Types

| Parameter Type | Description | | Examples of Actual Operands | |
|---|---|---|---|---|
| TIMER | Timer | 16 bits | T 15 | |
| COUNTER | Counter | 16 bits | C 16 | |
| BLOCK_FC | Function | 16 bits | FC 17 | |
| BLOCK_FB | Function block | 16 bits | FB 18 | |
| BLOCK_DB | Data block | 16 bits | DB 19 | |
| BLOCK_SDB | System data block | 16 bits | SDB 100 | |
| POINTER | DB pointer | 48 bits | P#M10.0<br>P#DB20.DBX22.2<br>MW 20<br>I 1.0 | (Pointer)<br>(Pointer)<br>(Address)<br>(Address) |
| ANY | ANY pointer | 80 bits | P#DB10.DBX0.0 WORD 20<br>or an arbitrary variable | |

# Basic Functions

This part of the book describes the functions of the STL programming language that represent a certain "basic functionality". These functions allow you to program a PLC as you would contactor or relay controls.

- The **binary logic operations** are used to simulate series and parallel circuits in a circuit diagram or to implement the AND and OR functions in electronic switching systems. Nesting functions make it possible to implement even complex binary logic operations.
- The **memory functions** retain the result of a logic operation (RLO) so that it can, for example, be checked and further processed at another point in the program.
- The **transfer functions** are the prerequisite for the handling of digital values. These functions are also required, for instance, to inform a timer of the time value.
- The **timers** are to programmable controllers what timing relays are to contactor controls and timers are to electronic switching systems. The timers integrated in the CPU allow you to program such values as wait and monitoring times.
- The **counters** are up and down counters that can count in the range from 0 to 999.

This part of the book describes the functions using the address areas for inputs, outputs, and memory bits. Inputs and outputs are the link to the process or plant. Memory bits correspond to auxiliary contactors that store binary states. The subsequent parts of the book deal with the remaining address areas which can be used in binary logic operations. Most importantly, these include the data bits in the global data blocks and the temporary and static local data bits.

Chapter 5, entitled "Memory Functions", contains a programming example for the binary logic operations and memory functions; Chapter 8, entitled "Counters", provides an example of the use of timers and counters. In each case, the example is in an FC function without block parameters.

**4 Binary Logic Operation**
AND, OR and Exclusive OR; scanning for signal state "1" and "0"; processing a binary logic operation; nesting functions

**5 Memory Functions**
Assign, Set and Reset; RS flip-flop; edge evaluation; example of conveyor belt control

**6 Transfer Functions**
Load and Transfer; accumulator functions

**7 Timers**
Starting 5 different types of timers; resetting, enabling and scanning a timer; time value

**8 Counters**
Setting a counter; up and down counting; resetting, enabling and scanning a counter; count value, example for a parts counter

# 4 Binary Logic Operation

This Chapter discusses the AND, OR and Exclusive OR functions as well as combinations of these functions. AND, OR and Exclusive OR are used to check the signal states of binary locations and link them with one another.

A binary location can be checked (checked) for signal state "1" or signal state "0". By negating the result of the logic operation and using nesting expressions, you can also program complex binary logic operations without saving the intermediate result.

The examples shown in this chapter are also on the diskette accompanying this book under the "Basic Functions" program in function block FB 104 or source file Chap_4.

## 4.1 Processing a Binary Logic Operation

Figure 4.1 shows, in broad outline, how a binary logic instruction is processed. An input module selects a sensor on the basis of the specified address, for instance the sensor at input I 1.2. The CPU checks the signal state (status) of that sensor, and links the result of the check (check result) with the result of the logic operation (RLO) saved from the preceding logic instruction. The result of this logic operation is saved and stored as the new RLO. The CPU then processes the next statement in the program, for instance storing the result of the logic operation in a specific memory location. The first check to follow the storing of the now "old" RLO is a new logic operation, in which the RLO is set to the check result.

### Status

The status of a bit is the same as its signal state, and can be "0" or "1". In SIMATIC S7, the signal state is "1" when voltage is present at the input (for instance 230 V AC or 24 V DC, depending on the module); if no voltage is present, the signal state of the input is "0".

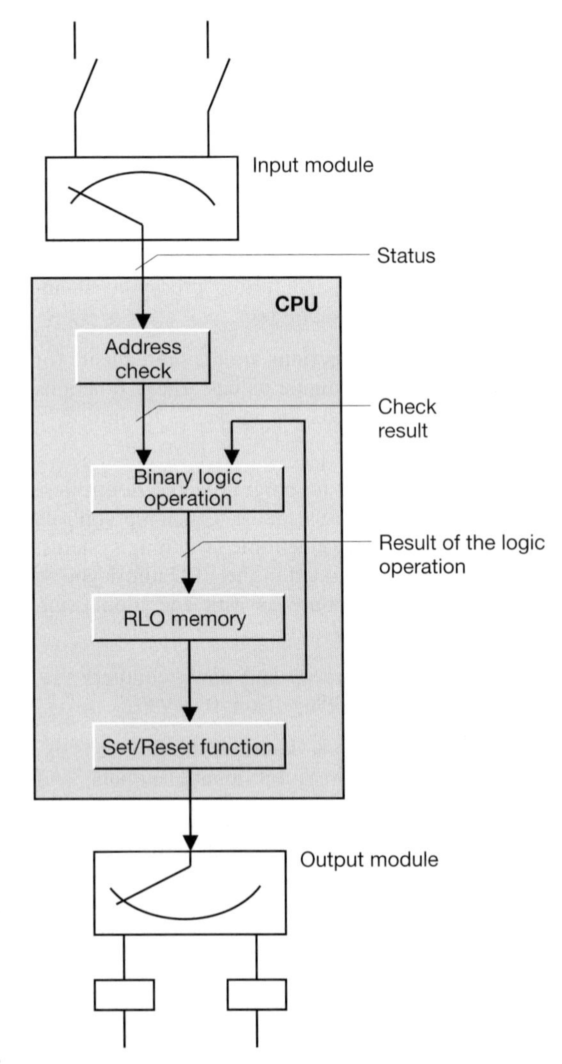


**Figure 4.1** How a PLC Works, Using as Example a Binary Logic Operation

A check statement queries the status of the bit. At the same time, it contains the rule of logic according to which the checked signal state is to be linked with the result of the logic operation stored in the processor. For example, the statement

```
A   I 17.1
```

checks input I 17.1 for signal state "1" and links the checked signal state according to AND; the statement

```
ON  M 20.5
```

checks memory bit M 20.5 for signal state "0" and links the checked signal state according to OR.

### Check result

Strictly speaking, the CPU does not link the signal state of the bit checked, but rather first forms a check result. In checks for signal state "1", the check result is identical to the signal state of the bit checked. In checks for signal state "0", the check result is the negated signal state of the bit checked.

### Result of the logic operation

The result of the logic operation (RLO) is the signal state in the CPU, which the CPU then uses for subsequent binary signal processing. The result of the logic operation is formed and modified by check statements. An RLO of "1" means that the condition of the binary logic operation was fulfilled; "0" means that the condition was not fulfilled. Bits are set or reset according to the result of the logic operation.

### Logic step

Just as one can define a sequence step in a sequential control system, it is also possible to define a logic step in a logic control system. In a logic step, a result of a logic operation is formed and evaluated (further processed). A logic step consists of check statements (also called scan statements) and conditional statements. The first check statement following a conditional statement is called the first check. In the following schematic of a logic control system, the logic step is emphasized:

```
...       ...
=  Q 4.0  Conditional statement
A  I 2.0  First check
A  I 2.1  Check statement
...       ...
A  I 1.7  Check statement
=  Q 5.1  Conditional statement
...       ...
=  Q 4.3  Conditional statement
O  I 2.6  First check
O  I 2.5  Check statement
...       ...
```

### First check

The first check statement following a conditional statement is called the first check, or first check. It has a special meaning because the CPU directly accepts the check result of this statement as the result of the logic operation. The "old" RLO is thus lost. The first check always represents the start of a logic operation. The rule of logic governing a first check (AND, OR, Exclusive OR) plays no role in this.

### Check statements

The result of the logic operation is formed with check statements. These statements check the signal state of a bit for "1" or "0" and link it according to AND, OR, or Exclusive OR. The CPU then saves the result of this logic operation as new RLO.

Figure 4.2. shows how checks for signal states "1" and "0" are programmed. The check for "1" takes the status of the bit checked as the check result for the next link. The check for "0" forms the check result from the negated status.

### Conditional statements

Conditional statements are statements whose execution depends on the result of the logic operation. They include statements for assigning, setting and resetting binary locations, for starting timers and counters, and so on.

The conditional statements (with very few exceptions) are executed when the RLO is "1", and not executed when the RLO is "0". They do not affect the RLO (with very few exceptions), and the RLO thus remains the same over several contiguous statements.

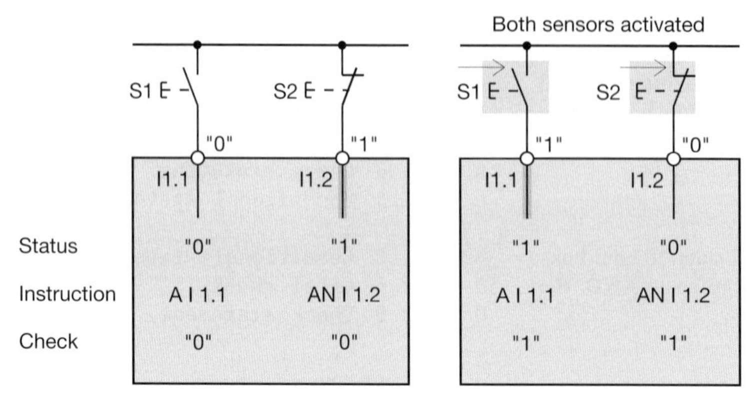


**Figure 4.2** Checking for Signal States "1" and "0"

### Intelligible programming

The rule of logic governing a first check is irrelevant, as the result of the check is taken directly as the result of the logic operation. For the purpose of intelligible programming, the rule of logic for a first check should be identical to the desired function. For example, the statement sequence

```
...
=   Q 15.3

O   I 18.5     First AND function
A   I 21.7
=   Q 15.4

A   I 18.4     Second AND function
A   I 21.6
=   Q 15.5
...
```

represents two AND functions, whereby the programming of the second AND function (in which both checks are programmed with AND) is to be preferred. In the case of individual check statements, such as

```
...
=   Q 10.0
A   I 20.1     Assign
=   Q 10.1     I 20.1 to Q 10.1
...
```

the AND is preferable.

## 4.2 Elementary Binary Logic Operations

STL uses the binary functions AND, OR and Exclusive OR. These functions are linked to the check for signal state "1" or with the check for signal state "0".

A *Bit address*
Check for "1" and combine according to AND

AN *Bit address*
Check for "0" and combine according to AND

O *Bit address*
Check for "1" and combine according to OR

ON *Bit address*
Check for "0" and combine according to OR

X *Bit address*
Check for "1" and combine according to Exclusive OR

XN *Bit address*
Check for "0" and combine according to Exclusive OR

The checks for "1" set the result of the check to "1" when the signal state of the bit is "1". The checks for "0" produce a check result of "1" when the signal state of the bit is "0". This corresponds to an input which leads to the relevant function when negated.

The CPU then combines the result of the check with the current RLO as per the specified function and forms the new RLO. When a binary logic operation immediately follows a memory function, the result of the check is entered in the RLO buffer without a logic operation being performed.

The number of binary functions and the scope of a binary function are theoretically arbitrary; in practice, however, the restriction is given by the length of a block or the size of the CPU's work memory.

### 4.2.1 AND Function

The AND function links two binary states with one another and returns an RLO of "1" when both states (both results of the check) are "1". When you program the AND function several times in succession, all check results must be "1" in order for the common result of the logic operation to be "1". In all other cases, the AND function returns an RLO of "0".

Figure 4.3 shows an example for AND. In network 1, the AND function has three inputs; these may be arbitrary bit addresses. All of these bits are checked for signal state "1", so that the signal state of the bits will be directly linked according to AND. If all bits checked are "1", the Assign statement sets the bit *Output1* to "1". In all other cases, the AND condition is not fulfilled and bit *Output1* is set to "0".

Network 2 shows an AND function with a negated input. The input is negated by checking it for "0". The check result of a bit checked for "0" is "1" if that bit is "0", that is, the AND condition in the example is fulfilled when bit *Input4* is "1" and bit *Input5* is "0".

### 4.2.2 OR Function

The OR function combines two binary states with one another and returns an RLO of "1" when one of these states (one of the check results) is "1". When you program the OR function several times in succession, only one check result need be "1" for the common result of the logic operation to be "1". If all check results are "0", the OR function returns an RLO of "0".

Figure 4.3 shows an example of an OR function. In network 3, the OR function has three inputs; these may be arbitrary bit addresses. All bits are checked for "1", so that the signal state of the bits is linked directly according to OR. If at least one of the bits checked is "1", the subsequent Assign statement sets bit *Output3* to "1". If all the bits checked are "0", the OR condition is not fulfilled and *Output3* is reset to "0".

Network 4 shows an OR function with negated input. The input is negated by a check for "0". The check result for a bit that was checked for "0" is "1" when that bit is "0", that is, the OR condition in the example is fulfilled when bit *Input4* is "1" or bit *Input5* is "0".

### 4.2.3 Exclusive OR Function

The Exclusive OR function links two binary states with one another and returns an RLO of "1" when the two states (the two check results) are not the same. The function returns an RLO of "0" when the two states (the two check results) are the same.

Figure 4.3 shows an example of an Exclusive OR function. In network 5, two inputs (arbitrary bit addresses) lead to the Exclusive OR function. Both inputs are checked for "1". If the signal state of only one of these bits is "1", the Exclusive OR condition is fulfilled and the Assign statement sets bit *Output5* to "1". If both bits are "1" or "0", *Output5* is reset to "0".

Network 6 shows an Exclusive OR function with a negated input. The input is negated by a check for signal state "0". The check result for a bit checked for "0" is "1" when that bit is "0", that is, the Exclusive OR condition in the example is fulfilled when both input bits have the same signal state.

You can also program the Exclusive OR function several times in succession, in which case the common RLO is "1" when an uneven number of the bits checked return a check result of "1".

### 4.2.4 Allowing for the Sensor Type

The binary functions AND, OR and Exclusive OR are described in preceding sections of this chapter as though normally open contacts were connected to the input modules (normally open contacts are sensors which return signal state "1" when activated). When implementing a control function, however, it is not always possible to use a normally open contact. In many cases, for

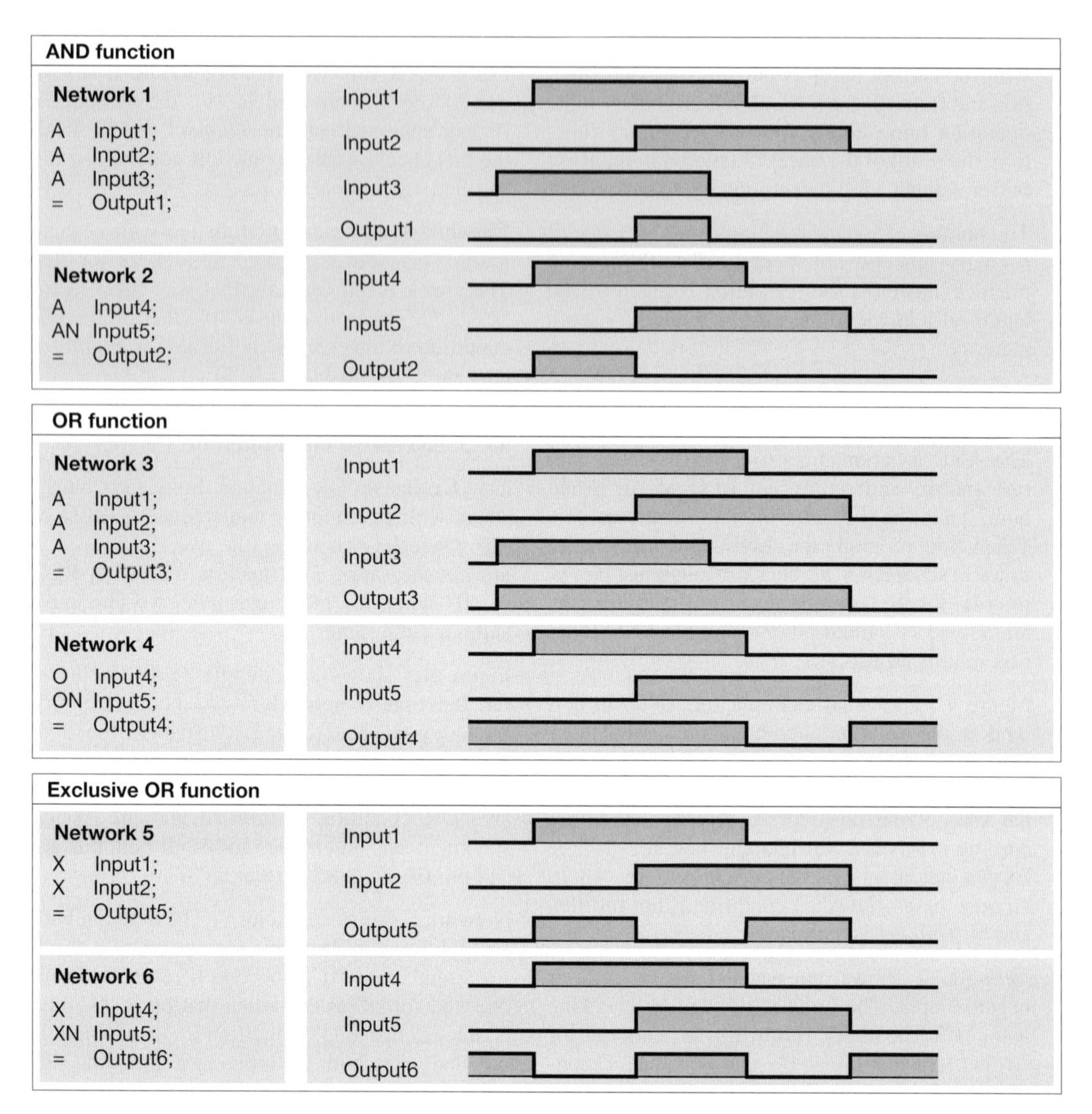


**Figure 4.3** Elementary Binary Functions

example in the case of closed circuits, the use of normally closed contacts is absolutely essential (a normally closed contact is a sensor which returns signal state "0" when activated).

If the sensor connected to an input is a normally open contact, the input carries signal state "1" when the sensor is activated. If the sensor is a normally closed contact, the input carries signal state "1" when the sensor is inactive. The CPU has no way of knowing whether an input is associated with a normally open or a normally closed contact; it can only differentiate between signal state "1" and signal state "0".

When developing the program, it is therefore necessary to take the sensor type into account. Before writing the program, you have to know whether the sensor is a normally closed contact or a normally open contact. Because the program is in part determined by the function of the sensor ("Sensor activated", "Sensor not activated"), it follows that you must check the input for signal state "1" or signal state "0", depending on the type of sensor used. In this way, you can also directly check inputs which are to execute various activities when "0" ("active when zero" inputs) and use the check result in subsequent links.

Figure 4.4 shows how to program in dependence on the sensor type. In the first case, two normally open contacts are connected to the programmable controller, in the second case, one normally open contact and one normally closed contact. In both cases, a contactor connected to an output is to pick up when both sensors are activated. When a normally open contact is activated, the signal state of the input is "1"; in order to fulfill the AND condition with check result "1", the input is checked for "1". When a normally closed contact is activated, the signal state of the input is "0". In order to fulfill the AND condition with a check result of "1" in this case, the input must be checked for signal state "0".

## 4.3 Negating the Result of the Logic Operation

NOT negates the result of the logic operation. You can use NOT at any location in the program, even within a logic operation. You can use NOT, for example, to negate the AND condition for an output (NAND function, Figure 4.5, network 7). Network 8 shows the negation of an OR function, which is called a NOR function.

You will find additional examples for NOT in Section 4.4.6, "Negating Nesting Expressions".

## 4.4 Compound Binary Logic Operations

Binary logic operations can be combined, for instance AND and OR functions can be programmed in any order. When such functions are programmed in arbitrary order, the CPU's handling of them is very difficult to duplicate. It is better, for instance, to illustrate the problem solution in the form of a function block diagram, then program it in STL. When compound binary logic operations are programmed, STL treats OR and Exclusive OR the same (they have the same priority). AND is executed "before" OR or Exclusive OR, and has a higher priority.

In order for the functions to be processed in the required order, it is sometimes necessary for the CPU to temporarily store the function value (the RLO that has been computed up to a certain point in the program). The nesting expressions

Case 1: Both sensors are normally-open contacts

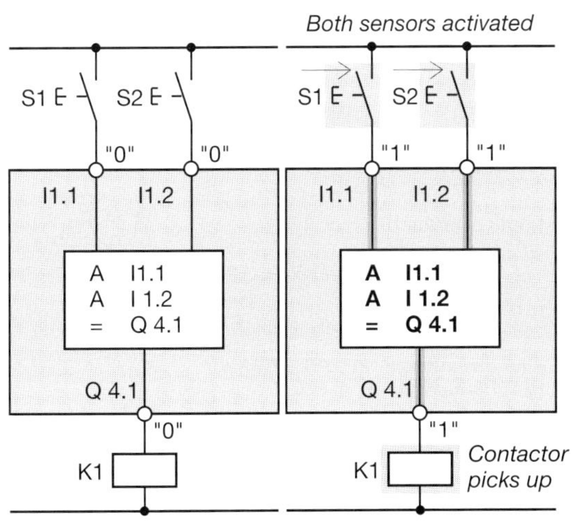


Case 2: One normally-open contact and one normally-closed contact

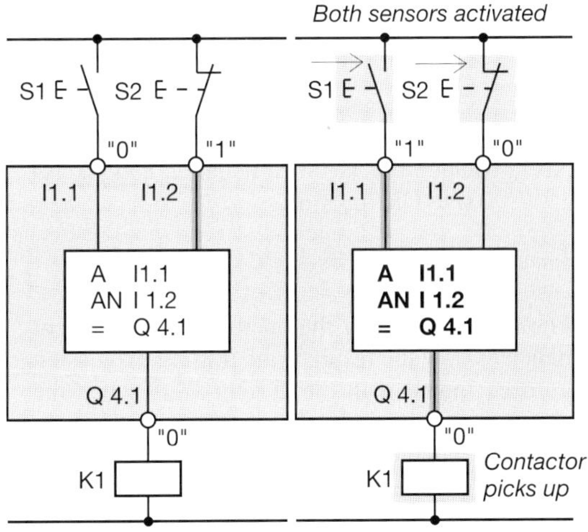


**Figure 4.4** Allowing for the Sensor Type

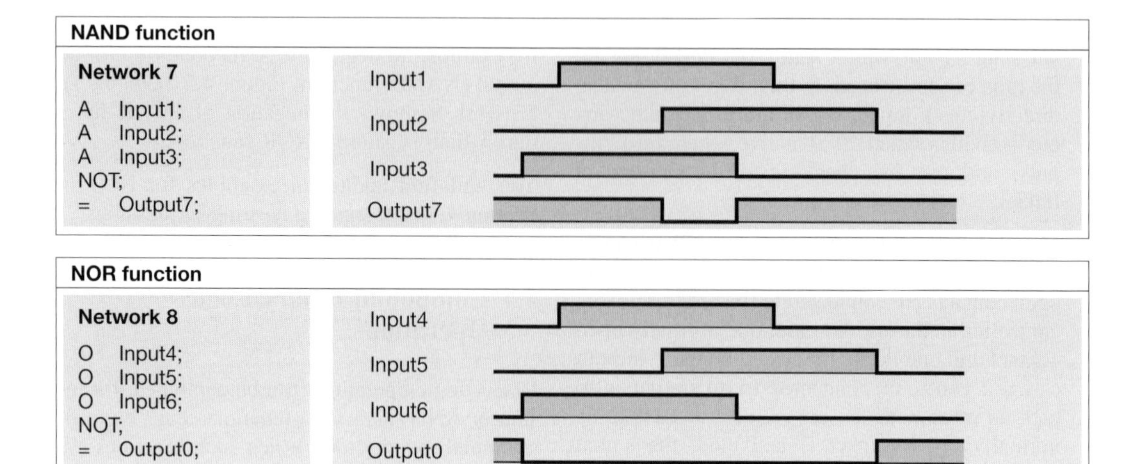


**Figure 4.5** Examples for the NOT Function

are provided for this purpose. As in the case of the notation used for Boolean algebra, the nesting expressions cause one function to be executed "before" another. The nesting expressions also include Exclusive OR. The STL programming language provides the following binary nesting expressions:

| | |
|---|---|
| O | ORing of AND functions |
| A( | Open bracket with AND function |
| O( | Open bracket with OR function |
| X( | Open bracket with Exclusive OR function |
| AN( | Open bracket with negation and AND function |
| ON( | Open bracket with negation and OR function |
| XN( | Open bracket with negation and Exclusive OR function |
| ) | Close bracket |

The rule of logic for the open bracket statement indicates how the result of the nesting expression is to be linked with the current RLO when the close bracket operation is encountered. Prior to this logic operation, the result of the nesting expression is negated if a negation character is specified.

### 4.4.1 Processing Nesting Expressions

In the STL programming language, the binary nesting expressions are used to define the order in which binary logic operations are processed. At runtime, the setting of brackets has the effect as the CPU processing the nesting expressions "first", that is, before executing the instructions outside the brackets.

When it encounters an open bracket statement, the CPU stores the current RLO internally, then processes the nesting expression, when it en-

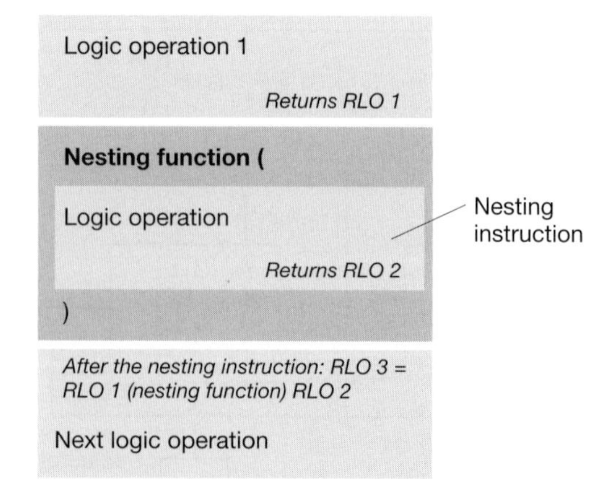


**Figure 4.6** Processing Nesting Expressions

counters the close bracket statement, it links the RLO from the nesting expression with the RLO it stored prior to processing the nesting expression as per the function given in the open bracket statement (Figure 4.6).

A check statement following a open bracket statement is always a first check because the CPU always regenerates the RLO within a nesting expression. A check statement following a close bracket statement is never a first check because, when a nesting expression is the first instruction in a logic operation, the CPU treats the RLO from the nesting statement like the result of a first check.

Nesting expressions can be nested, that is to say, you can program a nesting expression in a nesting expression (Figure 4.7). The nesting depth is seven, that is, you may begin a nesting expression seven times without first terminating one. Processing within the brackets is much as described above.

### Saving intermediate results with the aid of the nesting stack

Internally, the CPU sets up a nesting stack in order to process nesting functions. In this stack it stores:

- The result of the logic operation (RLO) preceding the bracket
- The binary result (BR) preceding the bracket
- The status bit (OR) (indicating whether an OR condition was already fulfilled) and
- The nesting function (with which function the nesting expression is to be linked).

The CPU sets the binary result BR following the *close bracket* statement the signal state it had prior to the nesting expression.

Within a nesting expression, you can not only program binary logic operations but all statements in the STL programming language. Care must be taken, however, that nesting expression be terminated with the *close bracket* statement. It is thus possible, for example, to program several logic steps or memory and comparison functions within a nesting expression.

## 4.4.2 Combining AND Functions According to OR

These logic operations, which are a combination of OR and AND functions, can be written in Boolean algebra without brackets. It is the rule that the AND functions are processed "first".

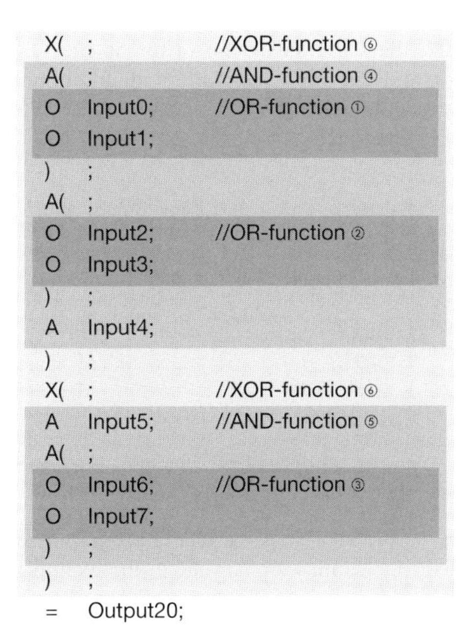


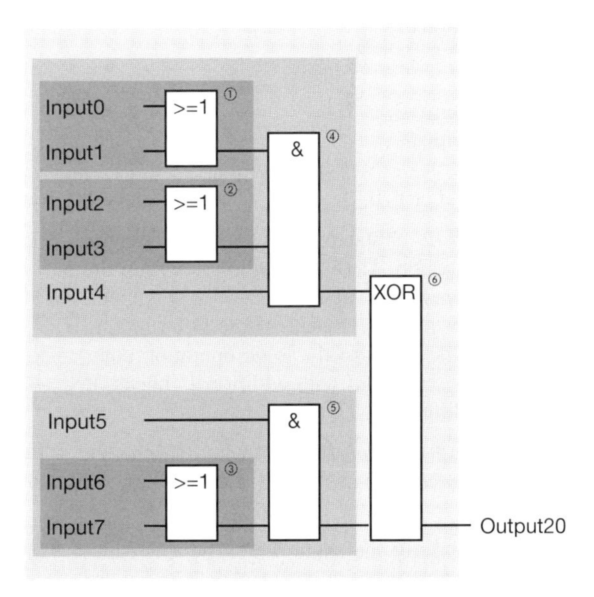


**Figure 4.7** Example for Nested Nesting Expressions

The results of the AND functions, together with additional OR checks, if any, are then linked with OR. Example:

```
A       Input0;
A       Input1;
O       ;
A       Input2;
A       Input3;
=       Output8;
```

In the example, a single O (for OR) is between the first and the second AND function. This operation makes "AND before OR" possible, and is always necessary when an AND function is placed "before" an OR function. The single "O" precedes the AND function, and is no longer necessary after the AND function.

In the example, *Output8* is set when {*Input0* and *Input1*} or {*Input2* and *Input3*} are "1".

### 4.4.3 Combining OR and Exclusive OR Functions According to AND

These compound logic operations comprising AND and OR functions must be written with brackets in Boolean algebra to indicate that the OR functions are to be processed "before" the AND functions. Example:

```
A(      ;
O       Input0;
O       Input1;
)       ;
A(      ;
O       Input2;
O       Input3;
)       ;
=       Output10;
```

The *open bracket* statement is "combined" with an AND function. The OR function is within the nesting expression. The *close bracket* statement, in this case, links the result of the OR function (generally the result of the logic operation computed in the brackets) with additional checks, if any, according to AND.

In the example, *Output10* is set when {*Input0* or *Input1*} and {*Input2* or *Input3*} are "1".

The ANDing of Exclusive OR functions is programmed exactly the same way. The OR functions in the example could be replaced by Exclusive OR, since the two functions have the same priority.

### 4.4.4 Combining AND Functions According to Exclusive OR

An AND function before an Exclusive OR function is written in brackets. With the aid of the brackets, the CPU saves the result of the AND function and can then combine it, possibly with additional checks, according to the rules governing Exclusive OR. Example:

```
X(      ;
A       Input0;
A       Input1;
)       ;
X(      ;
A       Input2;
A       Input3;
)       ;
=       Output12;
```

In the example, the first AND function need not be in brackets, as an AND function has a higher priority than an Exclusive OR. The brackets, however, make the program more readable.

*Output12* in the example is set either when {*Input0* and *Input1*} or {*Input2* and *Input3*} are "1".

### 4.4.5 Combining OR Functions and Exclusive OR Functions

An OR function before an Exclusive OR function is written in brackets. With the aid of the brackets, the CPU saves the result of the OR function and can link it, possibly with additional checks, according to the rules governing Exclusive OR. Example:

```
X(      ;
O       Input0;
O       Input1;
)       ;
X(      ;
O       Input2;
O       Input3;
)       ;
=       Output14;
```

In the example, *Output14* is set when one, and only one, of the two OR conditions is fulfilled.

An OR before an Exclusive OR is programmed in exactly the same way. An OR in the example can be replaced by an Exclusive OR and vice versa, since the two functions have the same priority.

### 4.4.6 Negating Nesting Expressions

Just as you can check a bit for signal state "0" (negate the status, as it were), you can also negate a nesting expression. This means that the CPU will post-process the result of the nesting expression in negated form. Negation is specified by an additional N in the *open bracket* statement. Example:

```
AN(     ;
O       Input0;
O       Input1;
)       ;
AN(     ;
X       Input2;
X       Input3;
)       ;
=       Output16;
```

In the example, *Output16* is set if neither the OR condition nor the Exclusive OR condition is fulfilled.

A second way to negate nesting expressions is through the use of the NOT (negation) statement. A NOT written before the *close bracket* statement negates the result of the nesting statement prior to further processing. Example:

```
A(      ;
O       Input0;
O       Input1;
NOT     ;
)       ;
A(      ;
X       Input2;
X       Input3;
NOT     ;
)       ;
=       Output17;
```

This logic operation has the same function as the preceding logic operation. Negation of the nesting expression is attained here within the brackets using NOT.

# 5 Memory Functions

This chapter describes the memory functions; these include Assign for dynamic bit control and Set and Reset for static control. The memory functions also include edge evaluations.

The memory functions are used in conjunction with binary logic operations in order to affect the signal states of bits with the help of the RLO generated in the CPU.

The examples in this chapter are also on the diskette which accompanies this book under the "Basic Functions" program in function block FB 105 or source file Chap_5.

## 5.1 Assign

= *Bit*
Assigns the result of the logic operation

The Assign statement "=" assigns the RLO in the processor directly to the bit specified in the statement. If the result of the logic operation is "1", the bit is set; if the result of the logic operation is "0", the bit is reset (Figure 5.1, network 1). If you want the bit to be set when the RLO is "0", you can negate the RLO prior to Assign with the NOT statement (network 2).

You will find additional examples for Assign in Chapter 4, "Binary Logic Operations".

### Simultaneous execution of multiple Assigns

You can also assign the result of the logic operation to several different bits by programming successive Assign statements specifying the relevant bits (Figure 5.1, network 3). All specified bits react the same, as the instructions used for bit control do not affect the RLO. The CPU does not generate a new RLO until it encounters the next check statement. If you want to use the signal state of an output in another logic operation, simply check that output with the appropriate check statement (network 4).

## 5.2 Set and Reset

S *Bit*
Sets the bit when the result of the logic operation is "1"

R *Bit*
Resets the bit when the result of the logic operation is "1"

The Set S and Reset R instructions are executed only when the result of the logic operation is "1". The Set instruction then sets the specified bit to "1", and the Reset instruction sets it to "0". RLO "0" has no effect on the Set or Reset instruction; when the RLO is "0", the bit specified in a Set or Reset instruction retains it current signal state (Figure 5.1, networks 5 and 6).

### Simultaneous execution of multiple memory functions

You can control multiple Set and Reset instructions, in any combination and together with Assigns, with the same RLO. Simply write successive statements specifying the relevant bits (Figure 5.1, network 7). As long as Set, Reset and Assign statements are being processed, the RLO does not change. The CPU does not generate a new RLO until it encounters the next check statement.

Here, too, you can use NOT to negate the RLO within the sequence of memory statements.

To ensure the clarity and readability of the program, you should use the Set and Reset statements for a specific bit in pairs, and only once for a given bit, and you should not then affect this same bit by programming an Assign statement for it.

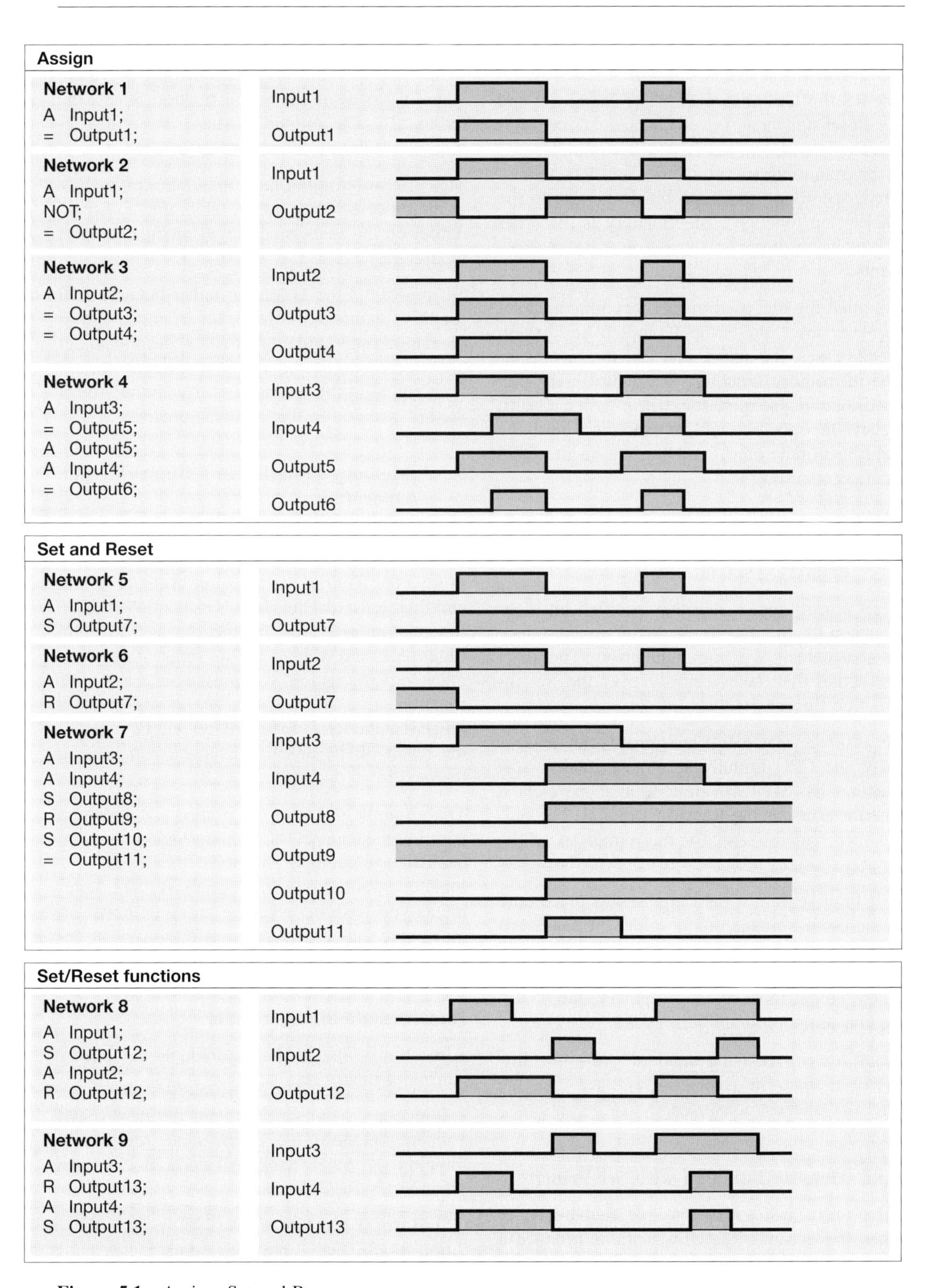


**Figure 5.1** Assign, Set and Reset

## 5.3 RS Flipflop Function

The RS flipflop function consists of one Set and one Reset statement; there is no special identifier in STL. The RS flipflop function is implemented by programming successive Set and Reset statements specifying the same bit. Important to the RS flipflop function's functionality is the order in which you program the Set and Reset statements.

Note that the bits used in memory functions are normally reset on startup (complete restart). In special cases, the signal states of the bits specified in memory functions are retained; this depends on the type of startup (for instance a warm restart), the specified bit (for example a bit in the static local data area), and settings in the CPU (such as retentivity).

### 5.3.1 Memory Functions with Reset Priority

Reset priority means that the specified bit is reset when the Set and Reset instructions produce a signal state of "1" "simultaneously". The Reset instruction then takes priority over the Set instruction (Figure 5.1, network 8).

Because the statements are processed sequentially, the CPU initially sets the bit because it executes the Set instruction first, then resets it when it executes the Reset instruction. The output then remains reset for the remainder of the program scan cycle.

This brief setting of the output takes place only in the process-image output table; the (external) output on the associated output module is not affected. The CPU does not transfer the process-image output table to the output modules until the end of the program scan cycle.

Reset priority is the "standard" form of this function, since the reset state is, as a rule, the safer or less hazardous state.

### 5.3.2 Memory Function with Set Priority

Set priority means that the specified bit is set when the Set and Reset instructions produce a signal state of "1" simultaneously. The Set instruction then takes priority over the Reset instruction (Figure 5.1, network 9).

When it processes the statement sequence, the CPU first sets the specified bit to "0"; then, when it processes the Set instruction, it sets the bit to "1". The output remains set for the remainder of the program scan cycle.

This brief resetting of the output takes place only in the process-image output table; the "external" output on the associated output module remains unaffected.

Set priority is the exception rather than the rule for this function. It is used, for example, to implement a fault signal latch when, despite an acknowledgment at the Reset input, the current fault signal at the Set input is to continue to set the bit specified in the memory function.

### 5.3.3 Memory Function in a Binary Logic Operation

In the STL programming language, you can use the memory functions very freely. It is possible to save the RLO at any location in the program, and then reuse it later. If you use an RS flipflop function in a nesting instruction, as shown in Figure 5.2, and want to use its signal state in another logic operation, you must re-check the RS function prior to the *close bracket* statement as otherwise the Reset input's RLO will be taken as the result of the nesting statement and used in the next logic operation.

This example does not use nesting statements to control the sequence of a binary logic operation, but rather to temporarily save the result of a logic operation. You can program any STL statements you wish between the brackets; however, make sure that you have the RLO you want before writing the *close bracket* statement.

#### Intermediate binary results

Almost any bits can be used for the temporary storing of binary results:

- Temporary local data bits are the best suited when you need the intermediate result only within the block itself. All code blocks have temporary local data areas.
- Static local data bits are available only in function blocks, and save their signal states until set again.
- Memory bits are available globally in a CPU-

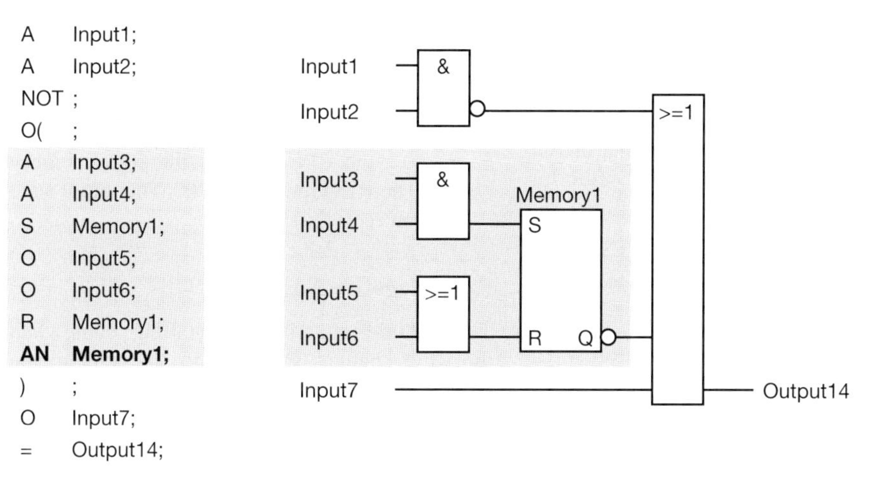


**Figure 5.2** Nesting Statement as Intermediate Bit Buffer

specific quantity; for clarity of programming, avoid multiple use of memory bits (the same bits for different tasks).

- Data bits in global data blocks are also available throughout the whole program, but before they can be used, the relevant data block must first be opened (even if this is only implied through the use of complete addressing).

Note: You can replace the "scratchpad memory" used in STEP 5 with temporary local data, which are available in every block.

## 5.4 Edge Evaluation

FP *Bit*
Positive (rising) edge

FN *Bit*
Negative (falling) edge

Edge evaluation is the detection of a change in a signal state, a signal edge. A positive (rising) edge is present when the signal goes from "0" to "1". The opposite is referred to as a negative (falling) edge.

In a relay logic diagram, the equivalent of edge evaluation is a pulse contact element. If this pulse contact element emits a pulse when the relay is switched on, this corresponds to a rising edge. Emission of a pulse when the relay is switched off corresponds to a falling edge.

The bit specified in the edge evaluation is referred to as "edge memory bit" (it need not necessarily be a memory bit). However, it must be a bit whose signal state is once again available in the next program scan cycle and which is otherwise not used in the program. Suitable bits are memory bits, data bits in global data blocks, and static local data bits in function blocks.

This edge memory bit stores the "old" RLO, which is the one the CPU used for the last edge evaluation. In each edge evaluation, the CPU compares the current RLO with the signal state of the edge memory bit. An edge is present when they have different signal states. In this case, the CPU updates the signal state of the edge memory bit by assigning the current RLO to it, and sets the RLO to "1" on either a positive or negative edge, depending on the instruction, following edge evaluation. If the CPU does not detect an edge, it sets the RLO to "0".

Signal state "1" following an edge evaluation thus means "edge detected". The signal state remains at "1" only briefly, as a rule for only one scan cycle. Because the CPU does not detect an edge on the next edge evaluation (when the edge evaluation's "input RLO" does not change), it sets the RLO back to "0" following the edge evaluation.

You can process the RLO directly after an edge evaluation, for example with a Set operation, or you can store it in a bit (a "pulse memory bit").

Use a pulse memory bit when the RLO from the edge evaluation is to processed at another location in the program; it is effectively the intermediate buffer for a detected edge. Bits suitable as pulse memory bits are memory bits, data bits in global data blocks, and temporary and static local data bits.

You can also further process the RLO after an edge evaluation directly with the following check statements.

Note the response of the edge evaluation when you switch on the CPU. If an edge is to be detected, the RLO prior to edge evaluation and the signal state of the edge memory bit must be equal. Under certain circumstances, the edge memory bit must be reset on startup (depending on the desired response and on the bit used).

The following examples illustrate how edge evaluation works. In simplified form, an input represents the RLO prior to edge evaluation and a memory bit (the "pulse memory bit") the RLO following edge evaluation. Of course, the edge evaluation may also be preceded and followed by a binary logic operation.

### 5.4.1 Positive Edge

The CPU detects a positive (rising) edge when the result of the logic operation changes from "0" to "1" prior to the edge evaluation. The procedures involved are shown in Figure 5.3, above; the consecutive number stands for successive scan cycles:

① The first time around, the signal state of both the input and the edge memory bit is "0". The pulse memory bit remains reset.

② The second time around, the signal state of the input should have changed from "0" to "1". The CPU detects the change by comparing the current RLO with the status of the edge memory bit. If the RLO is "1" and the edge memory bit "0", the edge memory bit is set to "1". The current RLO is also set to "1".

③ The next time around, the CPU finds that the there is no different between the state of the input and that of the edge memory bit, and therefore sets the current RLO to "0".

④ As long as there is no difference between the two states, the RLO remains at "0" and the edge memory bit remains set.

⑤ When the input once again has a signal state of "0", the CPU corrects the edge memory bit. The RLO remains at "0". The initial state is then reestablished.

### 5.4.2 Negative Edge

The CPU detects a negative (falling) edge when the RLO changes from "1" to "0" prior to the edge evaluation. The procedures involved are shown in Figure 5.3 below; the consecutive number stands for successive scan cycles:

❶ The first time around, the signal state of both the input and the edge memory bit is "0". The pulse memory bit remains reset.

❷ The second time around, the signal state of the input should have changed from "0" to "1". The CPU detects change by comparing the current RLO with the status of the edge memory bit. If the RLO is "1" and the edge memory bit "0", the edge memory bit is set to "1". Following edge evaluation, the RLO remains at "0".

❸ As long as there is no difference between the two states, the RLO remains at "0" and the edge memory bit remains set.

❹ When the input once again has a signal state of "0", the CPU corrects the status of the edge memory bit and sets the RLO to "1" following edge evaluation.

❺ In the next scan cycle, the signal state of the input and that of the edge memory bit are the same. The CPU therefore sets the RLO back to "0", thus reestablishing the original state.

### 5.4.3 Testing a Pulse Memory Bit

The signal states of the pulse memory bits are very difficult to monitor with the programming devices' test functions because they remain at "1" for only one scan cycle. It is for this reason that an output is also unsuitable as pulse memory bit, as the signal amplifiers on the output module or the actuators are not capable of duplicating the signal changes all that quickly.

With a "flying restart circuit", however, you can record the extremely brief signal states of the

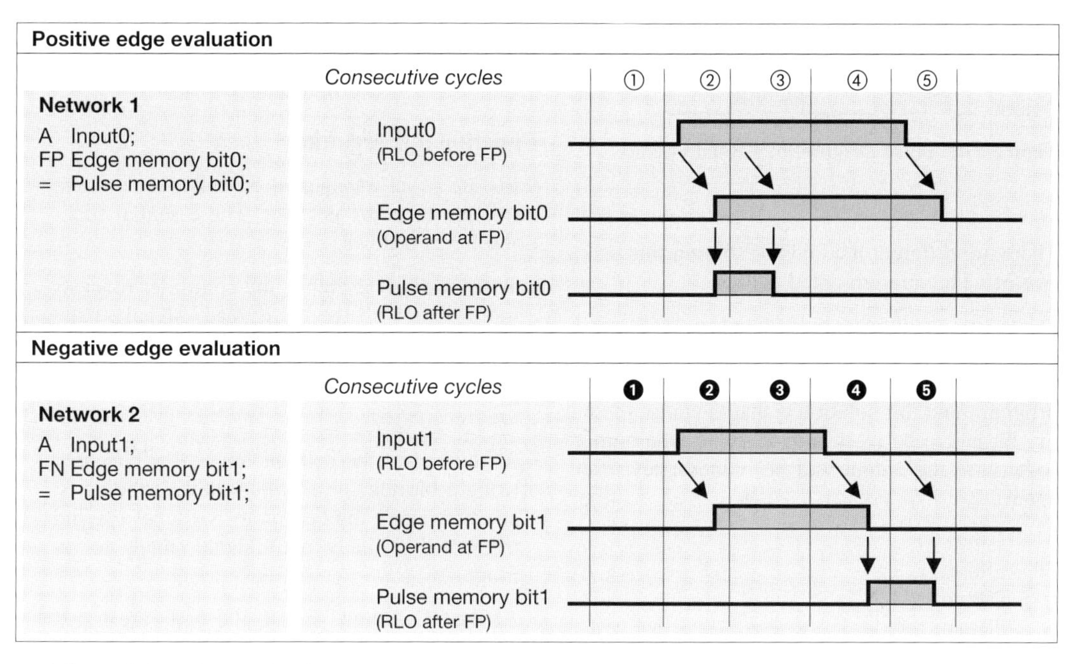


**Figure 5.3** Edge Evaluations

pulse memory bits in an RS flipflop. The pulse memory bit sets the RS flipflop, thus storing the "Edge Detected" signal. After you have evaluated this signal, you can reset the flipflop.

```
O    PMembit0;
O    PMembit1;
S    Flipflop2;
A    Input2;
R    Flipflop2;
```

### 5.4.4 Edge Evaluation in a Binary Logic Operation

Edge evaluation in a binary logic operation can be used to serve a practical purpose only when you use the signal state following edge evaluation (the "pulse") to control a memory, timer or counter function. Binary checks may lie between the edge evaluation and control of the relevant function.

```
O    Input3;
O    Input4;
FP   EMembit2
A    Input5;
S    Output15;
A    Input6;
FN   EMembit3;
R    Output15;
```

In the example, *Output15* is set at the instant at which the OR condition is fulfilled (when the bit in the OR statement goes from "0" to "1") and *Input5* is "1". *Output15* is reset on a falling edge at *Input6*.

An edge evaluation is effectively a first check, as the RLO generated by the edge evaluation can be post-processed. This also means that the logic operation up to the instant of the edge evaluation is regarded as "completed" (a fulfilled OR condition is not stored). Edge evaluation does not affect the processing of nesting instructions.

### 5.4.5 Binary Scaler

A binary scaler has one input and one output. If the signal at the binary scaler's input changes its state, for example from "0" to "1", the output changes its state as well (Figure 5.4). This (new) signal state is retained until the next, in our example positive, signal state change. Only then does the signal state of the output change again. This means that half the input frequency appears at the output of the binary scaler.

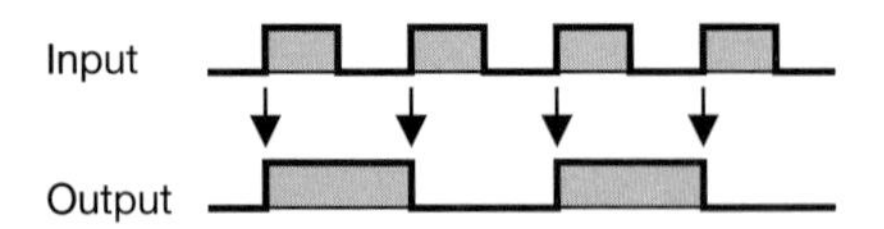


**Figure 5.4** Pulse Diagram of a Binary Scaler

There are different methods of solving this task, two of which are presented below.

In the first solution, a pulse memory bit is used to set the output if it was reset and to reset it if it was set. The important thing to remember when programming this solution is that the pulse memory bit has to be reset once it has set the output (otherwise the output will be immediately reset again).

```
A   Input_1;
FP  EMembit_1;
=   PMembit_1;
A   PMembit_1;
AN  Output_1;
S   Output_1;
R   PMembit_1;
A   PMembit_1;
A   Output_1;
R   Output_1;
```

The second solution uses a conditional jump JCN to evaluate the edge. When the CPU does not detect an edge, the RLO is "0" and the program scan is resumed at the jump label. In the case of a positive edge, the CPU does not execute the jump and executes the next two statements. If the output is reset, it is set; if it is set, it is reset. Although an Assign controls the output, the latter functions as a latch, as this program section is executed only when there is a positive edge.

```
    A     Input_2;
    FP    EMembit_2;
    JCN   M1;
    AN    Output_2;
    =     Output_2;
M1: . . . ;
```

## 5.5 Example of a Conveyor Belt Control System

A functionally extremely simple conveyor belt control system is used here as an example to show how binary logic operations and memory functions work in conjunction with inputs, outputs, and memory bits.

### Functional description

Parts are to be transported on conveyor system, one crate or pallet per belt. The essential functions are as follows:

- When the belt is empty, the controller requests more parts with the "readyload" signal (ready to load)
- The "Start" signal starts the belt, and the parts are transported
- At the end of the belt, an "end-of-belt" sensor (a light barrier, for example) detects the parts, at which point the belt motor switches off and triggers the "ready_rem" signal (ready to remove)
- At the "Continue" signal, the parts are transported further until the "end-belt" (end-of-belt) sensor no longer detects them.

The function block diagram for the conveyor belt control system is shown in Figure 5.5. The example is programmed with inputs, outputs and memory bits. it can be loaded anywhere in any block. In the example, a function without a function value was chosen as block.

In the section entitled "Block Parameters", the same example is programmed in a function block with block parameters; the function block can be called more than once (for more than one belt).

### Signals and symbols

A number of additional signals supplement the functionality of the conveyor belt control system:

- Basic_st
  Sets the controller to the basic state
- Man_on
  Switches on the belt without regard to any conditions
- /Stop
  Stops the belt as long as the "0" signal is present (an NC contact as sensor, "zero active")
- Light_barrier1
  The parts have reached the end of the belt

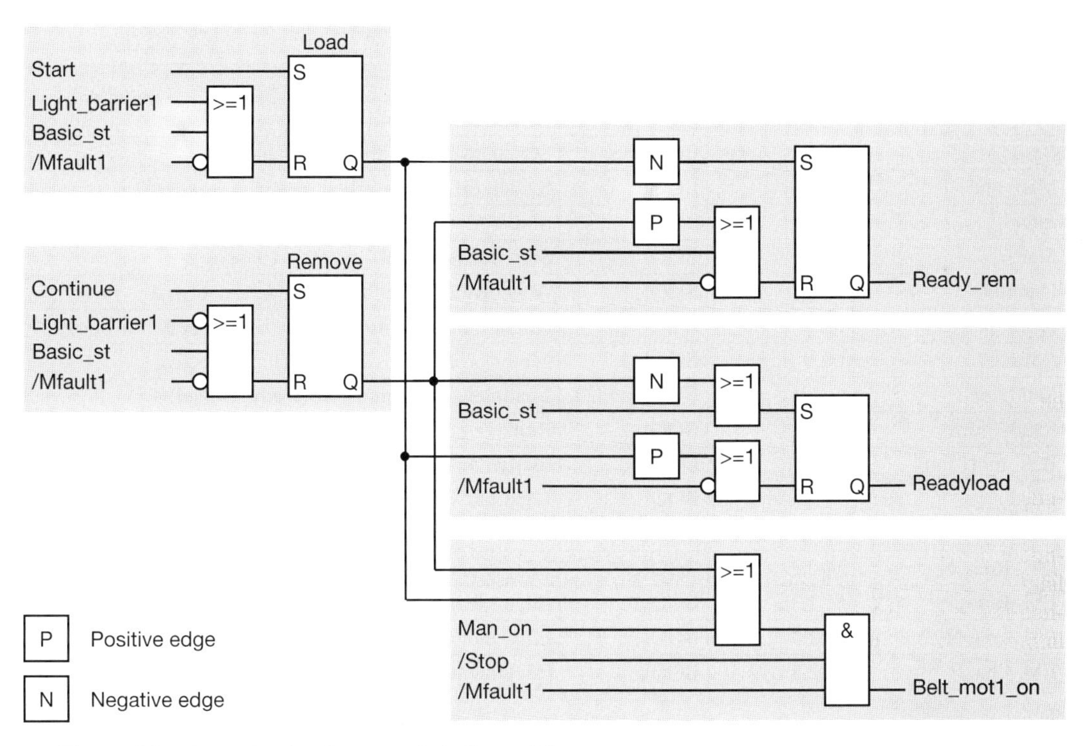


**Figure 5.5** Example of a Conveyor Belt Control System

- /Mfault
  Fault signal from the belt motor (e.g. motor protection switch); designed as "zero active" signal so that other malfunctions, such as a wire break, will also generate a fault signal

We want to use symbolic addressing, that is, the addresses are assigned names which we then use when writing the program. Prior to incremental program input or to compilation, we generate a symbol table (Table 5.1) that contains the inputs, outputs, memory bits and blocks.

## Program

The example is located in a function without block parameters. You can call this function, for instance, in organization block OB 1 as follows:

```
CALL Belt_control;
```

The example is in the form of source text with symbolic addressing. The global symbols can also be used without quotation marks when they contain no special characters. If a symbol contains a special character, it must be enclosed in quotation marks. The STL editor displays all global symbols in the compiled block in quotation marks.

The program is subdivided into networks to improve clarity and readability. The last network, with the title BLOCK END, is not absolutely necessary, but serves as a visual end of the block, a very useful feature for extremely long blocks.

**Table 5.1** Symbol Table for the Sample "Conveyor Belt Control System"

| Symbol | Address | Data Type | Comment |
|---|---|---|---|
| Belt_control | FC 11 | FC 11 | Belt control system |
| Basic_st | I 0.0 | BOOL | Set controllers to the basic state |
| Man_on | I 0.1 | BOOL | Switch on conveyor belt motor |
| /Stop | I 0.2 | BOOL | Stop conveyor belt motor (zero-active) |
| Start | I 0.3 | BOOL | Start conveyor belt |
| Continue | I 0.4 | BOOL | Acknowledgment that parts were removed |
| Light_barrier1 | I 1.0 | BOOL | (Light barrier) sensor signal "End of belt" for belt1 |
| /Mfault1 | I 2.0 | BOOL | Motor protection switch belt1, zero-active |
| Readyload | Q 4.0 | BOOL | Load new parts onto belt (ready to load) |
| Ready_rem | Q 4.1 | BOOL | Remove parts from belt (ready to remove) |
| Belt_mot1_on | Q 5.0 | BOOL | Switch on belt motor for belt1 |
| Load | M 2.0 | BOOL | Load parts command |
| Remove | M 2.1 | BOOL | Remove parts command |
| EM_Rem_N | M 2.2 | BOOL | Edge memory bit for negative edge of "remove" |
| EM_Rem_P | M 2.3 | BOOL | Edge memory bit for positive edge of "remove" |
| EM_Loa_N | M 2.4 | BOOL | Edge memory bit for negative edge of "load" |
| EM_Loa_P | M 2.5 | BOOL | Edge memory bit for positive edge of "load" |

```
FUNCTION Belt_control : VOID
TITLE = Control of a conveyor belt
//Example of binary logic operations and memory functions, without block
parameters
NAME    :  Belt1
AUTHOR  :  Berger
FAMILY  :  STL_Book
VERSION :  01.00
BEGIN
NETWORK
TITLE = Load parts
//This network generates the comman "Load" that initiates transport of parts
//to the end of the belt.
   A    Start;                //Start conveyor belt
   S    Load;
   O    Light_barrier1;       //Parts hve reached end of belt
   O    Basic_st;
   ON   "/Mfault1";           //Motor protection switch (zero active)
   R    Load;
```

```
NETWORK
TITLE = Parts ready to remove
//When parts have reached end of belt, they are ready to remove.
   A   Load;                  //When end of belt has been reached,
   FN  EM_Loa_N;              //"Load" is reset.
   S   Ready_rem;             //Parts are then "ready for removal"
   A   Remove;
   FP  EM_Rem_P;              //The parts are removed
   O   Basic_st;
   ON  "/Mfault1";
   R   Ready_rem;
NETWORK
TITLE = Remove parts
//The "Remove" command initiates removal of the parts from the belt.
   A   Continue;              //Switch belt back on
   S   Remove;
   ON  Ligh_barrier1;         //Parts leave the belt
   O   Basic_st;
   ON  "/Mfault1";            //Motor protection switch (zero active)
   R   Remove;
NETWORK
TITLE = Belt ready for loading
//The belt is ready for loading when the parts have left the belt.
   A   Remove;
   FN  EM_Rem_N:              //Parts have left the belt
   O   Basic_st;
   S   Readyload;             //Belt is empty
   A   Load;
   FP  EM_Loa_P;              //Belt is started
   ON  "/Mfault1";
   R   Readyload;
NETWORK
TITLE = Control belt motor
//The belt motor is switched on and off in this network.
   A(;
   O   Load;                  //Load parts onto belt
   O   Remove;                //Remove parts from belt
   O   Man_on;                //Start with "Man_on" (non-retentive)
   );
   A   "/Stop";               //Stop and motor fault prevent
   A   "/Mfault1";            //belt motor from running
   =   Belt_mot1_on;
NETWORK
TITLE = Block end
    BE;
END_FUNCTION
```

# 6 Transfer Functions (Move Functions)

This chapter describes functions which interchange data with the accumulators (registers). These include

- Load functions
  The load functions are used to fill the accumulators for subsequent digital post-processing, for instance compare, compute, and so on.
- Transfer functions
  The transfer functions transfer the digital results from accumulator 1 to memory areas in the CPU, for instance bit memory.
- Accumulator functions
  These functions transfer information from one accumulator to another or replace information in accumulator 1.

You also need load functions to specify initial values for timers and counters or to process current times and counts.

System functions SFC 20 BLKMOV and SFC 21 FILL are available for copying larger amounts of data in memory or to preset data areas. You need the load and transfer functions to address modules via the user data area; when you address modules via the system data area, you must use system functions to transfer data records. You can also use these system functions to parameterize the modules.

The examples in this chapter are also on the diskette accompanying the book under the "Basic Functions" program in function block FB 106 or source file Chap_6.

## 6.1 General Remarks on Loading and Transferring Data

The load and transfer functions enable the exchange of information between different areas of memory. This information exchange does not take place directly, but instead is "routed through" accumulator 1. An accumulator is a special register in the processor, and serves as "intermediate buffer".

When information is exchanged, the direction in which the information flows is indicated by the instruction used to transfer that information. The information flow from a memory area to accumulator 1 is called *loading*, the reverse direction of flow is called *transferring* (the contents of the accumulator are "transferred" to the memory area).

Loading and transferring are the prerequisites for the use of the *digital functions*, which manipulate a digital value (convert or shift, for example) or combine two digital values(for instance compare or add). In order to combine two digital values, two intermediate buffers are needed, namely accumulator 1 and accumulator 2. All CPUs are equipped with these two special registers. The S7-400 CPUs have two additional intermediate buffers, accumulator 3 and accumulator 4, which are used primarily in conjunction with arithmetic functions. The group of functions called the *accumulator functions* is used to copy the contents of one accumulator to another.

These associations are illustrated graphically in Figure 6.1. The load function transfers information from system memory, work memory and the I/O to accumulator 1, shifting the "old" (that is to say, current) contents of accumulator 1 over to accumulator 2. The digital functions manipulate the contents of accumulator 1 or combine the contents of accumulators 1 and 2 and write the result back into accumulator 1. The accumulator functions can access the contents of all accumulators. The source for transferring information to system memory, work memory or the I/O is always and only accumulator 1.

Each accumulator comprises 32 bits, while all memory areas are byte-oriented. Information can be exchanged between the memory areas and accumulator 1 by byte, word, or doubleword.

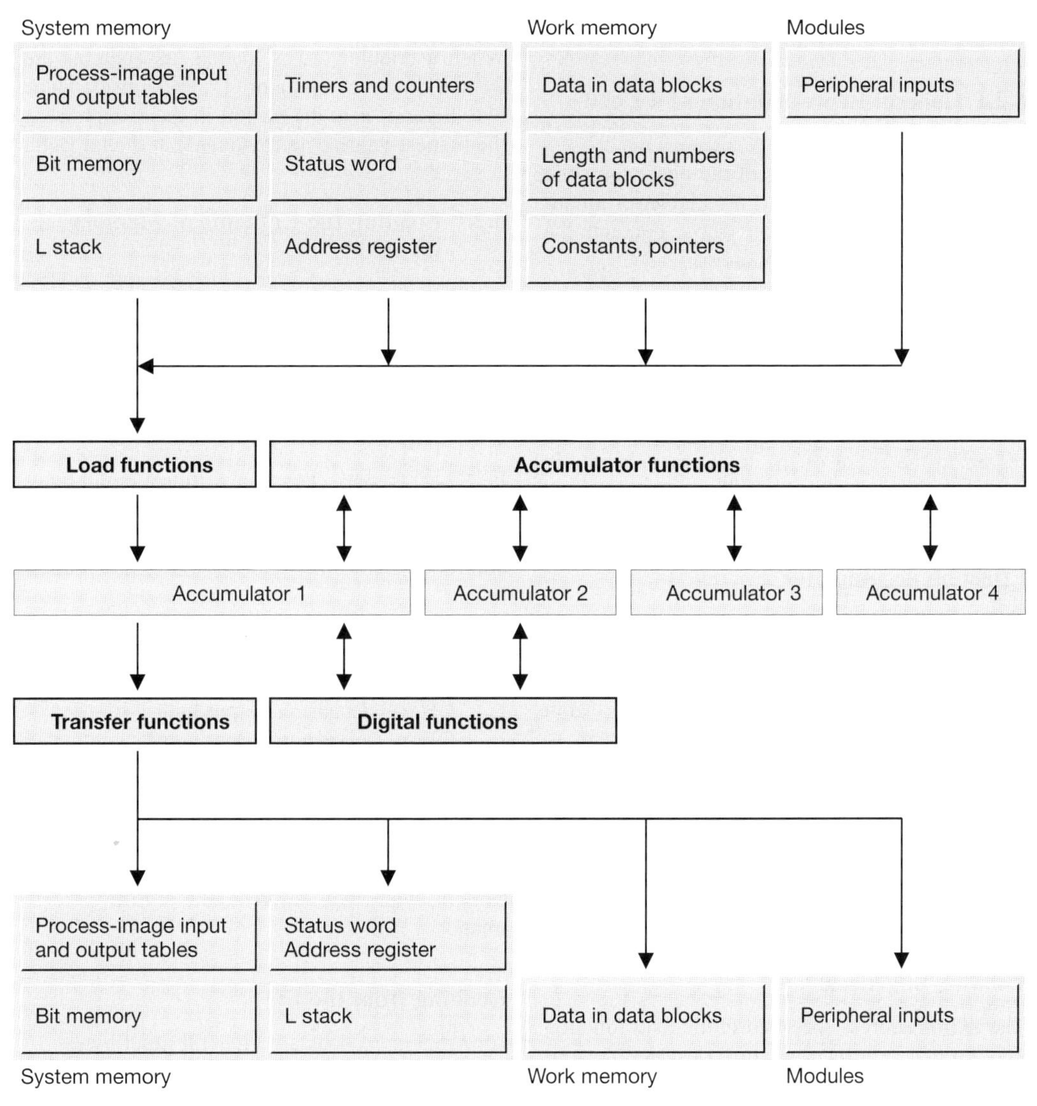


**Figure 6.1** Memory Areas for Loading and Transferring

In this chapter, the load and transfer functions are discussed in conjunction with the address areas for inputs, outputs, memory bits, I/O, and the loading of constants. The load and transfer functions can also be combined with the following address areas:

- Timers and counters
  (Chapter 7, "Timers", and Chapter 8, "Counters")
- Status word
  (Chapter 15, "Status Bits")
- Temporary local data
  (L stack, Section 18.1.5, "Temporary Local Data")
- Data addresses, lengths and numbers of data blocks
  (Section 18.2, "Block Functions for Data Blocks")
- Address registers, pointers
  (Chapter 25, "Indirect Addressing")
- Variable address
  (Chapter 26, "Direct Variable Access")

## 6.2 Load Functions

### 6.2.1 General Representation of a Load Function

The load function consists of the operation code L (for load) and a constant, a variable, or an address with address identifier whose contents the function loads into accumulator 1.

| | | |
|---|---|---|
| L | +1200 | Constant (immediate addressing) |
| L | IW 16 | Digital memory location (direct addressing) |
| L | Act_val | Variable (symbolic addressing) |

The CPU executes the load function without regard to the result of the logic operation or to the status bits. The load function affects neither the RLO nor the status bits.

#### Effect on accumulator 2

The load function also changes the contents of accumulator 2. While the value of the address, constant or variable specified in the load statement is loaded into accumulator 1, the current contents of accumulator 1 are transferred to accumulator 2. The load function transfers the entire contents of accumulator 1 to accumulator 2. The original contents of accumulator 2 are lost.

The load function does not affect the contents of accumulators 3 and 4 on the S7-400 CPUs.

#### Loading in general

The digital address specified in the load function may be that of a byte, a word, or a doubleword (Figure 6.2).

*Loading a byte*
When a byte is loaded, its contents are written right-justified into accumulator 1. The remaining bytes in the accumulator are padded with "0".

*Loading a word*
When a word is loaded, its contents are written right-justified into accumulator 1. The high-value byte of the word (n+1) is right-justified in the accumulator, the low-value byte of the word (byte n) is at its immediate left. The remaining bytes in the accumulator are padded with "0".

*Loading a doubleword*
When a doubleword is loaded, its contents are written into accumulator 1. The lowest-value byte (byte n) is at the far left in the accumulator, the highest-value byte (byte n+3) at the far right.

### 6.2.2 Loading the Contents of Memory Locations

#### Loading inputs

| | | |
|---|---|---|
| L | IB n | Loads an input byte |
| L | IW n | Loads an input word |
| L | ID n | Loads an input doubleword |

In S7-400 controllers, only those input bytes which address an input module may be loaded from the process-image input table. On S7-300 controllers, you may address the entire process image.

#### Loading outputs

| | | |
|---|---|---|
| L | QB n | Loads an output byte |
| L | QW n | Loads an output word |
| L | QD n | Loads an output doubleword |

On S7-400 controllers, only those bytes which address an output module may be loaded from the process-image output table. On S7-300 controllers, you may address the entire process image.

#### Loading from the I/O

| | | |
|---|---|---|
| L | PIB n | Loads a peripheral input byte |
| L | PIW n | Loads a peripheral input word |
| L | PID n | Loads a peripheral input doubleword |

When loading from the I/O area, the input modules are referenced as peripheral inputs (PIs). Only the existing modules may be referenced. Note that direct loading of an I/O module may produce a different value than the loading of inputs on a module with that same address, for whereas the signal states of the inputs are the same as they were at the start of the program scan cycle (when the CPU updated the process image), direct loading of the I/O modules loads the current value.

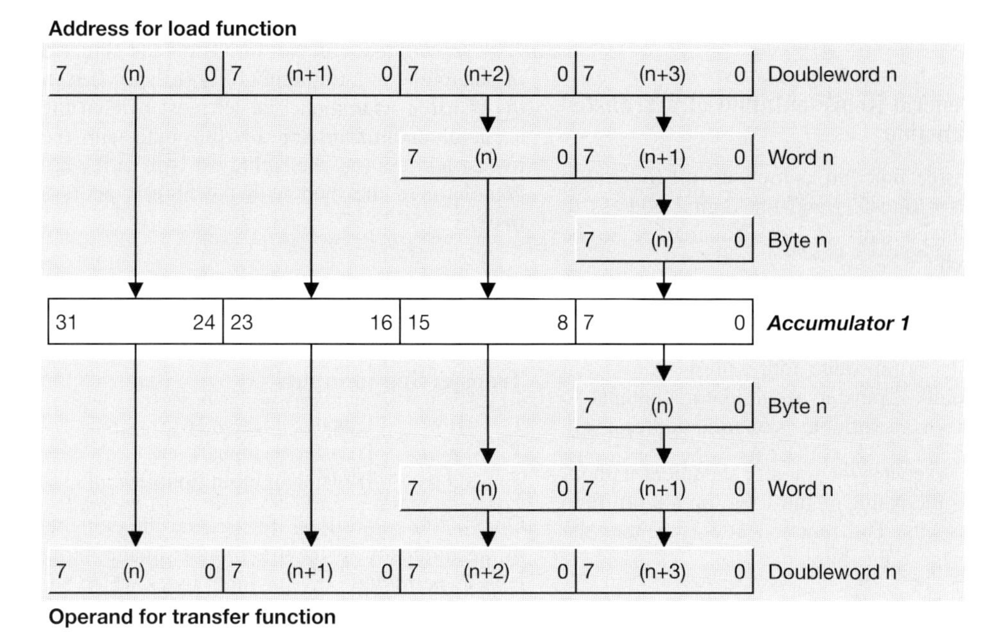


**Figure 6.2** Loading and Transferring Bytes, Words and Doublewords

### Loading bit memory

| | | |
|---|---|---|
| L | MB n | Loads a memory byte |
| L | MW n | Loads a memory word |
| L | MD n | Loads a memory doubleword |

Loading from the bit memory is always allowed, as this entire area in the CPU. Note, however, that different CPUs have different-sized bit memory areas.

## 6.2.3 Loading Constants

### Loading constants of elementary data type

You can load a constant, or fixed value, directly into the accumulator. For better readability, you can represent these constants in different formats. In the chapter 3 entitled "STL Programming Language" you will find an overview of the permissible formats. All constants which can be loaded into the accumulator are of elementary data type. The bit layout for constants (structure of the various data types) can be found in the chapter 24 entitled "Data Types".

| | | |
|---|---|---|
| L | B#16#F1 | Loads a 2-digit hexadecimal number |
| L | –1000 | Loads an INT number |
| L | 5.0 | Loads a REAL number |
| L | S5T#2s | Loads an S5 time value |
| L | TOD#8:30:00 | Loads a time of day |

### Loading pointers

Pointers are a special form of constant, and are used to calculate addresses. You can load the following pointers into the accumulator:

| | | |
|---|---|---|
| L | P#1.0 | Loads an area-internal counter |
| L | P#M2.1 | Loads an area-crossing counter |
| L | P#*name* | Loads the address of a local variable |

You cannot load a DB pointer or an ANY pointer into the accumulator, as these pointers exceed 32 bits. You will find further information on this topic in chapters 25 ("Indirect Addressing") and 26 ("Direct Variable Access").

## 6.3 Transfer Functions

### 6.3.1 General Representation of a Transfer Function

The transfer function comprises the operation code T (for transfer) and the digital address to which the contents of accumulator are to be transferred.

| | | |
|---|---|---|
| T | MW 120 | Transfer accumulator contents to the specified memory location (absolute addressing) |
| T | Setpoint | Transfer accumulator contents to a variable (symbolic addressing) |

The CPU executes the transfer statement without regard to the result of the logic operation or to the status bits. The function affects neither the RLO nor the status bits.

The transfer function transfers the contents of accumulator 1 by byte, word or doubleword to the specified address. The contents of accumulator 1 remain unchanged, making multiple transfers possible.

The transfer function may be used for accumulator 1 only. If you want to transfer a value from another accumulator, you must transfer that value to accumulator 1 using the accumulator functions, then transfer it to the desired address in memory.

#### Transferring in general

The digital address specified in the transfer statement may be that of a byte, a word, or a doubleword (Figure 6.2).

*Transferring a byte*
*Transferring* a byte transfers the rightmost byte in accumulator 1 to the byte specified in the transfer statement.

*Transferring a word*
*Transferring* a word transfers the contents of the rightmost word in accumulator 1 to the word specified in the transfer statement. The right byte of the word (low-order byte) is transferred to the byte with the higher address (n+1), the left word (high-order byte) to the byte with the lower address (n).

*Transferring a doubleword*
*Transferring* a doubleword transfers the contents of accumulator 1 to the doubleword specified in the transfer statement. The leftmost byte in the accumulator is transferred to the byte with the lowest address (n), the rightmost byte in the accumulator to the byte with the highest address (n+3).

### 6.3.2 Transferring to Various Memory Locations

#### Transferring to inputs

| | | |
|---|---|---|
| T | IB n | Transfer to input byte |
| T | IW n | Transfer to input word |
| T | ID n | Transfer to input doubleword |

On S7-400 controllers, transferring the contents of accumulator 1 to the process-image input table is permitted only for input bytes which address input modules. On S7-300 controllers, you may address the entire process image.

Transfers to inputs affect only the bits in the process image. Application: Presetting values for the purpose of testing or startup.

#### Transferring to outputs

| | | |
|---|---|---|
| T | QB n | Transfer to output byte |
| T | QW n | Transfer to output word |
| T | QD n | Transfer to output doubleword |

On S7-400 controllers, transfers to the process-image output table are permitted only for the output bytes which address an output module. On S7-300 controllers, you may address the entire process image.

#### Transferring to the I/O area

| | | |
|---|---|---|
| T | PQB n | Transfer to peripheral byte |
| T | PQW n | Transfer to a peripheral word |
| T | PQD n | Transfer to a peripheral doubleword |

When transferring the contents of accumulator 1 to the I/O area, the output modules are referenced as peripheral outputs (PQs). Only addresses on output modules may be specified. Transfers to I/O modules which have a process-image output table updates this table so that there is no difference between outputs and peripheral outputs with the same address.

**Transferring to bit memory**

| | | |
|---|---|---|
| T | MB n | Transfer to a memory byte |
| T | MW n | Transfer to a memory word |
| T | MD n | Transfer to a memory double-word |

Transfers to the memory areas are always permissible, since this entire area is located on the CPU. Note, however, that different CPUs have different-sized bit memory areas.

## 6.4 Accumulator Functions

The accumulator functions transfer values from one accumulator to another, or replace bytes in accumulator 1. Accumulator functions are executed without regard to the result of the logic operation or to the status bits. These functions affect neither the RLO nor the status bits.

### 6.4.1 Direct Transfers Between Accumulators

PUSH Shift accumulator contents "forward"

POP Shift accumulator contents "back"

ENT Shift accumulator contents "forward" (without accumulator 1)

LEAVE Shift accumulator contents "back" (without accumulator 1)

TAK Exchange contents of accumulators 1 and 2

CPUs with two accumulators (S7-300) need only the PUSH, POP and TAK operations; all five operations are available on CPUs with four accumulators (S7-400) (Figure 6.3).

**PUSH**

PUSH pushes the contents of accumulators 1 to 3 into the next higher accumulator (1 to 2, 2 to 3, and 3 to 4).. The contents of accumulator 1 remain unchanged.

You can use PUSH to enter the same value into more than one accumulator.

**POP**

POP transfers the contents of accumulators 4 to 2 into the next lower accumulator (4 to 3, 3 to 2, 2 to 1). The contents of accumulator 4 remain unchanged.

POP puts the values in accumulators 2 to 4 into accumulator 1, from whence they can be transferred to memory.

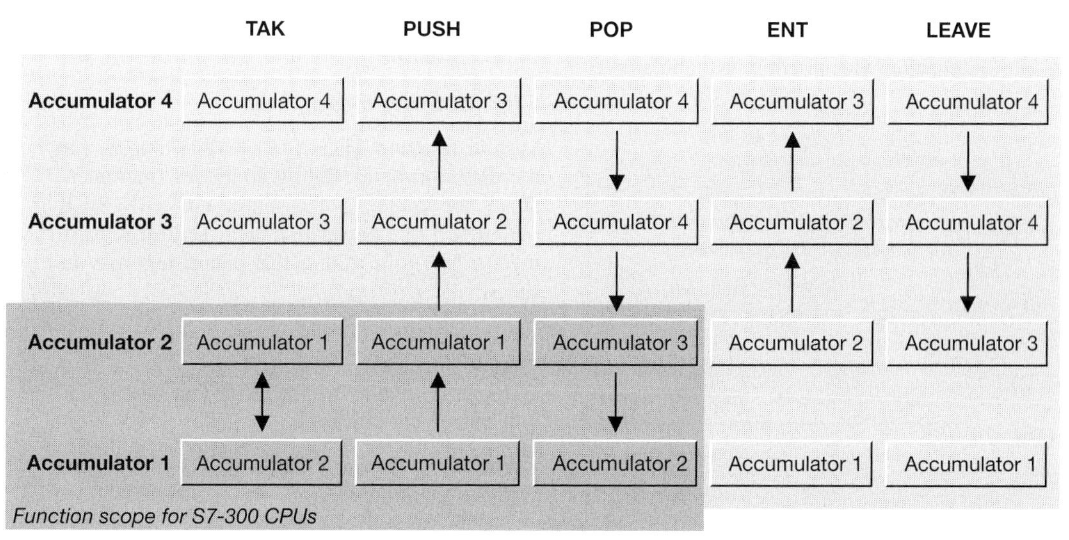


**Figure 6.3** Direct Transfers Between Accumulators on S7-300 and S7-400 CPUs

### TAK

TAK exchanges the contents of accumulators 1 and 2. The contents of accumulators 3 and 4 remain unchanged.

### ENT

ENT shifts the contents of accumulators 2 and 3 to the next higher accumulator. The contents of accumulators 1 and 2 remain unchanged.

If ENT is immediately followed by a load statement, the load shifts the contents of accumulators 1 to 3 "forward" (in a manner similar to that of PUSH); the new value is then in accumulator 1.

### LEAVE

LEAVE shifts the contents of accumulators 3 and 4 into the next lower accumulator. The contents of accumulators 4 and 1 remain unchanged.

The arithmetic functions include LEAVE functionality. Using LEAVE, you can also simulate the same functionality in other digital logic operations (in word logic, for example). When programmed after a digital logic operation, LEAVE places the contents of accumulators 3 and 4 into accumulators 2 and 3; the result of the digital logic operation remains unchanged in accumulator 1.

### 6.4.2 Exchange bytes in accumulator 1

CAW Exchange bytes in accumulator 1, low-order word

CAD Exchange bytes in entire accumulator 1

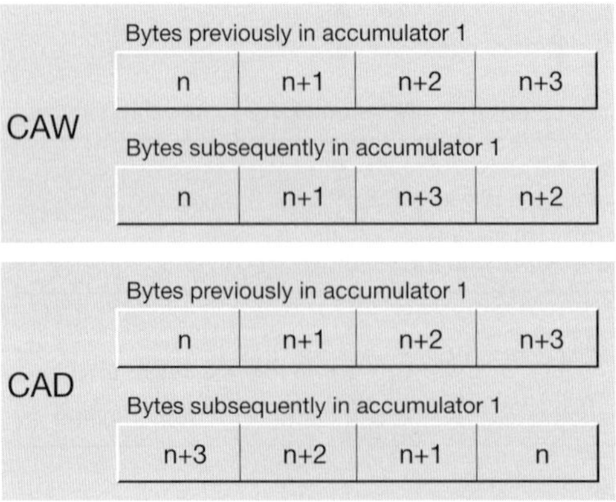

CAW

Bytes previously in accumulator 1

| n | n+1 | n+2 | n+3 |
|---|---|---|---|

Bytes subsequently in accumulator 1

| n | n+1 | n+3 | n+2 |
|---|---|---|---|

CAD

Bytes previously in accumulator 1

| n | n+1 | n+2 | n+3 |
|---|---|---|---|

Bytes subsequently in accumulator 1

| n+3 | n+2 | n+1 | n |
|---|---|---|---|

**Figure 6.4** Exchanging Bytes in Accumulator 1

CAW and CAD exchange the bytes in accumulator 1 by word or doubleword (Figure 6.4).

## 6.5 System Functions for Data Transfer

The following system functions are available for data transfer:

- SFC 20 BLKMOV
  Copy data area
- SFC 21 FILL
  Fill data area

Each of these system functions has two parameters of data type ANY (Table 6.1). Theoretically, each of these parameters may specify an arbitrary address, variable, or absolute memory location. If you specify a variable of complex data type, it must be a "complete" variable; components of a variable (such as individual array or structure components) are not permitted. For absolute addressing, use an ANY pointer; these pointers are discussed in detail in section 25.1, "Pointers".

If you use temporary local data as actual parameter in a block parameter of type ANY, the editor assumes that this actual parameter is of data type ANY. In this way, you can generate an ANY pointer in the temporary local data that can be modified at runtime, that is, you can set up a variable area. The "Message Frame " example in Chapter 26, "Direct Variable Access", shows how to use this "variable ANY pointer".

### 6.5.1 Copying a data area

System function SFC 20 BLKMOV copies , in the direction of ascending addresses (incrementally), the contents of a source area (SRCBLK parameter) to a destination area (DSTBLK parameter). The following actual parameters may be specified:

- Any variable from the address areas for inputs (I), outputs (Q), memory bits (M) or data blocks (variables from global data blocks and instance data blocks)
- Absolute-addressed data areas by specifying an ANY pointer
- Variables in temporary local data of data type ANY (special case)

You cannot use SFC 20 to copy timers or counters, or to copy information from or to the modules (I/O area) or system data blocks (SDBs).

In the case of inputs and outputs, the specified area is copied regardless of whether or not the addresses specified are on the input or output modules. You may also specify a variable or the address of an area in a data block in load memory (a data block programmed with the keyword UNLINKED).

Source and destination may not overlap. If source area and destination area are of different lengths, the shorter of the two determines the transfer.

Example: The variable Frame (a structured variable as user data type, for example) in data block "Rec_mailb" is to be copied to variable Frame1 (which is of the same data type as Frame) in data block "Buffer". The function value is to be entered in the variable Copyerror in data block "Evaluation".

```
CALL SFC 20 (
  SRCBLK  := Rec_mailb.Frame,
  RET_VAL := Evaluation.Copyerror,
  DSTBLK  := Buffer.Frame1);
```

### 6.5.2 Filling a data area

System function SFC 21 FILL copies a specified value (source area) to a memory area (destination area) until the destination area is completely filled. The transfer is in the direction of ascending addresses (incremental). The parameters may be assigned the following actual values:

- Any variables from the address areas for inputs (I), outputs (Q), memory bits (M), or data blocks (variables from global data blocks and instance data blocks)
- Absolute-addressed data areas by specifying an ANY pointer
- Variables in the temporary local data of data type ANY (special case)

You cannot use SFC 21 to copy timers or counters, or to copy information from or to the modules (I/O area) or system data blocks (SDB).

In the case of inputs and outputs, the specified area is copied without regard to the whether or not the addresses are those on input or output modules.

Source and destination may not overlap. The destination area is always completely filled, even when the source area is longer than the destination area or when the length of the destination area is not an integer multiple of the length of the source area.

Example: Data block DB 13 consists of 128 data bytes, all of which are to be set to the value of memory byte MB 80.

```
CALL SFC 21 (
  BVAL    := MB 80,
  RET_VAL := MW 32,
  BLK     := P#DB13.DBX0.0 BYTE 128);
```

**Table 6.1** Parameters for SFC 20 and 21

| SFC | Parameter | Declaration | Data Type | Contents, Description |
|---|---|---|---|---|
| 20 | SRCBLK | INPUT | ANY | Source area from which data are to be copied |
| | RET_VAL | OUTPUT | INT | Error information |
| | DSTBLK | OUTPUT | ANY | Destination to which data are to be copied |
| 21 | BVAL | INPUT | ANY | Source area to be copied |
| | RET_VAL | OUTPUT | INT | Error information |
| | BLK | OUTPUT | ANY | Destination to which the source area is to be copied (including multiple copies) |

# 7 Timer Functions

Timer functions are used to implement timing sequences. Some examples of timing sequences are waiting and monitoring times, measuring of a time interval, or the generation of pulses. The timer functions are stored in the system memory of the CPU; the number of timer functions is CPU-specific. On the S7-300, the timer functions are updated only during execution of the main program in organization block OB 1.

The following timer types are available:

- Pulse timer
- Extended pulse timer
- On-delay timer
- Retentive on-delay timer
- Off-delay timer

When you start a timer, you specify the dynamic response and the duration, the latter being the length of time the timer is to run; you can also reset or enable ("retrigger") timers. Binary logic operations are used to check timers ("timer running"). Load functions are used to transfer the current time value, in binary or BCD, to accumulator 1.

The examples shown in this chapter and the IEC timer calls are also on the diskette which accompanies the book under the "Basic Functions" program in function block FB 107 or source file Chap_7.

## 7.1 Programming a Timer

### 7.1.1 Starting a Timer

A timer is started (begins running) when the result of the logic operation (RLO) changes prior to the start instruction. Such a signal state change is always required to start a timer. In the case of an off-delay timer, the RLO must change from "1" to "0"; in all other cases, the timer begins running when the RLO goes from "0" to "1".

You can start any timer as one of five possible types (Figure 7.1). However, it is not a good idea to use a given timer as more than one type.

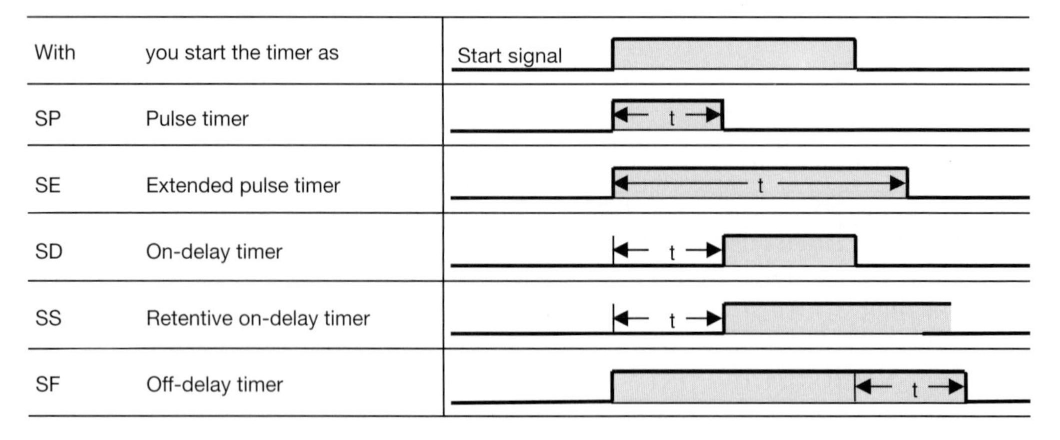


**Figure 7.1** Start Instructions for Timers

### 7.1.2 Specifying the Time

When a timer is started, it takes the value in accumulator 1 as its running time, or duration. How and when the value gets into the accumulator is of no consequence. To make your program more readable, the best way would be to load the running time directly into the accumulator before starting the timer, either as a constant (direct specification of the value) or as a variable (for example a memory word containing the value).

*Specifying the duration as a constant*

```
L    S5TIME#10s;    //Duration 10 s
L    S5T#1m10ms;    //Duration 1 min + 10 ms
```

The duration, or running time, is specified in minutes, seconds and milliseconds. The range extends from S5TIME#10ms to S5TIMEh46m30s (which corresponds to 9990 s). You may use either S5TIME# or S5T# to identify a constant.

*Specifying the duration as a variable*

```
L    S5T#10m;       //Duration 10 min
T    MW 20;         //Save duration
...  ;
L    MW 20;         //Load duration
```

*Structure of the duration*
Internally, the duration is composed of the time value and the time base. The duration is equal to time value x time base. The duration is the period of time during which the timer is active ("timer running"). The time value represents the number of timing periods the timer is to run. The time base specifies the timing period the CPU operating system is to use to decrement the timer (Figure 7.2).

You can also set up the duration directly in the word. The smaller the time base, the more accurately the actual duration processed. For example, if you want to implement a duration of 1 s, you can do so in one of three different ways:

Duration = $2001_{hex}$ Time base 1 s
Duration = $1010_{hex}$ Time base 100 ms
Duration = $0100_{hex}$ Time base 10 ms

The last of the three is the preferred method in this case.

When a timer is started, the CPU uses the programmed time value as the timer's running time. The operating system updates timers at fixed intervals and independently of the user program scan, that is, it decrements an active timer's time value as per the timing period indicated by the time base. When the value reaches zero, the timer is regarded as expired. The CPU then sets the timer status (signal state "0" or "1", depending on the type of timer involved) and drops all further activities until the timer is started again. When a time value of zero is specified, the timer remains active until the CPU processes it and discovers that the timer has expired.

Timers are updated asynchronously to the program scan. It is therefore possible that the status of the timer at the start of the cycle may differ from its status at the end of the cycle. If you use timer instructions at only one location in the program and in the suggested order (see below), there will be no errors due to asynchronous timer updating.

### 7.1.3 Resetting and Enabling a Timer

```
R     T n      Resets a timer
FR    T n      Enables a timer
```

A timer is reset when the RLO is "1" when the reset statement is encountered. As long as the RLO is "1", timer checks for "1" will return a

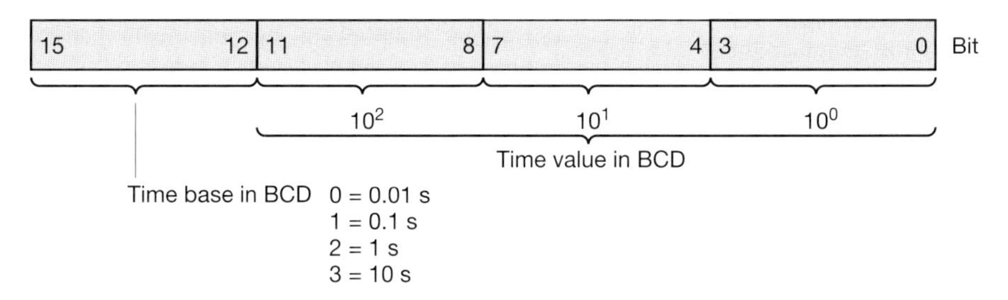


**Figure 7.2** Bit Assignments in the Duration

check result of "0" and timer checks for "0" will return a check result of "1". Resetting a timer sets the time value and the time base to zero.

The enable instruction is used to "retrigger" an active timer, that is, to restart it. This is possible only when the start instruction is processed while the RLO is at "1". The enabling of a timer requires a positive edge.

An enable instruction is not required to start or reset a timer, that is to say, it is not necessary to normal timer operation.

### 7.1.4 Checking a Timer

#### Checking the timer status

| | | |
|---|---|---|
| A | T n | Check for signal state "1" and combine according to AND |
| O | T n | Check for signal state "1" and combine according to OR |
| X | T n | Check for signal state "1" and combine according to Exclusive OR |
| AN | T n | Check for signal state "0" and combine according to AND |
| ON | T n | Check for signal state "0" and combine according to OR |
| XN | T n | Check for signal state "0" and combine according to Exclusive OR |

You can check a timer as you would an input, for instance, and further process the result. Depending on the type of timer, a check for signal state "1" produces different variations in the timing sequence (see the description of the dynamic response, below). As it does in the case of inputs, a check for signal state "0" returns precisely the reverse result as does the check for signal state "1".

#### Checking the time value

| | | |
|---|---|---|
| L | T n | Loads a binary time value |
| LC | T n | Loads a BCD time value |

Load functions L T and LC T check the specified time value and make it available in accumulator 1 in binary (L) or in binary-coded decimal (LC). The value loaded into the accumulator is the value current at the instant of the check (in the case of an active timer, the time value, in this case the value loaded into the accumulator, is counted down in the direction of zero).

*Loading a binary time value (direct load)*
The value specified in the timer instruction is in binary, and can be loaded into accumulator 1 in this form. The time base is lost in this case, and in its place in accumulator 1 is the value "0". The value in accumulator 1 therefore corresponds to a positive number in INT format, and can be further processed with, for example, compare functions. Please note that it is the *time value* that is in the accumulator, not the *duration*.

```
L    T 15;     //Load current time value
T    MW 34;    //and save
```

*Loading a BCD time value (coded load)*
You can also use a "coded load" instruction to load a binary value into accumulator 1. In this case, both the time value and the time base are available in binary-coded decimal (BCD). The contents of the accumulator are the same as when a time value is specified (see above), that is, the left-hand word (high-order word) in the accumulator contains zero. Example:

```
LC   T 16;     //Load current time value in BCD
T    MW 122;   //and save
```

### 7.1.5 Sequence of Timer Instructions

When you program a timer, you do not need to use all of the statements that are available for timers, but only those statements applicable to the timer you want to implement. Normally, this would include starting the timer with the specified duration, and binary checking of the timer.

In order for a timer to perform as described in the preceding sections, a certain order must be observed when programming timer operations. Table 7.1 shows the optimum order for all timer operations. Simply omit the statements that are not needed. If a timer is started and reset "simultaneously" in the statement sequence shown, the timer will start, but the subsequent reset statement will immediately reset it. When the timer is then checked, the fact that it was started will therefore go unnoticed.

Table 7.1 Sequence of Timer Operations

| Timer Operation | Examples: | | |
|---|---|---|---|
| Enable timer | A | I | 16.5 |
| | FR | T | 5 |
| Start timer | A | I | 17.5 |
| | L | S5T#1s | |
| | SP | T | 5 |
| Reset timer | A | I | 18.0 |
| | R | T | 5 |
| Digital timer check | L | T | 5 |
| | T | MW | 20 |
| | LC | T | 5 |
| | T | MW | 22 |
| Binary timer check | A | T | 5 |
| | = | Q | 2.0 |

## 7.2 Pulse Timers

### Starting a pulse timer

The diagram in Figure 7.3 describes the dynamic behavior of a timer that was started as a pulse timer, and its behavior when it is reset. The description applies when you keep to the sequence of operations shown above (the enable instruction is not necessary in this case).

① The timer is started when the RLO at its Start input changes from "0" to "1". It runs for the programmed duration as long as the RLO at the Start input remains at "1". Checks for signal state "1" return a check result of "1" as long as the timer is running.

② The timer stops if the RLO at its Start input goes to "0" before the time has elapsed. Checking the timer for signal state "1" returns a check result of "0". The time value shows the time remaining, which also shows at what point the timing sequence was prematurely interrupted.

### Resetting a pulse timer

③ RLO "1" at the Reset input of an active timer resets that timer. A check for signal state "1" then returns a check result of "0". The time value and the time base are also set to zero. A change in the RLO from "1" to "0" at the Reset input while RLO "1" is present at the Start input has no effect on the timer.

④ If a timer is not running, RLO "1" at its Reset input has no effect.

⑤ If the RLO at the Start input changes from "0" to "1" while the Reset signal is present, the timer is started but the subsequent reset instruction resets it immediately (indicated by a line in the diagram). If the prescribed sequence of timer operations is observed, the brief starting of the timer will have no effect on the check.

### Enabling a pulse timer

Comments relating to Figure 7.4:

❶ If the RLO goes from "0" to "1" at the Enable input, the timer, if active, will be restarted when the start instruction is processed (as long as it is processed when

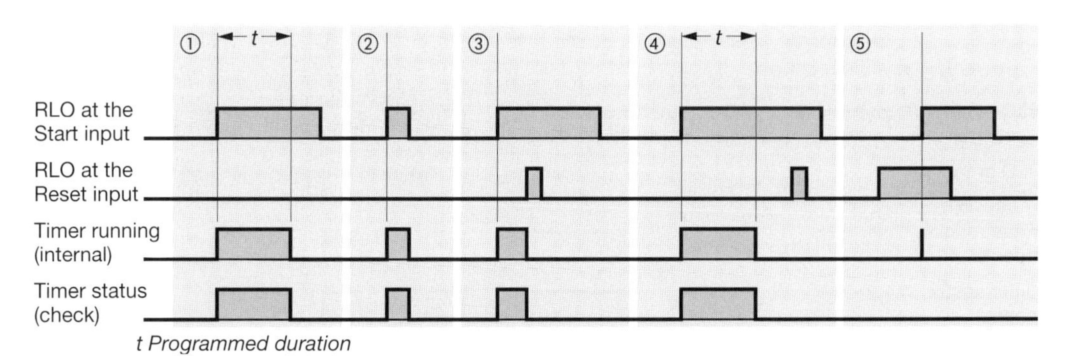


Figure 7.3 Dynamic Behavior when Starting and Resetting a Pulse Timer

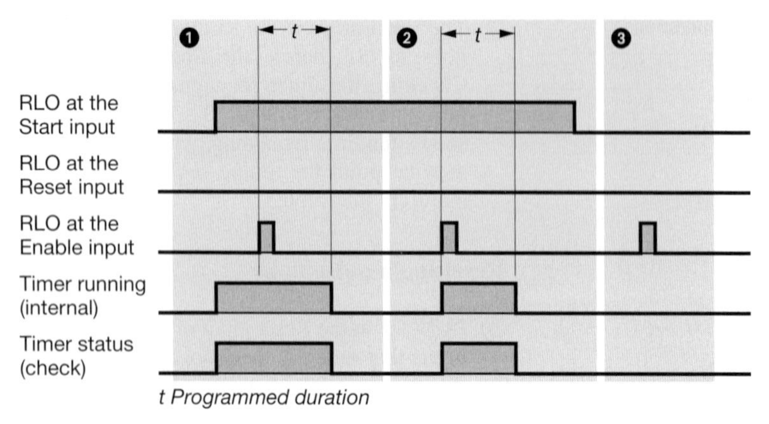


**Figure 7.4** Enabling a Pulse Timer

the RLO is "1"). The programmed duration is taken as the current time value for the restart. A change in the RLO from "1" to "0" at the Enable input has no effect.

❷ If the RLO at the Enable input changes from "0" to "1" and the Start instruction is processed with RLO "1", the timer, if inactive, is also started with the programmed duration as a pulse timer.

❸ A change in the RLO at the Enable input has no effect when the RLO is "0" at the Start input.

## 7.3 Extended Pulse Timers

### Starting an extended pulse timer

The diagram in Figure 7.5 describes a timer's performance after it is started as an extended pulse timer and when it is reset. The description applies when the sequence of operations shown above is observed (the Enable instruction is not required in this case).

①② When the RLO at the timer's Start input changes from "0" to "1", the timer is started. It runs for the programmed duration, without regard to changes in the RLO at the Start input. Checks for signal state "1" return a check result of "1" as long as the timer is running.

③ If the RLO at the Start input goes from "0" to "1" while the timer is running, the timer is restarted with the programmed time value (that is, the timer is "retriggered"). It can be restarted as often as required without the time having to expire first.

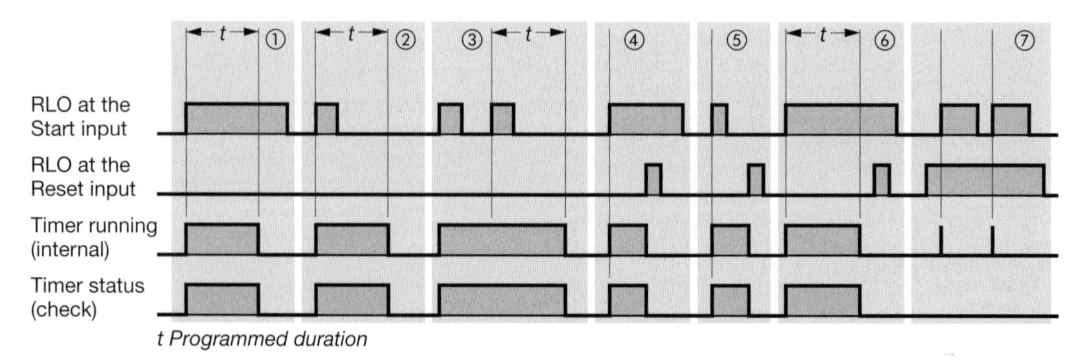


**Figure 7.5** Dynamic Behavior of an Extended Pulse Timer

### Resetting an extended pulse timer

④⑤ RLO "1" at an active timer's Reset input resets the timer. A check for signal state "1" then returns a check result of "0". The time value and the time base are also set to zero.

⑥ If the timer is not running, RLO "1" at the Reset input has no effect.

⑦ If the RLO at the Start input goes from "0" to "1" while the Reset signal is present, the timer is started but the subsequent reset resets it immediately (indicated by a line in the diagram). If the prescribed order for timer operations is observed, this brief starting of the timer will not affect a check.

### Enabling an extended pulse timer

Comments on Figure 7.6:

❶ When the RLO at an active timer's Enable input goes from "0" to "1", the timer is restarted when the start instruction is processed (as long as the RLO is "1"). The programmed duration is taken as the current time value for the restart. A change in the RLO from "1" to "0" at the Enable input has no effect.

❷ If the RLO at the Enable input of an inactive timer goes from "0" to "1" and the start instruction is executed with RLO "1", the timer is also started with the programmed duration as an extended pulse timer.

❸❹ If the RLO at the Start input is "0", a change in the RLO at the Enable input has no effect.

## 7.4 On-Delay Timers

### Starting an on-delay timer

The diagram in Figure 7.7 describes a timer's dynamic performance after it is started as an on-delay timer and when it is restarted. The description applies when the sequence of operations shown above is observed (the Enable is not necessary in this case).

① When the RLO at the timer's Start input changes from "0" to "1", the timer is started. The timer runs with the programmed duration as the time value. Checks for signal state "1" return a check result of "1" when the time expires without incident and the RLO at the Start input is still "1" (on delay).

② If the RLO at the Start input of a running timer goes from "1" to "0", the timer stops. In such cases, a check for signal state "1" always returns a check result of "0". The time value shows the time remaining, that is, the period by which the timer was prematurely interrupted.

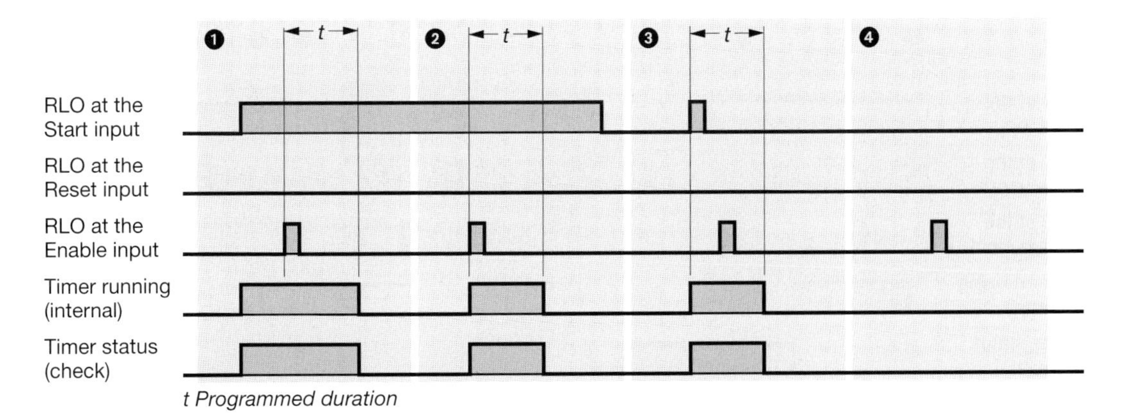


**Figure 7.6** Enabling an Extended Pulse Timer

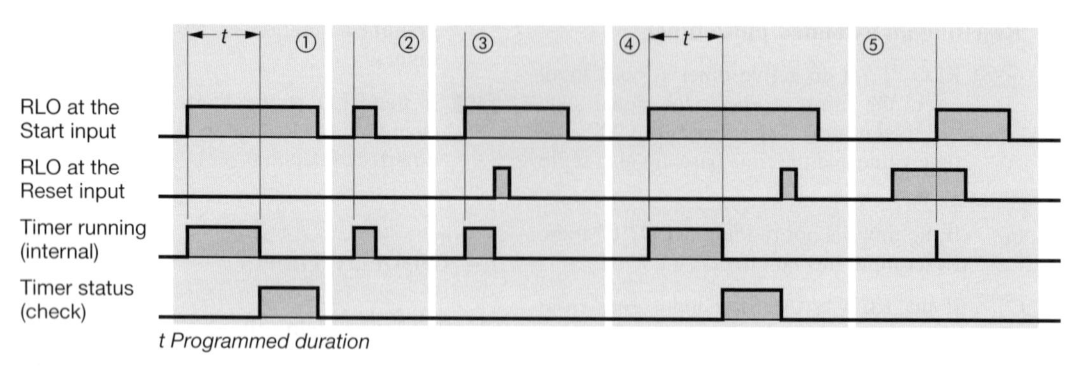


**Figure 7.7** Dynamic Behavior of an On-Delay Timer

### Resetting an on-delay timer

③ RLO "1" at an active timer's Reset input resets the timer. A check for signal state "1" returns a check result of "0" , even when the timer is not running and RLO "1" is still present at the Start input. Time value and time base are also set to zero. A change in the RLO at the Reset input from "1" to "0" while RLO "1" is still present at the Start input has no effect on the timer.

④ RLO "1" at the Reset input of a timer which is no longer running also resets the timer. A check for signal state "1" consequently returns a check result of "0".

⑤ If the RLO at the Start input changes from "0" to "1" while the Reset signal is still present, the timer is started but the subsequent reset resets it again immediately (indicated by a line in the diagram). If you are careful to observe the prescribed sequence of the timer operations, a check of the timer will not be affected by the brief start.

### Enabling an on-delay timer

Comments on Figure 7.8:

❶ If the RLO at a running timer's Enable input changes from "0" to "1", the timer is restarted when the start operation is processed (as long as it is done when the RLO is "1"). The programmed duration is taken as the current time value for the restart. A change in the RLO from "1" to "0" at the Enable input has no effect.

❷ If the RLO at the Enable input goes from "0" to "1" after the time has elapsed without incident, processing of the start operation does not affect the timer.

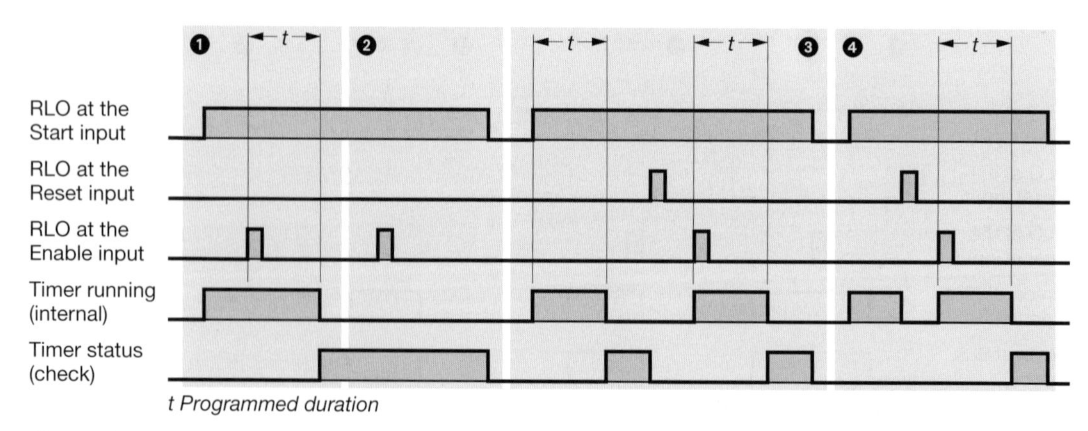


**Figure 7.8** Enabling an On-Delay Timer

❸❹ If the RLO goes from "0" to "1" at the Enable input of a timer that has been reset, and if the start operation is processed with RLO "1", the timer is restarted. The programmed duration is taken as the current time value for the restart.

If the RLO at the Start input is "0", a change in the RLO at the Enable input has no effect.

## 7.5 Retentive On-Delay Timers

### Starting a retentive on-delay timer

The diagram in Figure 7.9 describes a timer's dynamic performance after it is started as a retentive on-delay timer, and after it is restarted. The description applies if you have carefully observed the sequence of operations shown above (the Enable is not necessary in this case).

①② When the RLO at the timer's Start input changes from "0" to "1", the timer is started. It runs with the programmed duration without regard to any further changes in the RLO at the Start input. When the time has elapsed, checks for signal state "1" return a check result of "1", regardless of the RLO at the Start input. The check result is not "0" until the timer has been reset, regardless of what the RLO at the Start input is.

③ If the RLO at the Start input goes from "0" to "1" while the timer is running, the timer is restarted with the programmed time value (that is, the timer is "retriggered"). The timer may be restarted as often as required without the time having to elapse first.

### Resetting a retentive on-delay timer

④⑤ RLO "1" at the timer's Reset input resets the timer without regard to the RLO at the Start input. Checks for signal state "1" then return a check result of "0". Time value and time base are set to zero.

⑥ If the RLO at the Start input changes from "0" to "1" while the Reset signal is present, the timer is started, but the subsequent reset resets it again immediately (indicated by a line in the diagram). If you are careful to observe the prescribed sequence for timer operations, the brief start will have no effect on the check.

### Enabling a retentive on-delay timer

Comments on Figure 7.10:

❶ If the RLO at a running timer's Enable input changes from "0" to "1", the timer is restarted when the start instruction is processed (as long as this takes place while the RLO is "1"). The timer restarts with the programmed duration as the time value. A change in the RLO at the Enable input from "1" to "0" has no effect.

❷ When the RLO at the Enable input goes from "0" to "1" after the timer has expired without incident, the start instruction has no effect on the timer.

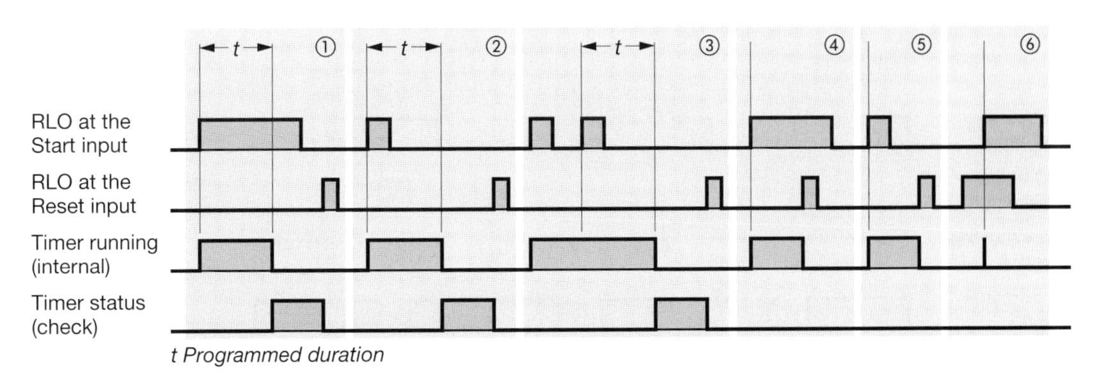


**Figure 7.9** Dynamic Behavior of a Retentive On-Delay Timer

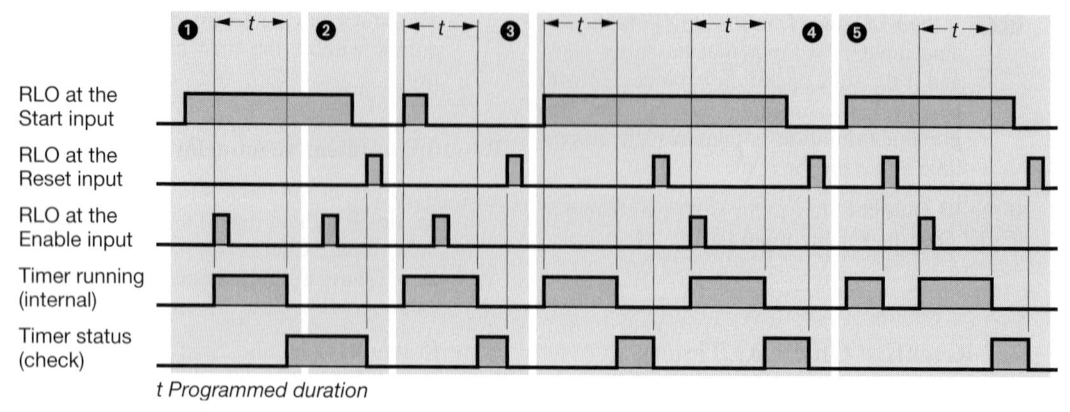


**Figure 7.10** Enabling a Retentive On-Delay Timer

❸ When the RLO at the Start input is "0", a change in the RLO at the Enable input has no effect.

❹❺ If the RLO at the Enable input of a restarted timer changes from "0" to "1", and if RLO "1" is still present at the Start input, the timer is restarted. The timer restarts with the programmed duration as the current time value.

## 7.6 Off-Delay Timers

### Starting an off-delay timer

The diagram in Figure 7.11 describes the dynamic performance of a timer after it is started as off-delay timer, and when it is restarted. The description applies when you have carefully observed the sequence for timer operations given above (the Enable is not required in this case).

①③ When the RLO at the timer's Start input goes from "0" to "1", the timer is started. It runs with the programmed duration as the time value. Checks for signal state "1" return a check result of "1" when the RLO at the Start input is "1" or when the timer is running (off delay).

② When the RLO at the Start input changes from "0" to "1" while the timer is running, the timer is reset. The timer is not restarted until there is a falling edge at the Start input.

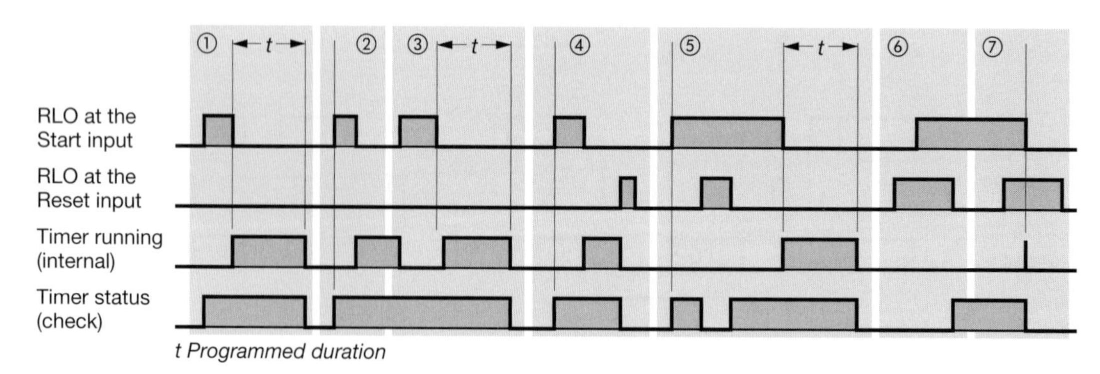


**Figure 7.11** Dynamic Behavior of an Off-Delay Timer

## Resetting an off-delay timer

④ RLO "1" at the Reset input of a running timer resets the timer. Checks for signal state "1" then return a check result of "0". Time value and time base are also set to zero.

⑤⑥ RLO "1" at the Start input and at the Reset input resets the timer's binary output (checks for signal state "1" then return a check result of "0"). If the RLO at the Reset input then changes back to "0", the timer's output goes back to "1".

⑦ If the RLO at the Start input changes from "1" to "0" while the Reset signal is present, the timer is started but the subsequent reset resets it again immediately (indicated by a line in the diagram). A check for signal state "1" then immediately returns a check result of "0".

## Enabling an off-delay timer

Comments on Figure 7.12:

❶❸ If the RLO at the Enable input of an inactive timer changes from "0" to "1", the timer is not affected by the execution of the start operation. A change in the RLO at the Enable input from "1" to "0" also has no effect.

❷ If the RLO at the Enable input of a running (active) timer changes from "0" to "1", the timer is restarted when the start operation is processed. The timer uses the programmed duration as the current time value for the restart.

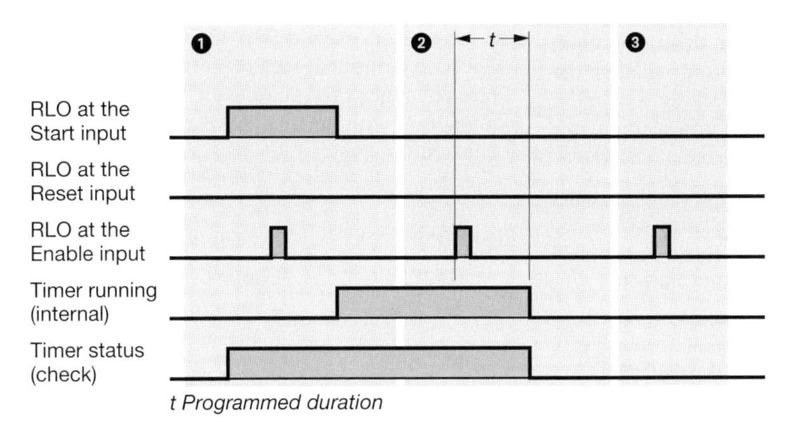


**Figure 7.12** Enabling an Off-Delay Timer

# 8 Counter Functions

Counter functions allow you to have counting tasks carried out directly by the central processor. The counters can count up and down, and the counting range extends over three decades (from 000 to 999).

The counting rate of these counters depends on your program's scan time! In order to count, the CPU must detect a signal state change in the input pulse, that is to say, an input pulse (or space (interpulse period)) must be present for at least one program scan cycle. The longer the program scan cycle, the lower the counting rate.

The counters described in this chapter are stored in the system memory of the CPU. You can set counters to an initial value, reset them, count up, and count down. You can find out whether the count is zero or not zero by checking a counter. The current count can be loaded into accumulator 1 in binary or in binary-coded decimal (BCD).

The examples shown in this chapter and the calls for IEC counters are on the diskette which accompanies the book under the "Basic Functions" program in FB 108 or source file Chap_8.

## 8.1 Setting and Resetting Counters

### Setting a counter

S  C n  Sets a counter

A counter is set when the RLO goes from "0" to "1" prior to the Set operation S. A positive edge is always required to set a counter.

"Set counter" means to load the counter with an initial value. The initial value with which it is to be loaded is in accumulator 1 (see below). The range extends from 0 to 999.

### Specifying the count

The "set counter" statement takes the value in accumulator 1 as count value. How and when that value got into accumulator 1 is of no concern. To make your program more readable, you should load the count value into the accumulator immediately before the Set statement, either in the form of a constant (direct specification of a count value) or a variable (such as a memory word containing the count value).

*Specifying a count in the form of a constant*

```
L    C#100;          //Count value 100
L    W#16#0100;      //Count value 100
```

A count comprises three decades, and may be in the range from 000 to 999. Only positive BCD values are permitted; the counters cannot process negative values. You may use C# or W#16# (in conjunction with decimal digits only) to identify a constant.

*Specifying a count in the form of a variable*

```
L    C#200;          //Count value 200
T    MW 56;          //Save count value
. . . ;
L    MW 56;          //Load count value
```

The Set operation expects there to be a count in accumulator 1 consisting of three right-justified decades. The meanings of the bits in the count (data type C#) are described in detail under "Data Types" in Chapter 24.

### Resetting a counter

R  C n  Resets a counter

A counter is reset when the RLO is "1" when the Reset statement is encountered. As long as RLO "1" is present, counter checks for "1" return a check result of "0" and counter checks for "0" return a counter result of "1". Resetting a counter sets the count value to "0".

## 8.2 Counting

### Counting up

CU C n Count up

A counter is counted up (incremented) when the RLO changes from "0" to "1" prior to the CU (count up) statement. Up counting always requires a positive signal edge.

Each positive edge preceding the CU operation increases the count value by one unit until the upper limit value of 999 is reached. A positive edge at the CU input then has no further effect. There is no carry.

### Counting down

CD C n Count down

A counter is counted down (decremented) when the RLO changes from "0" to "1" prior to the CD (count down) statement. Down counting always requires a positive signal edge.

Each positive edge preceding the CD statement decreases the count value by one unit until the lower limit value of 0 has been reached. A positive edge at the CD input then has no further effect. The count value does not go into the negative range.

## 8.3 Checking a Counter

### Binary counter check

| | | |
|---|---|---|
| A | C n | Check for signal state "1" and combine according to AND |
| O | C n | Check for signal state "1" and combine according to OR |
| X | C n | Check for signal state "1" and combine according to Exclusive OR |
| AN | C n | Check for signal state "0" and combine according to AND |
| ON | C n | Check for signal state "0" and combine according to OR |
| XN | C n | Check for signal state "0" and combine according to Exclusive OR |

You can check a logical counter combination as you would an input, for instance, and further combine the result of the check. Checks for signal state "1" return a check result of "1" when the count is greater than zero, and a check result of "0" when the count is zero.

### Direct loading of a count value

L C n Direct loading of a count value

The load function L C transfers the count specified in the counter function into accumulator 1 in the form of a binary number. This value is the value current at the instant of the check. The value now in accumulator 1 corresponds to a positive number in INT format, and can be further processed, for example with arithmetic functions. Example:

```
L    C 99;     //Load current count value
T    MW 76;    //and save
```

### Coded loading of a count value

LC C n Coded loading of a count value

The load function LC C transfers the count specified in the counter function to accumulator 1 in the form of a binary-coded decimal number. This value is the value current at the instant of the check. The count is subsequently available in the accumulator as a right-justified BCD number. It has the same structure as the specified count. Example:

```
LC  C 99;     //Load current count value in BCD
T   MW 50;    //and save
```

## 8.4 Enabling a Counter

FR C n Enable counter

When you enable a counter, you can set the counter and use it for counting without a positive signal edge having to precede the relevant operation. However, this is possible only when the relevant operation is processed while the RLO is "1".

The Enable is active when the RLO goes from "0" to "1" before the Enable instruction is encountered. A positive signal edge is always required to enable a counter.

A counter need not be enabled in order for it to be set, reset, or used for counting (that is to say, for normal operation of a counter).

*Note:* Enable affects setting, counting up and counting down simultaneously! A positive edge at the time of the Enable instruction causes all subsequent instructions (S, CU and CD) which have signal state "1" to be executed.

Examples of counter functions:

```
A    "Enable";
FR   "counter";

A    "Count up";
CU   "counter";

A    "Count down";
CD   "counter";

A    "Set";
L    C#020;
S    "counter";

A    "Reset";
R    "counter";

A    "Counter";
=    output;
```

Comments on Figure 8.1:

① The positive edge at the Set input sets the counter to the initial value of 20.

② A positive edge at the CU input increments the counter by one unit.

③ Because the RLO at the Set input is "1", an Enable increments the count by one unit.

④ The positive edge at the Reset input decrements the count by one unit.

⑤ The Enable causes the statements for up counting and down counting to be executed, as RLO "1" is present at both inputs.

⑥ The positive edge at the Set input sets the counter to the initial value of 20.

⑦ RLO "1" at the Reset input resets the counter. A check for signal state "1" returns a check result of "0".

⑧ Because RLO "1" is still present at the Set input, the Enable again sets the counter to 20. A check for signal state "1" returns a check result of "1".

## 8.5 Sequence of Counter Instructions

When programming a counter, you do not need all the statements available for it. You need program only those statements required for the timer in question. For example, all that is needed for a down counter is the setting of the initial value, down counting, and a binary check for "0".

For a counter to perform as described in the last several sections, you must observe a specific sequence when programming the counter instructions. Table 8.1 shows the optimum sequence for all counter instructions. Simply omit the unneeded statements when you write your program.

If a Reset is to have a "static" effect on the CU, CD and S statements and be independent of the

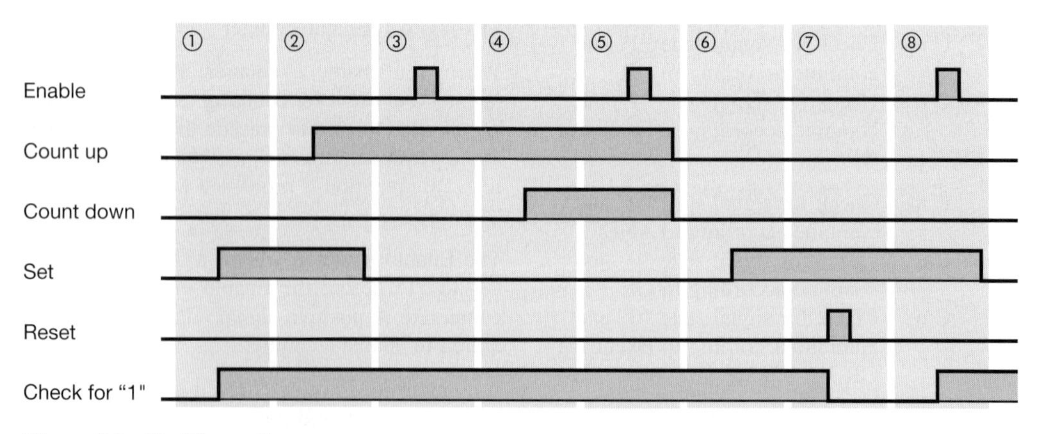


**Figure 8.1** Enabling a Counter

**Table 8.1** Sequence of Counter Instructions

| Counter Instruction | Examples | | |
|---|---|---|---|
| Enable counter | A | I | 22.0 |
| | FR | C | 17 |
| Count up | A | I | 22.1 |
| | CU | C | 17 |
| Count down | A | I | 22.2 |
| | CD | C | 17 |
| Set counter | A | I | 22.3 |
| | L | C#500 | |
| | S | C | 17 |
| Reset counter | A | I | 22.4 |
| | R | C | 17 |
| Digital check | L | C | 17 |
| | T | MW | 30 |
| | LC | C | 17 |
| | T | MW | 32 |
| Binary check | A | C | 17 |
| | = | Q | 13.0 |

result of the logic operation (RLO), you have to write the Reset statement for the counter in question after these statements and before the check statement for the counter. If the counter is then set and reset "simultaneously", it will still be assigned a value, but is then immediately reset by the Reset statement. The subsequent check therefore does not recognize the fact that the counter had been briefly set.

If the setting of a counter is to have a "static" effect on the counter statements and be independent of the RLO, the Set instruction for that counter must be programmed after the counting instructions. If a counter is set and reset "simultaneously", the counting instructions still affect the count, but the count is subsequently set to the programmed value, which it retains for the remainder of the program scan.

The sequence of statements for up and down counting is not relevant.

## 8.6 Parts Counter Example

The example illustrates the handling of timers and counters. It is programmed with inputs, outputs and memory bits so that it can be programmed at any point in any block. A function without block parameters has been used in the example.

### Functional description

Parts are transported on a conveyor belt. A light barrier detects and counts the parts. After a set number, the counter sends the signal "Finished". The counter is equipped with a monitoring circuit. If the signal state of the light barrier does not change within a specified time, the monitoring circuit emits a signal.

The "Set" input gives the counter its starting value (the number of parts to be counted). A positive edge at the light barrier decrements the counter by one unit. When the count reaches zero, the counter sends the "Finished" signal. Prerequisite is that the parts are lying individually (at intervals from one another) on the conveyor belt (Figure 8.2).

The "Set" input also sets the "Active" signal. The controller monitors a signal state change of the light barrier only in the active state. The "Active" signal is reset when counting is finished and the last item counted has exited the light barrier.

In the active state, a positive edge of the light barrier starts the timer with the time value "Duration1" ("Dura1") as retentive pulse timer. If the timer's Start input is processed with "0" in the next scan cycle, it nevertheless continues to run. A new positive signal edge "retriggers", that is, restarts, the timer. The next positive edge for restarting the timer is generated when the light barrier signals a negative edge. The timer is then started with time value "Duration2" ("Dura2"). If the light barrier is broken for a length of time exceeding "Dura1" or free for a period of time which exceeds "Dura2", the time elapses and the timer signals "Fault". The first "Active" signal starts the timer with the time value "Dura2".

The "Set" signal activates the counter and the monitoring circuit. The light barrier uses positive and negative signal edges to control the counter, the "Active" state, selection of the time value, and the starting (retriggering) of the watchdog timer.

Evaluation of the light barrier's positive and negative edge is required often, and temporary local data are suitable here as "scratchpad mem-

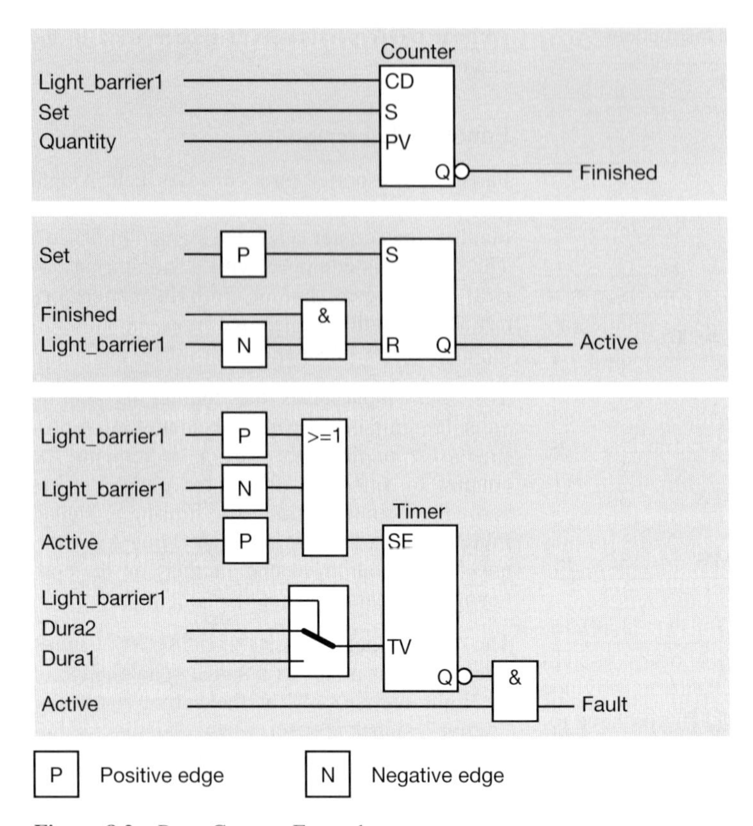


**Figure 8.2** Parts Counter Example

ory". Temporary local data are block-local variables, and are declared in the block (not in the symbol table). In the example, the edge evaluation's pulse memory bits are stored in temporary local data. (The edge memory bits also require their signal states in the next scan cycle, and must therefore not be temporary local data).

The program is located in a function without block parameters. You can call this function, in OB 1 for example, as follows:

```
CALL Counter_control;
```

The program is available as source text with symbolic addressing. The global symbols can also be used without quotation marks as long as they contain no special characters. If a special character is located in a symbol, then it must be enclosed in quotation marks. The editor shows all global symbols with quotation marks in the compiled block.

The program is subdivided into networks for better readability. The last network, which has the network title BLOCK END, is not absolutely necessary. However, it is a visual sign of the end of the block, which is very useful, particularly in the case of extremely long blocks.

On the diskette which accompanies the book you will find the symbol table in the "Conveyor Example" program under the object *Symbols*, the source program "Conveyor" in the *Sources* container, and the compiled program in the *Blocks* container, function FC 12.

```
FUNCTION "Counter_control" : VOID
TITLE = Parts counter with monitoring circuit
//Example of timer and counter functions
NAME    : CountDat
AUTHOR  : Berger
FAMILY  : STL_Book
Version : 01.00
VAR_TEMP
  PM_LB_P : BOOL;                   //Pulse positive edge light barrier
  PM_LB_N : BOOL;                   //Pulse negative edge light barrier
END_VAR
BEGIN
NETWORK
TITLE = Counter_control
    A    Light_barrier1;            //When light barrier is tripped,
    CD   "Count"                    //decrement counter by 1
    A    Set;
    L    Quantity;                  //Preset count with "Quantity"
    S    "Count";
    AN   "Count";                   //When count reaches zero,
    =    Finished;                  //output "Finished" signal
NETWORK
TITLE = Activate monitor
    A    Light_barrier1;
    FP   EM_LB_P;                   //Generate pulse memory bit
    =    PM_LB_P;                   //on positive edge of light barrier
    A    Light_barrier1;
    FN   EM_LB_N;                   //Generate pulse memory bit
    =    PM_LB_N;                   //on negative edge of light barrier
    A    Set;
    FP   EM_ST_P;
    S    Active;                    //Activate monitoring circuit
    A    Finished;
    A    PM_LB_N;
    R    Active;                    //Deactivate monitoring circuit
NETWORK
TITLE = Monitoring circuit
    L    Dura1;                     //If light barrier is "1"
    A    Light_barrier1;            //jump JC to D1 is executed and the
    JC   D1;                        //accumulator contains "Dura1" otherwise
    L    Dura2;                     //the accumulator contains "Dura2"
D1: A    Active;
    FP   EM_Ac_P;                   //If there is a positive edge at "active"
    O    PM_LB_P;                   //or a positive edge at the light barrier,
    O    PM_LB_N;                   //or a negative edge at the light barrier,
    SE   "Monitor";                 //the timer is started or retriggered
    AN   "Monitor";
    A    Active;                    //If time elapses while "active",
    =    "Fault";                   //"Fault" is signaled
NETWORK
TITLE = Block End
    BE;
END_FUNCTION
```

# Digital Functions

The digital functions process digital values predominantly of data type INT, DINT and REAL, and thus extend the functionality of the PLC.

- The **comparison functions** form a binary result from the comparison of two values. They take into account the data types INT, DINT and REAL.
- You use the **arithmetic functions** to make calculations in your program. All basic arithmetic operations in data types INT, DINT and REAL are available.
- The **math functions** extend the calculation options beyond the basic arithmetic operations to include such additions as trigonometric functions.
- Before and after performing calculations, you match the digital values to the required data type using the **conversion functions**.
- The **shift functions** make it possible to align the contents of the accumulator by shifting them to the right or left. It is always possible to scan the last bit shifted.
- **Word logic** is used to mask digital values by targeting individual bits and setting them to "1" or "0".

Digital logic operations work mainly with values in data blocks. These can be global data blocks or instance data blocks if static local data are used. Section 18.2, which is entitled "Block Functions for Data Blocks", deals with the use of data blocks and the addressing options for data.

With the exception of the accumulators, the temporary local data are extremely well suited for storing temporary results.

**9 Comparison Functions**
Compare for equal to, not equal to, greater than, greater than or equal to, less than, and less than or equal to; comparison function in a binary logic operation

**10 Arithmetic Functions**
Basic arithmetic operations; chain calculations; constant addition; decrementation and incrementation

**11 Mathematical Functions**
Trigonometric functions; inverse trigonometric functions; squaring, square-root extraction, exponentiation and logarithms

**12 Conversion functions**
Conversion from INT/DINT to BCD and vice versa; conversion from DINT to REAL and vice versa with different methods of rounding; one's complement, negation and absolute-value generation

**13 Shift Functions**
Shifting to left and right, by word and doubleword, shifting according to the rules for signs; rotating to left and right and through accumulator 1; shifting and rotating with a constant or with the contents of accumulator 2

**14 Word Logic**
AND, OR, Exclusive OR; logic operations by word and doubleword, with a constant or with the contents of accumulator 2

# 9 Comparison Functions

The comparison functions compare two digital values, one of which is in accumulator 1, the other in accumulator 2. As the result of the comparison, the comparison function sets the result of the logic operation (RLO) and status bits CC0 and CC1. The result can be post-processed with binary logic operations, memory functions, or jump statements. Table 9.1 provides an overview of the available comparison functions.

In Chapter 15, "Status Bits", you will learn how the comparison functions set status bits CC0 and CC1.

The examples in this chapter are also on the diskette which accompanies the book under the "Digital Functions" program in function block FB 109 or source file Chap_9.

## 9.1 General Representation of a Comparison Function

The CPU executes a comparison function, without regard to conditions, according to the schematic shown in Figure 9.1. A comparison function does not alter the contents of the accumulators.

Table 9.2 shows an example for the various data types. The comparison instruction carries out the comparison according to the specified characteristic, without regard to the contents of the accumulators. In the case of data type INT, the CPU compares only the right-hand (low-order) words of the accumulators; the contents of the left-hand (high-order) words are not taken into account. In comparisons involving data type REAL, a check is made to make sure that the accumulators contains valid REAL numbers. If they do not, the CPU sets the RLO to "0" and status bits CC0, CC1, OV and OS to "1".

## 9.2 Description of the Comparison Functions

### Comparison for equal to

The "comparison for equal to" instruction interprets the contents of the accumulators in accordance with the data type specified in the instruction and checks to see if the two values are equal. The RLO is "1" following the operation in the following cases:

- Data type INT
  If the contents of the low-order word of accumulator 2 is equal to the contents of the low-order word of accumulator 1
- Data type DINT
  If the contents of accumulator 2 are equal to the contents of accumulator 1
- Data type REAL
  If the contents of accumulator 2 are equal to

**Table 9.1** Overview of Comparison Functions

| Comparison Function | Comparison According To Data Type | | |
|---|---|---|---|
| | INT | DINT | REAL |
| Comparison for equal to | ==I | ==D | ==R |
| Comparison for not equal to | <>I | <>D | <>R |
| Comparison for greater than | >I | >D | >R |
| Comparison for greater than or equal to | >=I | >=D | >=R |
| Comparison for less than | <I | <D | <R |
| Comparison for less than or equal to | <=I | <=D | <=R |

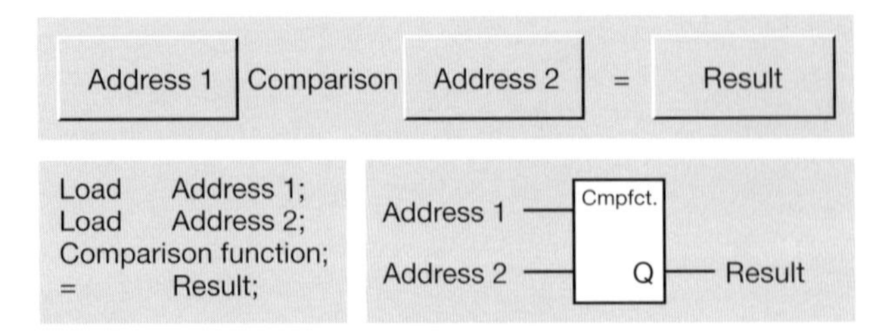


**Figure 9.1** Processing a Comparison Function

the contents of accumulator 1, on condition that both accumulators contains valid REAL numbers.

If two REAL numbers are equal but invalid, the "equal to" condition is not fulfilled (RLO = "0").

## Comparison for not equal to

The "comparison for not equal to" instruction interprets the contents of the accumulators in accordance with the data type specified in the comparison instruction and checks to see whether the two values differ. The RLO is "1" following the comparison operation in the following cases:

- Data type INT
  If the contents of the low-order word of accumulator 2 are not equal to the contents of the low-order word of accumulator 1.
- Data type DINT
  If the contents of accumulator 2 are not equal to the contents of accumulator 1.
- Data type REAL
  If the contents of accumulator 2 are not equal to the contents of accumulator 1, on condition that both accumulators contain valid REAL numbers.

  If two REAL numbers are not equal but at least one of them is invalid, the "not equal to" condition is not fulfilled (RLO = "0").

## Comparison for greater than

The "comparison for greater than" instruction interprets the contents of the accumulators in accordance with the data type specified in the comparison instruction and checks to see whether the value in accumulator 2 is greater than the value in accumulator 1. Following the operation, the RLO is "1" in the following instances:

- Data type INT
  If the contents of the low-order word of accumulator 2 are greater than the contents of the low-order word of accumulator 1.

**Table 9.2** Examples of Comparison Functions

| | | |
|---|---|---|
| **Comparison According to INT** | Memory bit M99.0 is reset if the value in memory word MW 92 is equal to 120, otherwise it is not. | L MW 92;<br>L 120;<br>==I ;<br>R M 99.0; |
| **Comparison According to DINT** | The variable "CompResult" in data block "Global_DB" is set if variable "CompVal1" is less than "CompVal2"; otherwise it is reset. | L "Global_DB".CompVal1;<br>L "Global_DB".CompVal2;<br><D ;<br>= "Global_DB".CompResult; |
| **Comparison According to REAL** | If variable #Actval is greater than or equal to variable #Calibra, #Recali is set, otherwise not. | L #Actval;<br>L #Calibra;<br>>=R ;<br>S #Recali; |

- Data type DINT
  If the contents of accumulator 2 are greater than the contents of accumulator 1.
- Data type REAL
  If the contents of accumulator 2 are greater than the contents of accumulator 1, on condition that both accumulators contain valid REAL numbers.

### Comparison for greater than or equal to

The "comparison for greater than or equal to" instruction interprets the contents of the accumulators in accordance with the data type specified in the comparison statement and checks to see whether the value in accumulator 2 is greater than or equal to the value in accumulator 1. Following the comparison, the RLO is "1" in the following instances:

- Data type INT
  If the contents of the low-order word of accumulator 2 is greater than the contents of the low-order word of accumulator 1 or if the bit patterns of the two words are equal
- Data type DINT
  If the contents of accumulator 2 are greater than the contents of accumulator 1 or if the bit patterns in the two accumulators are equal
- Data type REAL
  If the contents of accumulator 2 are greater than the contents of accumulator 1 or if the contents of the two accumulators are equal on condition that both accumulators contain valid REAL numbers.

### Comparison for less than

The "comparison for less than" instruction interprets the contents of the accumulators in accordance with the data type specified in the comparison operation and checks whether the value in accumulator 2 is less than the value in accumulator 1. Following the comparison, the RLO is "1" in the following instances:

- Data type INT
  If the contents of the low-order word of accumulator 2 are less than the contents of the low-order word of accumulator 1.
- Data type DINT
  If the contents of accumulator 2 are less than the contents of accumulator 1.
- Data type REAL
  If the contents of accumulator 2 are less than the contents of accumulator 1 on condition that both accumulators contain valid REAL numbers.

### Comparison for less than or equal to

The "comparison for less than or equal to" operation interprets the contents of the accumulators in accordance with the data type specified in the comparison instruction and checks whether the value in accumulator 2 is less than or equal to the value in accumulator 1. Following the comparison operation, the RLO is "1" in the following instances:

- Data type INT
  If the contents of the low-order word of accumulator 2 are less than the contents of the low-order word of accumulator 1 or if the bit patterns of the two words are equal
- Data type DINT
  If the contents of accumulator 2 are less than the contents of accumulator 1 or if the bit patterns of the values in the two accumulators are equal
- Data type REAL
  If the contents of accumulator 2 are less than the contents of accumulator 1 or if the contents of the two accumulators are equal on condition that both contain valid REAL numbers.

## 9.3 Comparison Function in a Logic Operation

Because the comparison function returns a binary RLO, this function can be used in conjunction with other binary functions. The comparison function sets status bit /FC, that is to say, in binary logic operations, a comparison function is always a first check.

## Comparison at the start of a logic operation

At the start of a logic operation, a comparison function is always a first check. The RLO returned by the comparison function can be directly combined using binary checks.

```
L       MW 120;
L       512;
>I      ;
A       Input1;
=       Output1;
```

In the example, *Output1* is set if the comparison condition is fulfilled and *Input1* has a signal state of "1".

## Comparison within a logic operation

A comparison function within a binary logic operation must be enclosed, as the comparison function begins a new logic step (first check).

```
O       Input2;
O(      ;
L       MW 122;
L       200;
<=I     ;
)       ;
O       Input3;
=       Output2;
```

In the example, *Output2* is set if *Input2* or *Input3* has a signal state of "1" or if the compare condition is fulfilled.

## Multiple comparisons

Because a comparison function does not alter the contents of the accumulators, multiple successive comparisons are possible in STL.

```
L       MW 124;
L       1200;
>I      ;
JC      GREA;
==I     ;
JC      EQUA;
```

In the example, two comparison functions are applied to the same accumulator contents. The first comparison generates RLO = "1" if MW 124 is greater than 1200, so that the jump to GREA is executed. Without reloading the accumulators, the second comparison function compares for equal to and generates a new RLO.

The comparison function sets the status bits based on the relationship between the values compared, that is, independently of the condition on which the comparison is based. You can make use of this fact by checking the status bits with the relevant jump functions. The example above can also be programmed as follows:

```
L       MW 124;
L       1200;
>I      ;
JP      GREA;
JZ      EQUA;
```

In this example, the comparison is evaluated on the basis of status bits CC0 and CC1. The comparison itself, in this case "greater than", does not affect the setting of the status bits; a different comparison, for example for "less than", could also have been used. JP scans to see whether the first comparison value is greater than the second, JZ to see whether they are equal.

# 10 Arithmetic Functions

The arithmetic functions link two digital values in accumulators 1 and 2 in accordance with one of the basic arithmetic operations. The result is placed in accumulator 1. Status bits CC0, CC1, OV and OS provide information about the result and the progress of the calculation (see Chapter 15, "Status Bits"). Table 10.1 provides an overview of the available arithmetic functions.

In addition to applying the basic arithmetic operations to values in accumulator 2, you can also add constants directly to the contents of accumulator 1 or modify the contents of accumulator 1 by a fixed amount.

The examples in this chapter are also on the diskette which accompanies the book under the "Digital Functions" program in function block FB 110 or source file Chap_10.

## 10.1 General Representation of an Arithmetic Function

The arithmetic functions link the contents of accumulators 1 and 2 as per the schematic shown in Figure 10.1 and place the result in accumulator 1. These functions execute without regard to any conditions.

Table 10.2 shows an example for each of the different data types. An arithmetic function performs the calculation in accordance with the specified arithmetic operator, without regard to the contents of the accumulators. In the case of data type INT, only the low-order accumulator words are linked; the high-order words are ignored. In the case of data type REAL, the accumulators are checked to make sure that both contain valid REAL numbers.

**Table 10.1** Overview of the Arithmetic Functions

| Arithmetic Functions | With Data Type INT | DINT | REAL |
|---|---|---|---|
| Addition | +I | +D | +R |
| Subtraction | –I | –D | –R |
| Multiplication | *I | *D | *R |
| Division with quotient as result | /I | /D | /R |
| Division with remainder as result | - | MOD | – |

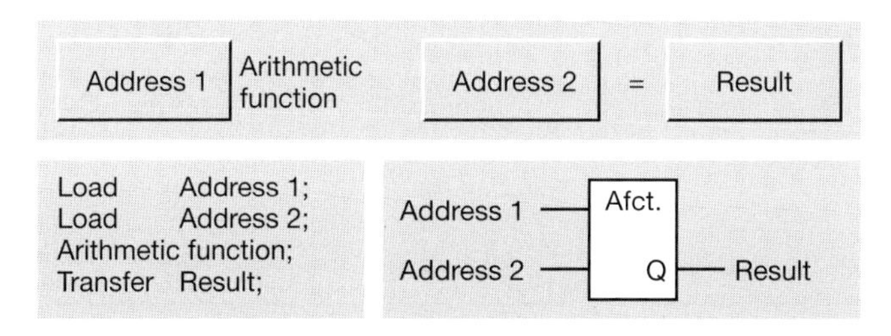


**Figure 10.1** Processing an Arithmetic Function

**Table 10.2** Examples for Arithmetic Functions

| | | |
|---|---|---|
| **INT** | The value in memory word MW 100 is divided by 250; the integer result is stored in memory word MW 102. | L MW 100;<br>L 250;<br>/I ;<br>T MW 102; |
| **DINT** | The values in variables "CalcVal1" and "CalcVal2" are added and the result stored in variable "CalcResult". All variables are in data block "Global_DB". | L "Global_DB".CalcVal1;<br>L "Global_DB".CalcVal2;<br>+D ;<br>T "Global_DB".CalcResult; |
| **REAL** | Variables #Actval and #Factor are multiplied; the product is transferred to variable #Display. | L #Actval;<br>L #Factor;<br>*R ;<br>T #Display; |

On the S7-300 CPUs, execution of an arithmetic function does not alter the contents of accumulator 2; on S7-400 CPUs, the contents of accumulator 2 are overwritten by the contents of accumulator 3. The contents of accumulator 4 are then "shifted over" to accumulator 3 (Figure 10.2).

You must pay particular attention to this aspect of the arithmetic functions when writing programs or blocks containing these functions which must be executable in both S7-300 and S7-400 systems.

## 10.2 Calculating with Data Type INT

### INT addition

The +I function interprets the values in the low-order words of accumulators 1 and 2 as numbers of data type INT. It adds the two numbers and stores the sum in accumulator 1. After the calculation has been performed, status bits CC0 and CC1 indicate whether the sum is negative, zero, or positive. Status bits OV and OS flag any range violations.

The high-order word of accumulator 1 remains unchanged.

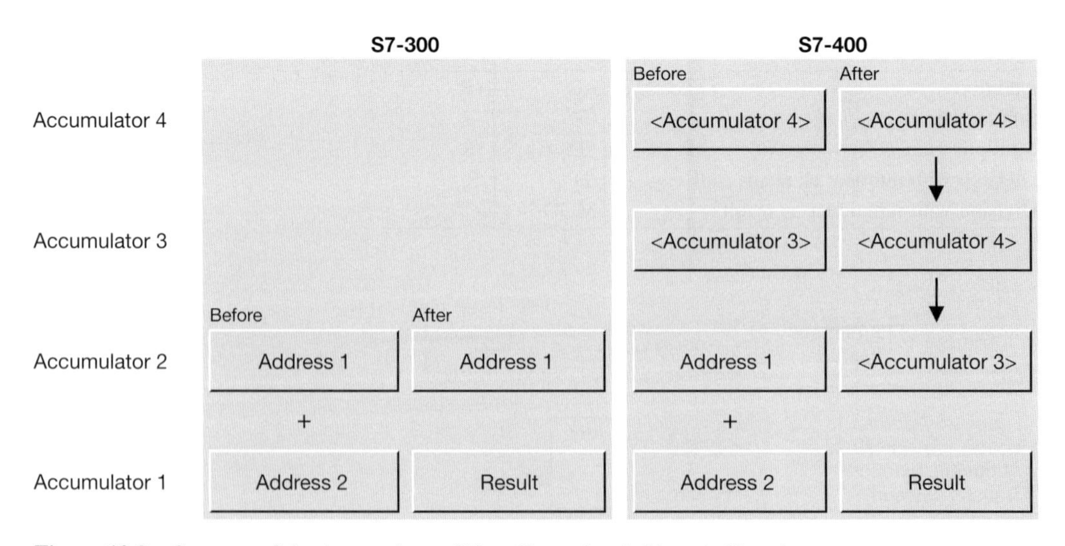


**Figure 10.2** Contents of the Accumulators When Executing Arithmetic Functions

### INT subtraction

The –I function interprets the values in the low-order words of accumulators 1 and 2 as numbers of data type INT. It subtracts the value in accumulator 1 from the value in accumulator 2 and stores the difference in accumulator 1. After the calculation has been performed, status bits CC0 and CC1 indicate whether the difference is negative, zero, or positive. Status bits OV and OS flag any range violations.

The high-order word of accumulator 1 remains unchanged.

### INT multiplication

The *I function interprets the values in the low-order words of accumulators 1 and 2 as numbers of data type INT. It multiplies the two numbers and stores the product as a number of data type DINT in accumulator 1. After the calculation has been performed, status bits CC0 and CC1 indicate whether the product is negative, zero, or positive. Status bits OV and OS flag any INT range violations.

Following execution of the *I function, the product is available as a DINT number in accumulator 1.

### INT division

The /I function interprets the values in the low-order words of accumulators 1 and 2 as numbers of data type INT. It divides the value in accumulator 2 (dividend) by the value in accumulator 1 (divisor) and returns two results: the quotient and the remainder, both of which are of data type INT (Figure 10.3).

After the function has executed, the low-order word of accumulator 1 contains the quotient. The quotient is the integer result of the division operation. The quotient is zero when the dividend is zero and the divisor is not zero or when the dividend is smaller than the divisor. The quotient is negative if the divisor is negative.

After the /I function has executed, the high-order word contains the remainder of the division (not the places after the decimal point!). If the dividend is negative, the remainder is also negative.

After the calculation has been performed, status bits CC0 and CC1 indicate whether the quotient is negative, zero, or positive. Status bits OV and OS flag any range violations.

Division by zero returns a quotient of zero and a remainder of zero, and sets status bits CC0, CC1, OV and OS to "1".

## 10.3 Calculating with Data Type DINT

### DINT addition

The +D function interprets the values in accumulators 1 and 2 as numbers of data type DINT. It adds the two numbers and stores the sum in accumulator 1. Following execution of the function, status bits CC0 and CC1 indicate whether the sum is negative, zero, or positive. Status bits OV and OS flag any range violations.

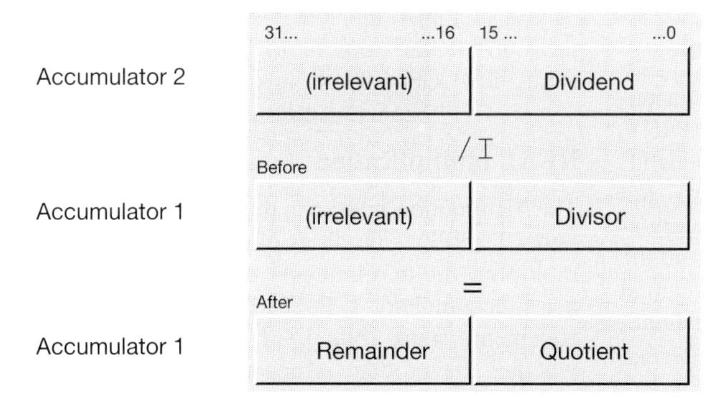


**Figure 10.3** Results Returned by the Arithmetic Function /I

### DINT subtraction

The –D function interprets the values in accumulators 1 and 2 as numbers of data type DINT. It subtracts the value in accumulator 1 from the value in accumulator 2 and stores the difference in accumulator 1. Following execution, status bits CC0 and CC1 indicate whether the difference is negative, zero, or positive. Status bits OV and OS flag any range violations.

### DINT multiplication

The *D function interprets the values in accumulators 1 and 2 as numbers of data type DINT. It multiplies the two numbers and stores the product in accumulator 1. Following execution of the function, status bits CC0 and CC1 indicate whether the product is negative, zero, or positive. Status bits OV and OS flag any range violations.

### DINT division with quotient as result

The /D function interprets the values in accumulators 1 and 2 as numbers of data type DINT. It divides the value in accumulator 2 (dividend) by the value in accumulator 1 (divisor) and stores the quotient in accumulator 1. The quotient is the integer result of the division. It is zero if the dividend is zero and the divisor is not zero or when the dividend is smaller than the divisor. The quotient is negative if the divisor is negative.

Following execution of the function, status bits CC0 and CC1 indicate whether the quotient is negative, zero, or positive. Status bits OV and OS flag any range violations.

Division by zero returns a quotient of zero and sets status bits CC0, CC1, OV and OS to "1".

### DINT division with remainder as result

The MOD function interprets the values in accumulators 1 and 2 as numbers of data type DINT. It divides the value in accumulator 2 (dividend) by the value in accumulator 1 (divisor) and stores the remainder of the division in accumulator 1. The remainder is what is left over from the division; it does not correspond to the decimal places. If the dividend is negative, the remainder is also negative.

Following execution of the function, status bits CC0 and CC1 indicate whether the remainder is negative, zero, or positive. Status bits OV and OS flag any range violations.

Division by zero returns a remainder of zero and sets status bits CC0, CC1, OV and OS to "1".

## 10.4 Calculating with Data Type REAL

### REAL addition

The +R function interprets the values in accumulators 1 and 2 as numbers of data type REAL. It adds the two numbers and stores the sum in accumulator 1. Following execution of the function, status bits CC0 and CC1 indicate whether the sum is negative, zero, or positive. Status bits OV and OS flag any range violations.

In the case of an impermissible calculation (one of the input values is an invalid REAL number or you tried to add $+\infty$ and $-\infty$), +R returns an invalid value in accumulator 1 and sets status bits CC0, CC1, OV and OS to "1".

### REAL subtraction

The –R function interprets the values in accumulators 1 and 2 as numbers of data type REAL. It subtracts the number in accumulator 1 from the number in accumulator 2 and stores the difference in accumulator 1. Following execution of the function, status bits CC0 and CC1 indicate whether the difference is negative, zero, or positive. Status bits OV and OS flag any range violations.

In the case of an impermissible calculation (one of the input values is an invalid REAL number or you attempted to subtract $+\infty$ and $+\infty$), –R returns an invalid value in accumulator 1 and sets status bits CC0, CC1, OV and OS to "1".

### REAL multiplication

The *R function interprets the values in accumulators 1 and 2 as numbers of data type REAL. It multiplies the two numbers and stores the product in accumulator 1. Following execution of the statement, status bits CC0 and CC1 indicate whether the product is negative, zero, or positive. Status bits OV and OS flag any range violations.

In the case of an impermissible calculation (one of the input values is an invalid REAL number or you attempted to multiply ∞ and 0), *R returns an invalid value in accumulator 1 and sets status bits CC0, CC1, OV and OS to "1".

### REAL division

The /R function interprets the values in accumulators 1 and 2 as numbers of data type REAL. It divides the number in accumulator 2 (dividend) by the number in accumulator 1 (divisor) and stores the quotient in accumulator 1. Following execution of the function, status bits CC0 and CC1 indicate whether the quotient is negative, zero, or positive. Status bits OV and OS flag any range violations.

In the case of an impermissible calculation (one of the input values is an invalid REAL number or you attempted to divide ∞ by ∞ or 0 by 0), /R returns an invalid value in accumulator 1 and sets status bits CC0, CC1, OV and OS to "1".

## 10.5 Successive Arithmetic Functions

You can program one arithmetic function immediately behind another, in which case the result of the first function is post-processed by the second, the accumulators serving as temporary storage.

Please note that S7-300 CPUs and S7-400 CPUs handle successive arithmetic functions differently (the S7-300 CPUs have only 2 accumulators while the S7-400 CPUs have 4).

### Chain calculations on an S7-300

A chain calculation is performed by following an arithmetic function with the loading and subsequent linking of the next value.

Example: Result1 := Value1 + Value2 – Value3

```
L     Value1;
L     Value2;
+I    ;         //Value1 + Value2
L     Value3;
-I    ;         //Sum - Value3
T     Result1;
```

On CPUs with two accumulators, the first value loaded remains unchanged in accumulator 2 during execution of the arithmetic function, and can be reused without having to be reloaded.

Example: Result2 := Value5 + 2 × Value6

```
L     Value6;
L     Value5;
+R    ;         //Value5 + Value6
+R    ;         //Sum + Value6
T     Result2;
```

Example: Result3 := Value7 × Value8$^2$

```
L     Value8;
L     Value7;
*D    ;         //Value7 * Value8
*D    ;         //Product * Value8
T     Result3;
```

### Chain calculations on an S7-400

A chain calculation is performed by following an arithmetic function with the loading and subsequent linking of the next value. On CPUs with four accumulators, the value in accumulator 3 "shifts over" to accumulator 3 following execution of the arithmetic function. Beforehand, you can store an intermediate result in accumulator 3 (in the case of a dot-before-line calculation, for instance) with an ENT instruction (see Section 6.4, "Accumulator Functions').

Example:
Result4 := (Value1 + Value2) × (Value3 – Value4)

```
L     Value1;
L     Value2;
+I    ;
L     Value3;
ENT   ;
L     Value4;
-I    ;
*I    ;
T     Result4;
```

First, the sum of *Value1* and *Value2* is computed. While *Value3* is being loaded into accumulator 1, this sum is moved over to accumulator 2. From there, the ENT instruction copies it into accumulator 3. After *Value4* is loaded, the contents of *Value3* are in accumulator 2. When the two values are subtracted, the sum is "fetched back" into accumulator 2 from accumulator 3. The sum and the difference can now be multiplied.

## 10.6 Adding Constants to Accumulator 1

| | | |
|---|---|---|
| + | B#16#bb | Adds a byte constant |
| + | ±w | Adds a word constant |
| + | L#±d | Adds a doubleword constant |

The "Add Constant" instruction adds the constant specified in the instruction to the contents of accumulator 1. You may specify this constant as a hexadecimal byte constant or as a decimal word or decimal doubleword constant. If you want to add a word constant using DINT, precede the constant with #L. If a decimal constant exceeds the permissible INT range, the calculation automatically becomes DINT.

You may write a decimal number with a minus sign, thus making it possible to subtract constants. Before a byte constant is added, it is expanded into a signed INT number.

As does a calculation with data type INT, the addition of a byte constant or word constant affects only the low-order word of accumulator 1; there is no carry to the high-order word. If the INT value range is exceeded, bit 15 (which is the sign bit) is overwritten. The addition of a doubleword constant affects all 32 bits of accumulator 1, corresponding to a DINT calculation.

Execution of these statements is independent of any conditions, and affects neither the RLO nor – in contrast to arithmetic functions – the status bits.

Examples for adding constants:

```
L   AddValue1;
+   B#16#21;
T   AddResult1;
```

The value of variable *AddValue1* is increased by 33 and transferred to variable *AddResult1*.

```
L   AddValue2;
+   -33;
T   AddResult2;
```

The value of variable *AddValue2* is decreased by 33 and stored in variable *AddResult2*.

```
L   AddValue3;
+   L#-1;
T   AddResult3;
```

The value of variable *AddValue3* is decreased by 1 and stored in variable *AddResult3*. The subtraction is as for a DINT calculation.

## 10.7 Decrementing and Incrementing

| | | |
|---|---|---|
| DEC | n | Decrement |
| INC | n | Increment |

The Decrement and Increment statements alter the value in accumulator 1. That value is decreased (decremented) or increased (incremented) by the number of units specified in the statement parameter. The parameter may assume a value between 0 and 255.

Only the value in the low-order byte of the accumulator is altered. There is no carry to the high-order byte. The calculation is carried out in "modulo 256", that is, when incrementation produces a value which exceeds 255, the "count" begins again from the beginning, or if decrementation produces a value which falls below 0, the "count" begins again at 255.

The Decrement and Increment instructions are executed without regard to the RLO. They are always executed when encountered, and affect neither the RLO nor the status bits.

Examples:

```
L     IncValue;
INC   5;
T     IncValue;
```

The value of variable *IncValue* is incremented by 5.

```
L     DecValue;
DEC   7;
T     DecValue;
```

The value of variable *DecValue* is decremented by 7.

# 11 Math Functions

A math function takes the number in accumulator 1 as input value for the function, and stores the result in accumulator 1. The numbers to be processed must be of data type REAL. The following math functions are available:

- Sine, cosine, tangent
- Arc sine, arc cosine, arc tangent
- Squaring, square-root extraction
- Exponential function to base e, natural logarithm

Depending on the result, a math function sets status bits CC0, CC1, OV and OS as described in Chapter 15, "Status Bits".

The examples in this chapter are also on the diskette which accompanies the book under the "Digital Functions" program in function block FB 111 or source file Chap_11.

## 11.1 Processing a Math Function

A math function alters only the contents of accumulator 1; the contents of all other accumulators remain unchanged. Figure 11.1 shows three examples of math functions. A math function computes in accordance with the rules governing REAL numbers, even when absolute addressing is used and no data types are declared.

If accumulator 1 contains an invalid REAL number at the time the function is executed, the math function returns an invalid REAL number and sets status bits CC0, CC1, OV and OS to '1".

## 11.2 Trigonometric Functions

The trigonometric functions

- SIN Sine,
- COS Cosine and
- TAN Tangent

assume an angle in radian measure in form of a REAL number in accumulator 1.

Two units are normally used for the size of an angle: degrees from 0° to 360° and radian measure from 0 to $2\pi$ (where $\pi$ = +3.141593e+00). Both can be converted proportionally. For example, the radian measure for a 90° angle is $\pi/2$, or +1.570796e+00. With values greater than $2\pi$ (+6.283185e+00), $2\pi$ or a multiple thereof is subtracted until the input value for the trigonometric function is less than $2\pi$.

**Table 11.1** Examples of Math Functions

| | | |
|---|---|---|
| Sine | The value in memory doubleword MD 110 contains an angle in radian measure. The sine of this angle is generated and stored in memory doubleword MD 104. | `L MD 110;`<br>`SIN ;`<br>`T MD 104;` |
| Square root | The square root of the value in variable "MathVal1" is generated and stored in the variable "MathRoot". | `L "Global_DB".MathVal1;`<br>`SQRT ;`<br>`T "Global_DB".MathRoot;` |
| Exponent | Variable #Result contains the power of e and #Exponent. | `L #Exponent;`<br>`EXP ;`<br>`T #Result;` |

Example: Computing the idle power
$P_s = U \cdot I \cdot \sin(\varphi)$

```
L    PHI;
SIN  ;
L    Current;
*R   ;
L    Voltage;
*R   ;
T    I_Power;
```

## 11.3 Arc Functions

The arc functions (inverse trigonometric functions)

- ASIN Arc sine,
- ACOS Arc cosine and
- ATAN Arc tangent

are the inverse functions of the corresponding trigonometric functions. They assume a REAL number in a specific range in accumulator 1, and return an angle in radian measure (Table 11.2).

If the permissible range is exceeded, the arc function returns an invalid REAL number and sets status bits CC0, CC1, OV and OS to "1".

Example: In a right-angled triangle, one of the short sides of the triangle and the hypotenuse form an aspect ratio of 0.343. How big is the angle between them in degrees?

Arcsin (0.343) returns the angle in radian measure; multiplication with factor $360/2\pi$ (= 57.2958) gives you the angle in degrees (approx. 20°).

```
L    0.343;
ASIN ;
L    57.2958;
*R   ;
T    Angle_Degree;
```

**Table 11.2** Value Ranges for Arc Functions

| Function | Permissible Value Range | Value Returned |
|---|---|---|
| ASIN | –1 to +1 | $-\pi/2$ to $+\pi/2$ |
| ACOS | –1 to +1 | 0 to $\pi$ |
| ATAN | Entire range | $-\pi/2$ to $+\pi/2$ |

## 11.4 Other Math Functions

The following math functions are also available:

- SQR Squaring,
- SQRT Square-root extraction,
- EXP Exponential function to base e and
- LN Compute natural logarithm (logarithm to base e).

### Squaring

The SQR function squares the value in accumulator 1.

Example: Computing the volume of a cylinder $V = r^2 \cdot \pi \cdot h$

```
L    Radius;
SQR  ;
L    Height;
*R   ;
L    3.141592;
*R   ;
T    Volume;
```

### Square-root extraction

The SQRT function extracts the square root of the value in accumulator 1. If the value in accumulator 1 is less than zero, SQRT sets status bits CC0, CC1, OV and OS to "1" and returns an invalid REAL number. If accumulator 1 contains –0 (minus zero), –0 is returned.

Example: $c = \sqrt{a^2 + b^2}$

```
L    #a;
SQR  ;
L    #b;
SQR  ;
+R   ;
SQRT ;
T    #c;
```

(If b or c is declared as a local variable, it must be preceded by # if the compiler is to recognize it as a local variable; if b or c is a global variable, it must be enclosed in quotation marks.)

### Exponentiation to base e

The EXP function computes the power from base e (= 2.718282e+00) and the value in accumulator 1 ($e^{Accu1}$).

Example: Any power can be computed with the formula

$a^b = e^{b \ln a}$

```
L     Value_a;
LN    ;
L     Value_b;
*R    ;
EXP   ;
T     Power;
```

### Computing the natural logarithm

The LN function computes the natural logarithm to base e (= 2.718282e+00) from the number in accumulator 1. If accumulator 1 contains a value less than or equal to zero, LN sets status bits CC0, CC1, OV and OS to "1" and returns an invalid REAL number.

The natural logarithm is the inverse of the exponential function: If y = $e^x$ then x = ln (y).

Example: Computing a logarithm to base 10 and to any other base.

The basic formula is

$$\log_b a = \frac{\log_n a}{\log_n b}$$

where $b$ or $n$ is any base. If you make n = e, you can compute a logarithm to any base using the natural logarithm:

$$\log_b a = \frac{\ln a}{\ln b}$$

In the special case for base 10, the formula is:

$$\lg a = \frac{\ln a}{\ln 10} = 0.4342945 \cdot \ln a$$

# 12 Conversion Functions

The conversion functions convert the data type of the value in accumulator 1. Figure 12.1 provides an overview of the data type conversions described in this chapter.

You will find details on the bits of the data formats in Chapter 24, "Data Types", and information on how the conversion functions set the status bits in Chapter 15, "Status Bits".

The examples in this chapter are also on the diskette which accompanies the book under the "Digital Functions" program in FB 112 or source file Chap_12.

## 12.1 Processing a Conversion Function

The conversion functions affect only accumulator 1. Some functions affect only the low-order word (bits 0 to 15), others the entire accumulator. The conversion functions do not change the contents of any other accumulator.

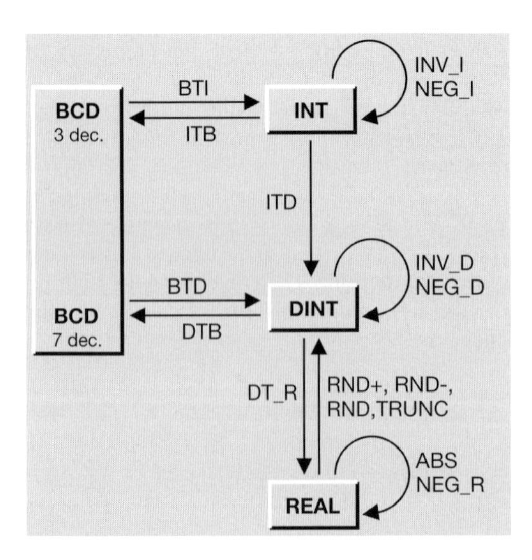


**Figure 12.1** Overview of the Conversion Functions

Table 12.1 shows an example for each data type. A conversion function converts as per the specified characteristic even if absolute addressing is used and no data types have been declared for the operands.

### Successive conversion functions

You can subject the contents of accumulator 1 to several successive conversions and so carry out conversions in stages without having to temporarily store the converted values.

Example:

```
L    BCD_Number;
BTI  ;                //BCD to INT
ITD  ;                //INT to DINT
DTR  ;                //DINT to REAL
T    REAL_Number;
```

This example converts a BCD number with 3 decades to a REAL number.

## 12.2 Converting INT and DINT Numbers

The following functions are provided for converting INT and DINT numbers:

- ITD  Converts INT to DINT
- ITB  Converts INT to BCD
- DTB  Converts DINT to BCD
- DTR  Converts DINT to REAL

### Converting INT to DINT

The ITD statement interprets the value in the low-order word of accumulator 1 (bits 0 to 15) as a number of data type INT and transfers the signal state of bit 15 (the sign) to the high-order word, that is, bits 16 to 31.

**Table 12.1** Examples for Conversion Functions

| | | |
|---|---|---|
| **Converting INT Numbers** | The value in memory doubleword MW 120 is interpreted as an INT number and stored in memory word MW 122 as a BCD number. | L MW 120;<br>ITB ;<br>T MW 122; |
| **Converting DINT Numbers** | The value in variable "ConvertDINT" is interpreted as a DINT number and stored as a REAL number in the variable "ConvertREAL". | L "Global_DB".<br>ConvertINT;<br>ITR ;<br>T "Global_DB".<br>ConvertREAL; |
| **Converting REAL Numbers** | The absolute value is generated from the variable #Display. | L #Display;<br>ABS ;<br>T #Display; |

The conversion of INT to DINT sets no status bits.

### Converting INT to BCD

The ITB statement interprets the value in the low-order word of accumulator 1 (bits 0 to 15) as a number of data type INT and converts it to a 3-decade BCD number. The three decades are right-justified in accumulator 1 and represent the value of the decimal number. The sign is in bits 12 to 15. If all bits are "0", the sign is positive; if all bits are "1", it is negative. The contents of the high-order word (bits 16 to 31) remain unchanged.

If the INT number is too large to be converted to BCD (999), the ITB statement sets status bits OV and OS. The conversion is not performed in this case.

### Converting DINT to BCD

The DTB statement interprets the value in accumulator 1 as a number of data type DINT and converts it to a 7-decade BCD number. The seven decades are right-justified in accumulator 1 and represent the value of the decimal number. The sign is in bits 28 to 31. If all bits are "0", the sign is positive; if all bits are "1", it is negative.

If the DINT number is too large to be converted to BCD (> 9 999 999), status bits OV and OS are set and the conversion is not carried out.

### Converting DINT to REAL

The DTR statement interprets the contents of accumulator 1 as a number in DINT format and converts it to a number in REAL format.

Since a number in DINT format has a higher accuracy than a number in REAL format, rounding may take place during conversion, but only to the next whole number (as per the RND statement).

DTR sets no status bits.

## 12.3 Converting BCD Numbers

The following functions are available for converting BCD numbers:

- BTI Converts BCD to INT
- BTD Converts BCD to DINT

### Converting BCD to INT

The BTI statement interprets the value in the low-order word of accumulator 1 (bits 0 to 15) as a 3-decade BCD number. The three decades are right-justified in the accumulator and represent the value of the decimal number. The sign is in bits 12 to 15. If all bits are "0", the sign is positive; if all bits are "1", it is negative. During conversion, only the signal state of bit 15 is taken into account. The contents of the high-order word of accumulator 1 (bits 16 to 31) remain unchanged.

If there is a pseudo tetrad in the BCD number (numerical value 10 to 15 or A to F in hexadecimal), the CPU reports a programming error and calls organization block 121 (synchronous errors). If OB 121 has not been programmed, the CPU goes to STOP.

The BTI statement sets no status bits.

### Converting BCD to DINT

The BTD statement interprets the value in accumulator 1 as a 7-decade BCD number. The seven decades are right-justified in the accumulator and represent the value of the decimal number. Bits 28 to 31 contain the sign. If these bits are all "0", the sign is positive; if they are all "1", the sign is negative. During conversion, only the signal state of bit 31 is taken into account.

If the BCD number contains a pseudo tetrad (numerical value 10 to 15 or A to F in hexadecimal), the CPU reports a programming error and calls organization block OB 121 (synchronous errors). If OB 121 is not available, the CPU goes to STOP.

The BTD statement sets no status bits.

## 12.4 Converting REAL Numbers

There are several statements for converting a number in REAL format to DINT format (conversion of a fractional value to an integer value). They differ from one another in the way they perform rounding.

- RND+ With rounding to the next higher integer
- RND– With rounding to the next lower integer
- RND With rounding to the next integer
- TRUNC No rounding

### Rounding to the next higher integer number

The RND+ statement interprets the contents of accumulator 1 as a number in REAL format and converts it to a number in DINT format.

The RND+ statement returns an integer greater than or equal to the number to be converted.

If the value in accumulator 1 exceeds or falls short of the permissible range for a DINT number or if it is not a REAL number, RND+ sets status bits OV and OS. The conversion is not carried out.

### Rounding to the next lower integer

The RND– statement interprets the contents of accumulator 1 as a number in RAL format and converts it to a number in DINT format.

The RND– statement returns an integer less than or equal to the number to be converted.

If the value in accumulator 1 exceeds or falls short of the permissible range for a DINT number or if it is not a REAL number, RND– sets status bits OV and OS. The conversion is not carried out.

### Rounding to the next integer

The RND statement interprets the contents of accumulator 1 as a number in REAL format and converts it to a number in DINT format.

The RND statement returns the next higher or next lower integer, whichever is closest to the true result. If the result lies exactly between an even and an odd number, the even number takes priority.

If the value in accumulator 1 exceeds or falls short of the permissible range for a DINT number or if it is not a REAL number, RND sets status bits OV and OS. The conversion is not carried out.

### No rounding

The TRUNC statement interprets the contents of accumulator 1 as a number in REAL format and converts it to a number in DINT format.

The TRUNC statement returns the integer component of the number to be converted; the fractional component is "truncated".

If the value in accumulator 1 exceeds or falls short of the permissible range for a DINT number or if it is not a REAL number, TRUNC sets status bits OV and OS. The conversion is not carried out.

### Summary of conversions from REAL to DINT

Table 12.2 shows the different effects of the functions for converting RAL to DINT. For the example, the range –1 to +1 has been chosen.

**Table 12.2** Rounding Modes for the Conversion of REAL Numbers

| Input Value | | Result | | | |
|---|---|---|---|---|---|
| REAL | DW#16# | RND | RND+ | RND– | TRUNC |
| 1.0000001 | 3F80 0001 | 1 | 2 | 1 | 1 |
| 1.00000000 | 3F80 0000 | 1 | 1 | 1 | 1 |
| 0.99999995 | 3F7F FFFF | 1 | 1 | 0 | 0 |
| 0.50000005 | 3F00 0001 | 1 | 1 | 0 | 0 |
| 0.50000000 | 3F00 0000 | 0 | 1 | 0 | 0 |
| 0.49999996 | 3EFF FFFF | 0 | 1 | 0 | 0 |
| 5.877476E–39 | 0080 0000 | 0 | 1 | 0 | 0 |
| 0.0 | 0000 0000 | 0 | 0 | 0 | 0 |
| –5.877476E–39 | 8080 0000 | 0 | 0 | –1 | 0 |
| –0.49999996 | BEFF FFFF | 0 | 0 | –1 | 0 |
| –0.50000000 | BF00 0000 | 0 | 0 | –1 | 0 |
| –0.50000005 | BF00 0001 | –1 | 0 | –1 | 0 |
| –0.99999995 | BF7F FFFF | –1 | 0 | –1 | 0 |
| –1.00000000 | BF80 0000 | –1 | –1 | –1 | –1 |
| –1.0000001 | BF80 0001 | –1 | –1 | –2 | –1 |

## 12.5 Other Conversion Functions

Other available conversion functions are one's complement generation, negation functions, and generation of an absolute REAL number.

- INVI One's complement INT
- INVD One's complement DINT
- NEGI Negation of an INT number (two's complement)
- NEGD Negation of a DINT number (two's complement)
- NEGR Negation of a REAL number
- ABS Generation of an absolute REAL number

### One's complement INT

The INVI statement negates the value in the low-value word of accumulator 1 (bits 0 to 15) bit for bit. It replaces the zeroes with ones and vice versa. The contents of the high-order word (bits 16 to 31) remain unchanged.

The INVI statement sets no status bits.

### One's complement DINT

The INVD statement negates the value in accumulator 1 bit for bit. It replaces the zeroes with ones and vice versa.

The INVD statement sets no status bits.

### Negation INT

The NEGI statement interprets the value in the low-order word of accumulator 1 (bits 0 to 15) as an INT number and changes the sign by generating the two's complement. This operation is identical to multiplication with –1. The high-order word of accumulator 1 (bits 16 to 31) remains unchanged.

The NEGI statement sets status bits CC0, CC1, OV and OS.

### Negation DINT

The NEGD statement interprets the value in accumulator 1 as DINT number and changes the sign by generating the two's complement. This operation is identical to multiplication with –1.

The NEGD statement sets status bits CC0, CC1, OV and OS.

## Negation REAL

The NEGR statement interprets the value in accumulator 1 as a REAL number and multiplies this number by –1 (it changes the sign of the mantissa, even if the number in the accumulator is not a valid REAL number).

The NEGR statement sets no status bits.

## Absolute-value generation REAL

The ABS statement interprets the value in accumulator 1 as a REAL number and generates the absolute value of this number (it sets the sign of the mantissa to "0", even if the number in the accumulator is not a valid REAL number).

The ABS statement sets no status bits.

# 13 Shift Functions

The shift functions shift the contents of accumulator 1 bit by bit to the left or right. Table 13.1 provides an overview of the available shift functions.

The examples in this chapter are also on the diskette accompanying the book under the "Digital Functions" program in function block FB 113 or source file Chap_13.

## 13.1 Processing a Shift Function

The contents of accumulator 1 are shifted bit by bit to the left or right; depending on the function, the accumulator contains either a word or a doubleword. The bits that are shifted out are either lost (Shift operations) or are added at the other side of the word or doubleword (Rotate operations). The shift functions have no effect on the other accumulators.

The shift functions are executed without regard to any conditions. They affect only the contents of accumulator 1. The result of the logic operation (RLO) is not affected. The shift functions set status bit CC0 to "0" and status bit CC1 to the signal state of the last bit shifted (Figure 13.1). The status bits are evaluated with binary check or jump instructions as described in Chapter 15, "Status Bits", and Chapter 16, "Jump Functions".

Table 13.2 shows several examples of shift functions. A word shift changes only the low-order word of accumulator 1; the contents of the high-order word are not affected. Rotation through status bit CC1 shifts the contents of the accumulator by one bit position.

### Successive shift functions

Shift functions can be applied as often as required to the contents of the accumulator.

Example:

```
L    Value1;
SSD  4;
SLD  2;
T    Result1;
```

In the example, a value is shifted, with sign, to the right by (effectively) 2 positions, whereby the two right bit positions are reset to "0".

**Table 13.1** Overview of the Shift Functions

| Shift Functions | Word | | Doubleword | |
|---|---|---|---|---|
| | with no. of shifts as a parameter | with no. of shifts in accum2 | with no. of shifts as a parameter | with no. of shifts in accum2 |
| Shift left | SLW n | SLW | SLD n | SLD |
| Shift right | SRW n | SRW | SRD n | SRD |
| Shift with sign | SSI n | SSI | SSD n | SSD |
| Rotate left | – | – | RLD n | RLD |
| Rotate right | – | – | RRD n | RRD |
| Rotate left through CC1 | – | – | RLDA[1] | – |
| Rotate right through CC1 | – | – | RRDA[1] | – |

[1] Without parameter, as only one bit is shifted

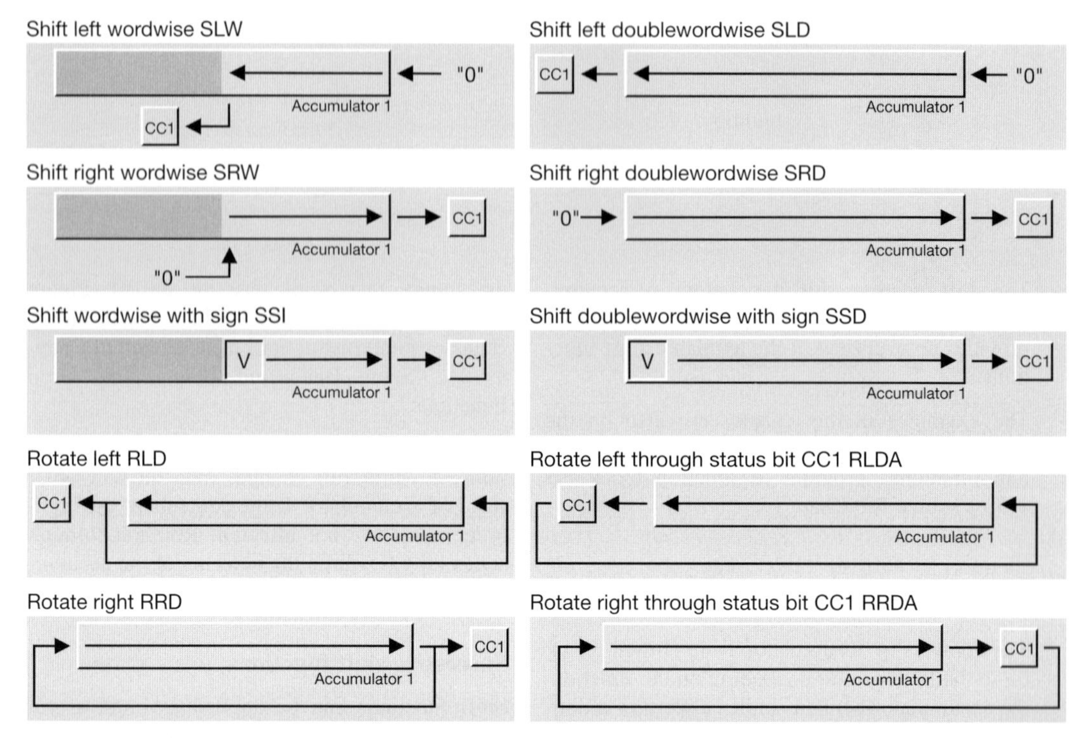


**Figure 13.1** How Shift Functions Work

## 13.2 Shifting

### Shift left word

SLW n — Shift left word by n bits
SLW — Shift left word by the number of bits in accumulator 2

Shift function SLW shifts bits 0 to 15 of accumulator 1 bit by bit to the left. The bit positions freed by the shift operation are padded with zeroes. The high-order word of accumulator 1 remains unchanged; there is no carry to bit 16.

The number of shifts may be specified as a parameter in the SLW statement or loaded as a positive number in INT format right-justified into accumulator 2. If the number of shifts is = 0, the statement is not executed (no operation, or NOP); if it is greater than 15, the low-order word

**Table 13.2** Examples for Shift Functions

| | | |
|---|---|---|
| **Shifting word variables** | The value in memory word MW 130 is shifted 4 positions to the left and stored in memory word MW 132. Here, the number of shifts appears as a parameter in the shift operation. | L MW 130;<br>SLW 4;<br>T MW 132; |
| **Shifting doubleword variables** | The value in variable "ShiftOn" is shifted right by "ShiftPos" positions and stored in "ShiftOff". Here, the number of shifts is in accumulator 2. | L "Global_DB".ShiftPos;<br>L "Global_DB".ShiftOn;<br>SRD ;<br>T "Global_DB".ShiftOff; |
| **Shifting with sign** | The variable #Actval is shifted, with sign, 2 positions to the right and transferred to the variable #Display. | L #Actval;<br>SSI 2;<br>T #Display; |

of accumulator 1 contains zero following execution of the SLW statement.

If the contents of accumulator 1 (low-order word) are interpreted as a number in INT format, a shift to the left would be equivalent to multiplication with a power to base 2. The exponent is then the number of shifts.

### Shift left doubleword

| | | |
|---|---|---|
| SLD | n | Shift left doubleword by n bits |
| SLD | | Shift left doubleword by the number of bits in accumulator 2 |

Shift function SLD shifts the entire accumulator bit by bit to the left. The bit positions freed by the shift operation are padded with zeroes.

The number of shifts may be specified as a parameter in the SLD statement or as a positive number in INT format right-justified in accumulator 2. If the number of shifts is = 0, the statement is not executed (no operation, or NOP); if it exceeds 31, accumulator 1 contains zero following execution of the SLD statement.

If the contents of accumulator 2 is interpreted as a number in DINT format, it would correspond to multiplication with a power to base 2, with the exponent as the number of shifts.

### Shift right word

| | | |
|---|---|---|
| SRW | n | Shift right word by n bits |
| SRW | | Shift right word by the number of bits in accumulator 2 |

Shift function SRW shifts bits 0 to 15 of accumulator 1 bit by bit to the right. The bit positions freed by the shift operation are padded with zeroes. Bits 16 to 31 are not affected.

The number of shifts may be specified as a parameter in the SRW operation or as a positive number in INT format right-justified in accumulator 2. If the number of shifts is = 0, the statement is not executed (no operation, or NOP); if it exceeds 15, the low-order word of accumulator 1 contains zero following execution of the SRW statement.

If the contents of accumulator 1 (low-order word) is interpreted as a number in INT format, a shift to the right is equivalent to division by a power to base 2, with the exponent as the number of shifts. Because the freed bits are padded with zeroes, this applies only to positive numbers. The result of such a division corresponds to the rounded integer component.

### Shift right doubleword

| | | |
|---|---|---|
| SRD | n | Shift right doubleword by n bits |
| SRD | | Shift right doubleword by the number of bits in accumulator 2 |

Shift function SRD shifts the entire accumulator bit by bit to the right. The bit positions freed by the shift operation are padded with zeroes.

The number of shifts may be specified as a parameter in the SRD statement or as a positive number in INT format right-justified in accumulator 2. If the number of shifts is = 0, the statement is not executed (no operation, or NOP); if it exceeds 31, accumulator 1 contains zero following execution of the SRD statement.

If the contents of accumulator 1 are interpreted as a number in DINT format, a shift to the right is equivalent to division with a power to base 2. The exponent is the number of shifts. Because the freed bits are padded with zeroes, this applies only to a positive number. The result of such a division corresponds to the rounded integer component.

### Shift word with sign

| | | |
|---|---|---|
| SSI | n | Shift word with sign by n bits |
| SSI | | Shift word with sign by the number of bits in accumulator 2 |

Shift function SSI shifts bits 0 to 15 of accumulator 1 bit by bit to the right. The bit positions freed by the shift operation are padded with the signal state of bit 15 (which is the sign of an INT number), that is to say, with "0" in the case of a positive number and "1" in the case of a negative number. Bits 16 to 31 are not affected.

The number of shifts may be specified as a parameter in the SSI statement or as a positive number in INT format right-justified in accumulator 2. If the number of shifts is = 0, the operation is not executed (no operation, or NOP); if it exceeds 15, the sign is in all bit positions of the low-order word of accumulator 1 following execution of the statement.

If the contents of accumulator 1 (low-order word) are interpreted as a number in INT format, a shift to the right is equivalent to division by a power to base 2, with the exponent as the number of shifts. The result of such a division corresponds to the rounded integer component.

### Shift doubleword with sign

SSD n Shift doubleword with sign by n bits
SSD Shift doubleword with sign by the number of bits in accumulator 2

Shift function SSD shifts the entire accumulator 1 bit by bit to the right. The bit positions freed by the shift operation are padded with the signal state of bit 31 (which is the sign of a DINT number), that is to say, with "0" in the case of a positive number and with "1" in the case of a negative number.

The number of shifts may be specified as a parameter in the SSD statement or as a positive number in INT format right-justified in accumulator 2. If the number of shifts is = 0, the statement is not executed (no operation, or NOP); if it exceeds 31, the sign is in all bit positions of accumulator 1 following execution of the statement.

If the contents of accumulator 1 are interpreted as a number in DINT format, a shift to the right is equivalent to division by the power to base 2, with the exponent as the number of shifts. The result of such a division corresponds to the rounded integer component.

## 13.3 Rotating

### Rotate left

RLD n Rotate left by n bits
RLD Rotate left by the number of bits in accumulator 2

Shift function TLD shifts the entire accumulator 1 bit by bit to the left. The bit positions freed by the shift operation are padded with the contents of the bit positions that were shifted out.

The number of shifts may be specified as a parameter in the RLD statement or as a positive number in INT format right-justified in accumulator 2. If the number of shifts is = 0, the statement is not executed (no operation, or NOP); if it exceeds 32, the contents of accumulator 1 remain unchanged and status bit CC1 assumes the signal state of the last bit shifted (bit 0). If the number of shifts is 33, the accumulator is shifted by one bit position, if 34 by two bit positions, and so on (the shift is executed modulo 32).

### Rotate right

RRD n Rotate right by n bits
RRD Rotate right by the number of bits in accumulator 2

Shift function RRD shifts the entire accumulator 1 bit by bit to the right. The bit positions freed by the shift operation are padded with the values of the bit positions that were shifted out.

The number of shifts may be specified as a parameter in the RRD statement or as a positive number in INT format right-justified in accumulator 2. If the number of shifts is = 0, the statement is not executed (no operation, or NOP); if it exceeds 32, the contents of accumulator 1 remain unchanged and status bit CC1 assumes the signal state of the last bit shifted (bit 31). If the number of shifts is 33, the contents of the accumulator are shifted by one position, if 34 by two positions, and so on (the shift is executed modulo 32).

### Rotate left through CC1

RLDA Rotate left through status bit 1 by one position

The RLDA function shifts the entire contents of accumulator 1 one bit to the left. The bit position freed by the shift (bit 0) assumes the signal state of status bit CC1. Status bit CC1 assumes the signal state of the bit that was shifted out (bit 31); status bit CC0 is set to "0".

### Rotate right through CC1

RRDA Rotate right through status bit CC1 by one position

The RRDA function shifts the entire contents of accumulator 1 one bit to the right. The bit position freed by the shift (bit 31) assumes the signal state of status bit CC1. Status bit CC1 assumes the signal state of the bit that was shifted out (bit 0); status bit CC0 is set to "0".

# 14 Word Logic

Word logic instructions link the value in accumulator 1 bit by bit with a constant or with the contents of accumulator 2 and store the result in accumulator 1. The logic operation can be performed on a word or a doubleword.

The following word logic instructions are available:

- AND
- OR
- Exclusive OR

Chapter 15, "Status Bits", provides information on the status bits set by these instructions.

The examples in this chapter are also on the diskette accompanying the book under the "Digital Functions" program in function block FB 114 or source file Chap_14.

## 14.1 Processing a Word Logic Instruction

A word logic instruction links the contents of accumulator 1 as shown in Figure 14.1. The operands are linked bit by bit with AND, OR, or Exclusive OR. The result is stored in accumulator 1 for further processing. The contents of accumulator 2 remain unchanged.

Word logic instructions execute without regard to any conditions. They do not affect the RLO.

### Word logic with the contents of accumulator 2

The actual word logic instruction is preceded by two load operations, one for each of the two values to be linked. When the word logic instruction has executed, the result is in accumulator 1.

Example:

```
L     MW 142;      //Address 1
L     MW 144;      //Address 2
AW    ;            //Logic operation
T     MW 146;      //Result
```

### Word logic with a constant

The operand to be linked is loaded into accumulator 1 and then linked with the value specified as a constant in the instruction. Following execution, the result of the word logic operation is in accumulator 1.

Example:

```
L     MW 148;
AW    W#16#807F;
T     MW 150;
L     MD 152;
OD    DW#16#8000_F000;
T     MD 156;
```

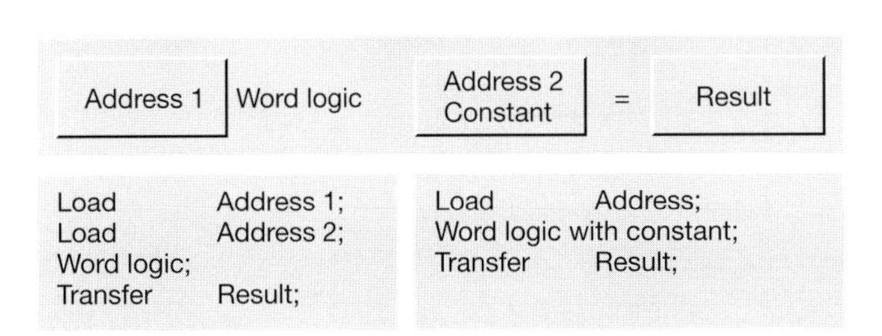


**Figure 14.1** Processing a Word Logic Instruction

**Table 14.1** Generating the Result of Word Logic Instructions

| Contents of accumulator 2 or bit in the constant | 0 | 0 | 1 | 1 |
|---|---|---|---|---|
| Contents of accumulator 1 | 0 | 1 | 0 | 1 |
| Result of AW, AD | 0 | 0 | 0 | 1 |
| Result of OW, OD | 0 | 1 | 1 | 1 |
| Result of XOW, XOD | 0 | 1 | 1 | 0 |

In the example above, the logic operation is performed on a word; in the example below, it is performed on the entire accumulator, that is, on a doubleword.

## Generating the result of a word logic instruction

A word logic instruction generates the result bit by bit, exactly as described in Chapter 4, "Binary Logic Operations" (Table 14.1). The operation links bit 0 of accumulator 1 with bit 0 of accumulator 2 or the constant specified in the instruction; the result is stored in bit 0 of the accumulator. The same logic is used on bit 1, bit 2, and, up to and including bit 15 (word instructions) or 31 (doubleword instructions).

## Performing word logic instructions on words

The logic instructions for words affect only the low-order word (bits 0 to 15) of the two accumulators. The high-order word (bits 16 to 31) remains unchanged (Figure 14.2).

## Successive word logic instructions

Following execution of a word logic instruction, you can proceed immediately to the next (load the operands and execute the word logic operation or execute the word logic operation using a constant) without having to store the intermediate result (in the local data area, for instance). The accumulators serve here as temporary stores.

Examples:

```
L     Value1;
L     Value2;
AW    ;
L     Value3;
OW    ;
T     Result1;
```

The result of the AW instruction is in accumulator 1, and is moved to accumulator 2 when

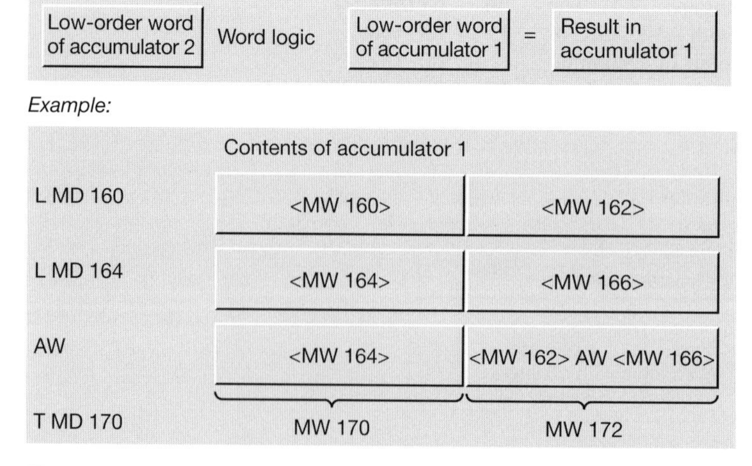


**Figure 14.2** Performing Word Logic Instructions on Words

*Value3* is loaded. The two values can now be linked with OW.

```
L      Value4;
L      Value5;
XOW    ;
AW     W#16#FFF0;
T      Result2;
```

The result of the XOW instruction is in accumulator 1. Bits 0 to 3 of accumulator 1 are set to "0" with the AW statement.

Table 14.2 shows one example for each of the different word logic instructions.

## 14.2 Description of the Word Logic Instructions

### Digital AND operation

| | |
|---|---|
| AW | AND operation (word) with accum1 and accum2 |
| AW W#16# | AND operation (word) with constant and accum1 |
| AD | AND operation (doubleword) with accum2 and accum2 |
| AD DW#16# | AND operation (doubleword) with constant and accum1 |

The digital AND operation links the bits of the value in accumulator 1 with the corresponding bits of the value in accumulator 2 or the constant according to AND. A bit in the result word is "1" only when the corresponding bits in both of the values being ANDed (combined according to logic AND) are also "1".

Since those bits in accumulator 2 or the constant which are "0" also set the corresponding bits in the result to "0", regardless of their signal state in accumulator 1, one also says of these bits that they are "masked". This so-called "masking" is the main purpose of the digital AND operation.

### Digital OR operation

| | |
|---|---|
| OW | OR operation (word) with accum2 and accum1 |
| OW W#16# | OR operation (word) with constant and accum1 |
| OD | OR operation (doubleword) with accum2 and accum1 |
| OD DW#16# | OR operation (doubleword) with constant and accum1 |

The digital OR operation links the bits of the value in accumulator 1 with the corresponding bits of the value in accumulator 2 or the constants according to OR. A bit in the result word is "0" only when the corresponding bits in both of the values being ORed (combine according to logic OR) are also "0".

Since those bits in accumulator 2 or the constant which are "1" also set the corresponding bits in the result to "1", regardless of their signal state in accumulator 1, one also says of these bits that they are "masked". This so-called "masking" is the main purpose of the digital OR operation.

**Table 14.2** Examples for Word Logic Instructions

| | | |
|---|---|---|
| **AND logic** | The four high-order bits of memory word MW 138 are set to "0"; the result is stored in memory word MW 140. | `L MW 138;`<br>`AW W#16#0FFF;`<br>`T MW 140;` |
| **OR logic** | Variables "WLogicVal1" and "WLogicVal2" are linked bit for bit according to OR and the resultstored in "WLogicReslt". | `L "Global_DB".WLogicVal1;`<br>`L "Global_DB".WLogicVal2;`<br>`OD ;`<br>`T "Global_DB".WLogicReslt;` |
| **Exclusive OR** | The value generated by linking variables #Input and #Mask with Exclusive OR is in variable #Buffer. | `L #Input;`<br>`L #Mask`<br>`XOW ;`<br>`T #Buffer;` |

**Digital Exclusive OR operation**

| | |
|---|---|
| XOW | Exclusive OR operation (word) with accum2 and accum1 |
| XOW W#16# | Exclusive OR operation (word) with constant and accum1 |
| XOD | Exclusive OR operation (doubleword) with accum2 and accum1 |
| XOD DW#16# | Exclusive OR operation (doubleword) with constant and accum1 |

The digital Exclusive OR operation links the bits of the value in accumulator 1 with the corresponding bits of the value in accumulator 2 or the constants according to Exclusive OR. A bit in the result word is "1" only when precisely one of the corresponding bits being linked is "1". If a bit in accumulator 2 has signal state "1", the result will contain at this position the reverse signal state that the bit previously had in accumulator 1.

In the result, only those bits are "1" that had different signal states in the two accumulators or in accumulator 1 and the constant prior to execution of the Exclusive OR instruction. Ascertaining these bits or "negation" of the signal states of individual bits is the primary purpose of the digital Exclusive OR operation.

# Program Flow Control

STL provides you with a variety of possibilities for controlling the flow of the program. You can exit linear program execution within a block or you can structure the program with parameterizable block calls. You can influence program execution depending on values calculated at runtime, or depending on process parameters, or according to your plant status.

The **status bits** provide information on the result of an arithmetic or math function and on errors (for example, number range violation in a calculation). You can incorporate the signal state of the status bits direct into your program using binary logic combinations.

- You can use the **jump functions** to branch within your program either unconditionally or dependent on the status bits, the RLO or the binary result. With STL, you can execute the jumps with calculated jump width (jump distributor) or you can easily implement program loops (loop jump).
- A further method of influencing program execution is provided by the **Master Control Relay** (MCR). Originally developed for relay contactor controls, STL offers a software version of this program control method.
- STL provides the **block functions** as a means for you to structure your program. You can use functions and function blocks again and again by defining **block parameters**.

For details of how to program blocks (with the keywords for source-oriented programming), see Section 3.3 "Programming Blocks". Chapter 18 "Block Functions" and 19 "Block Parameters" expand on this topic.

Chapter 26 "Direct Variable Access" contains further information on the block parameters, such as how they are stored in memory and how they can be used in conjunction with complex data types.

**15 Status Bits**
Status bits RLO, BR, CC0, CC1 and overflow; checking the status bits; status word; EN/ENO

**16 Jump Functions**
Unconditional jump; jump conditional on the RLO, BR, CC0, CC1 and overflow; jump distributor; loop jump

**17 Master Control Relay**
MCR-dependence; MCR range; MCR zone

**18 Block functions**
Block types, block call, block end; static local data; handling data blocks, data block register, handling data operands

**19 Block Parameters**
Parameter declaration; formal parameters, actual parameters; passing parameters on to called blocks; Examples: Conveyor belt, parts counter and feed

# 15 Status Bits

The status bits are binary flags (indicator bits). The CPU uses them for controlling the binary logic operations and sets them in digital processing. You can check these status bits (for example, as a result check in calculations) or you can influence specific bits. The status bits are combined into a word, the status word.

The examples in this chapter can also be found on the diskette accompanying the book under the "Program Flow Control"program in function block FB 115 or source file Chap_15.

## 15.1 Description of the Status Bits

Table 15.1 shows the status bits available with STL. The first column shows the bit number in the status word. The CPU uses the binary flags for controlling the binary functions; the digital flags indicate primarily results of arithmetic and math functions.

### First check

The /FC status bit steers the binary logic operation within a logic control system. A binary logic step always starts with /FC = "0" and a binary check instruction, the first check, as shown in the description of binary logic operations. The first check sets /FC = "1". A binary logic step ends with a binary value assignment or with a conditional jump or a block change. These set /FC = "0". The next binary check is then the start of a new binary logic combination.

**Table 15.1** Status Bits

| Bit | Binary Flags | |
|---|---|---|
| 0 | /FC | First check |
| 1 | RLO | Result of logic operation |
| 2 | STA | Status |
| 3 | OR | OR status bit |
| 8 | BR | Binary result |
| | **Digital Flags** | |
| 4 | OS | Stored overflow |
| 5 | OV | Overflow |
| 6 | CC0 | CC0 (condition code) status bit |
| 7 | CC1 | CC1 (condition code) status bit |

### Result of logic operation (RLO)

The RLO status bit is the intermediate buffer in binary logic operations. In first check, the CPU transfers the check result to the RLO, combines the check result with the stored RLO at each subsequent check, and stores the result, in turn, in the RLO (as described in Chapter 4 "Binary logic Operations"). You can also set, reset or negate the RLO direct or store it in the BR. Memory, timer and counter functions are controlled using the RLO and certain jump functions are executed.

### Status

The STA status bit corresponds to the signal state of the addressed binary operand or of the checked condition in the case of binary logic operations (A, AN, O, ON, X, XN).In the case of memory functions (S, R, =), the value of STA is the same value as the written value or (if no write operation takes place, for example, in the case of RLO = "0"or MCR active), STA corresponds to the value of the addressed (and unmodified) binary operand. With edge evaluations FP or FN, the value of the RLO prior to the edge evaluation is stored in STA. All other binary statements set STA = "1"; also the binary flag-dependent jumps JC, JCN, JBI, JNBI (Exception: CLR sets STA = "0").

The STA status bit has no effect on the processing of STL statements. It is displayed in the programming device test functions (such as program status) so that you can use it to trace binary logic sequences or for troubleshooting.

### OR status bit

The OR status bit stores the result of a fulfilled AND operation ("1") and indicates to a subsequent OR operation that the result is already fixed (in conjunction with the O statement in an AND before OR operation). All other binary statements reset the OR status bit.

Table 15.2 (under "Binary Flags") uses the example of a binary logic step to show how the binary flags are affected. The binary logic step starts with the first check following a memory function and ends with the last memory function prior to a check.

### Overflow

The OV status bit indicates a number range overflow or the use of invalid REAL numbers. The following functions influence the OV status bit: Arithmetic functions, math functions, some conversion functions, REAL comparison functions.

You can evaluate the OV status bit with check statements or with JO jump statement.

### Stored overflow

The OS status bit stores an OV status bit setting: When the CPU sets the OV status bit, it also always sets the OS status bit. However, while the next properly executed operation resets OV, OS

**Table 15.2** Example of Influencing the Status Bit

| STL statements | | Binary Flags: /FC | RLO | STA | OR | Remark | |
|---|---|---|---|---|---|---|---|
| ... | | | | | | | |
| = | M 10.0 | 0 | x | x | | | |
| A | I 4.0 | 1 | 1 | 1 | 0 | I 4.0 has "1" | |
| AN | I 4.1 | 1 | 1 | 0 | 0 | I 4.1 has "0" | |
| O | | 1 | 1 | 1 | 1 | | *The shaded area is a binary logic-step* |
| O | I 4.2 | 1 | 1 | 0 | 0 | I 4.2 has "0" | |
| ON | I 4.3 | 1 | 1 | 1 | 0 | I 4.3 has "1" | |
| = | Q 8.0 | 0 | 1 | 1 | 0 | Q 8.0 to "1" | |
| R | Q 8.1 | 0 | 1 | 0 | 0 | Q 8.1 to "0" | |
| S | Q 8.2 | 0 | 1 | 1 | 0 | Q 8.2 to "1" | |
| A | I 5.0 | 1 | x | x | | | |
| ... | | | | | | | |

| STL statements | | Digital Flags: CC0 | CC1 | OV | OS | Remark | |
|---|---|---|---|---|---|---|---|
| ... | | | | | | | |
| T | MW 10 | x | x | x | x | | |
| L | +12 | x | x | x | x | | |
| L | +15 | x | x | x | x | | |
| -I | | 1 | 0 | 0 | 0 | Result negative | |
| L | +20000 | 1 | 0 | 0 | 0 | | |
| *I | | 0 | 1 | 1 | 1 | Overflow | OV and OS to "1" |
| L | +20 | 0 | 1 | 1 | 1 | | |
| +I | | 0 | 1 | 0 | 1 | OV becomes "0" | OS remains "1" |
| T | MW 30 | 0 | 1 | 0 | 1 | | |
| L | MW 40 | 1 | 1 | 0 | 1 | | |
| ... | | | | | | | |

remains set. This provides you with the opportunity of evaluating, even at a later point in your program, a number range overflow or an operation with an invalid REAL number.

You can evaluate the OS status bit with check statements or with the JOS jump statement. JOS or a block change reset the OS status bit.

#### CC0 and CC1 status bits (condition code bits)

The CC0 and CC1 status bits provide information on the result of a comparison function, an arithmetic or math function, a word logic operation or on the shifted out bit in the case of a shift function.

You can evaluate all digital flags with jump func tions and check statements (see below in this chapter). Table 15.2 shows an example of setting digital flags in the lower section "Digital Flags".

#### Binary result

The BR status bit helps in the implementation of the EN/ENO mechanism for block calls (in conjunction with graphical languages). Section 15.4 "Using the Binary Result" shows you how STEP 7 uses the binary result. You can also set or reset the BR status bit yourself and check it with binary checks or with jump statements.

#### Status word

The status word contains all status bits. You can load it into accumulator 1 or write it out of accumulator 1 with a value.

```
L STW;  //Load the status word
        //. . .
T STW;  //Transfer to the status word
```

See Chapter 6 "Transfer Functions" for a description of the load and transfer statements. Table 15.1 contains the assignment of the status word with the status bits.

You can use the status word to check the status bits or to set them according to your wishes. In this way, you can store a current status word or begin a program section with a specific assignment of status bits.

Please note that the S7-300-CPUs do not load status bits /FC, STA and OR into the accumulator; the accumulator contains "0" at these locations.

## 15.2 Setting the Status Bits and the Binary Flags

The digital functions affect the CC0, CC1, OV and OS status bits as shown in Table 15.3. There are special STL statements for influencing the RLO and BR status bits.

#### Status bits with INT and DINT calculation

The arithmetic functions with data formats INT and DINT set all digital flags (status bits). A result of zero sets CC0 and CC1 to "0". CC0 = "0" and CC1 = "1" indicates a positive result, CC0 = "1" and CC1 = "0" indicates a negative result. A number range overflow sets OV and OS (please note the other meaning of CC0 and CC1 in the case of overflow). Division by zero is indicated by "1" at all digital status bits.

#### Status bits with REAL calculation

The arithmetic functions with data format REAL and the math functions set all digital status bits. A result of zero sets CC0 and CC1 to "0". CC0 = "0" and CC1 = "1" indicates a positive result, CC0 = "1" and CC1 = "0" indicates a negative result. A number range overflow sets OV and OS (please note the other meaning of CC0 and CC1 in the case of overflow). An invalid REAL number is indicated with "1" at all digital status bits.

A REAL number is referred to as "denormalized" if it is represented with reduced accuracy. the exponent is then zero; the absolute value of a denormalized REAL number is less than $1.175494 \times 10^{-38}$ (see also the chapter 24 on "Data Types"). S7-300 CPUs treat denormalized REAL numbers as equal to zero.

#### Status bits with the conversion functions

Of the conversion functions, the two's complements affect all digital status bits. In addition, the following conversion functions set status bits

**Table 15.3** Setting the Status Bits

| **INT Calculation** | | | | | |
|---|---|---|---|---|---|
| The result is: | | CC0 | CC1 | OV | OS |
| < –32 768 | +I, –I | 0 | 1 | 1 | 1 |
| < –32 768 | *I | 1 | 0 | 1 | 1 |
| –32 768 to –1 | | 1 | 0 | 0 | – |
| 0 | | 0 | 0 | 0 | – |
| +1 to +32 767 | | 0 | 1 | 0 | – |
| > +32 767 | +I, –I | 1 | 0 | 1 | 1 |
| > +32 767 | *I | 0 | 1 | 1 | 1 |
| 32 768 | /I | 0 | 1 | 1 | 1 |
| (–) 65 536 | | 0 | 0 | 1 | 1 |
| Division by zero | | 1 | 1 | 1 | 1 |

| **INT Calculation** | | | | | |
|---|---|---|---|---|---|
| The result is: | | CC0 | CC1 | OV | OS |
| < –2 147 483 648 | +D, –D | 0 | 1 | 1 | 1 |
| < –2 147 483 648 | *D | 1 | 0 | 1 | 1 |
| –2 147 483 648 to –1 | | 1 | 0 | 0 | – |
| 0 | | 0 | 0 | 0 | – |
| +1 to +2 147 483 647 | | 0 | 1 | 0 | – |
| > +2 147 483 647 | +D, –D | 1 | 0 | 1 | 1 |
| > +2 147 483 647 | *D | 0 | 1 | 1 | 1 |
| 2 147 483 648 | /D | 0 | 1 | 1 | 1 |
| (–) 4 294 967 296 | | 0 | 0 | 1 | 1 |
| Division by zero | /D, MOD | 1 | 1 | 1 | 1 |

| **REAL Calculation** | | | | |
|---|---|---|---|---|
| The result is: | CC0 | CC1 | OV | OS |
| + infinite (division by zero) | 0 | 1 | 1 | 1 |
| + normalized | 0 | 1 | 0 | – |
| ± denormalized | 0 | 0 | 1 | 1 |
| ± zero | 0 | 0 | 0 | – |
| – normalized | 1 | 0 | 0 | – |
| – infinite (division by zero) | 1 | 0 | 1 | 1 |
| ± invalid REAL number | 1 | 1 | 1 | 1 |

| **Comparison** | | | | |
|---|---|---|---|---|
| The result is: | CC0 | CC1 | OV | OS |
| equal to | 0 | 0 | 0 | – |
| greater than | 0 | 1 | 0 | – |
| less than | 1 | 0 | 0 | – |
| invalid REAL number | 1 | 1 | 1 | 1 |

| **Conversion NEG_I** | | | | |
|---|---|---|---|---|
| The result is: | CC0 | CC1 | OV | OS |
| +1 to +32 767 | 0 | 1 | 0 | – |
| 0 | 0 | 0 | 0 | – |
| –1 to –32 767 | 1 | 0 | 0 | – |
| (–) 32 768 | 1 | 0 | 1 | 1 |

| **Conversion NEG_D** | | | | |
|---|---|---|---|---|
| The result is: | CC0 | CC1 | OV | OS |
| +1 to +2 147 483 647 | 0 | 1 | 0 | – |
| 0 | 0 | 0 | 0 | – |
| –1 to –2 147 483 647 | 1 | 0 | 0 | – |
| (–) 2 147 483 648 | 1 | 0 | 1 | 1 |

| **Shift function** | | | | |
|---|---|---|---|---|
| The shifted out bit is: | CC0 | CC1 | OV | OS |
| “0” | 0 | 0 | 0 | – |
| “1” | 0 | 1 | 0 | – |
| with number of shifts 0 | – | – | – | – |

| **Word logic** | | | | |
|---|---|---|---|---|
| The result is: | CC0 | CC1 | OV | OS |
| zero | 0 | 0 | 0 | – |
| not zero | 0 | 1 | 0 | – |

OV and OS in the event of an error (number range overflow or invalid REAL number):

- ITB and DTB: Conversion of INT to BCD
- RND+, RND–, RND, TRUNC: Conversion of REAL to DINT

### Status bits with comparison functions

The comparison functions set the CC0 and CC1 status bits. The flags are set independently of the executed comparison function; it depends only on the relation between the two values involved in the comparison function. A REAL comparison checks for valid REAL numbers.

### Status bits with word logic operations and shift functions

Word logic operations and shift functions set the CC0 and CC1 status bits. OV is reset.

### Setting and resetting the RLO

SET sets the RLO to "1" and CLR sets it to "0". In parallel with this, the STA status bit is also set to "1" or to "0". Both statements are executed unconditionally.

SET and CLR also reset the OR and /FC status bits, that is, after SET or CLR a new logic operation starts with the next scan (check).

You can program an absolute set or reset of binary operands with SET:

```
SET  ;
S    M 8.0; //Memory bit is set
R    M 8.1; //Memory bit is reset

CLR  ;      //Reset edge memory bit
S    C 1;   //for 'Set counter'
```

Direct setting and resetting of the RLO is also useful in conjunction with timers and counters. To start a timer or counter, you require a change of the RLO from "0" to "1" (please note that you also require a positive edge for enabling). In program sections with predominantly digital logic operations, the RLO is generally not defined, for example, following the jump functions for evaluating the digital flags (status bits). Here you can use SET and CLR for defined setting or resetting of the RLO or for programming an RLO change.

See Chapter 4 "Binary Logic Operations" for details of how to negate the RLO with NOT.

### Setting and resetting the BR

With SAVE you can save the RLO in the binary result. SAVE transfers the signal state from the RLO to the BR status bit. SAVE operates unconditionally and does not affect any other status bits.

```
SET  ;
SAVE ;    //Set BR to '1'
. . .
AN   OV;
SAVE ;    //Set BR to '0' on overflow
. . .
```

## 15.3 Evaluating the Status Bit

The status bits RLO and BR and all digital flags can be evaluated with binary checks and jump functions. It is also possible to further process all status bits after loading the status word into the accumulator.

### Evaluation with binary checks

You can use all checks described in Chapter 4 "Binary Logic Operations" to check the digital flags and the binary result (see next page). The principle of operation is the same as for checking an input, for example.

A – Check for fulfilled condition and logic AND
O – Check for fulfilled condition and logic OR
X – Check for fulfilled condition and logic exclusive OR

AN – Check for unfulfilled condition and logic AND
ON – Check for unfulfilled condition and logic OR
XN – Check for unfulfilled condition and logic exclusive OR

| | | |
|---|---|---|
| >0 | Result greater than zero | [(CC0=0) & (CC1=1)] |
| >=0 | Result greater than or equal to zero | [(CC0=0)] |
| <0 | Result less than zero | [(CC0=1) & (CC1=0)] |
| <=0 | Result less than or equal to zero | [(CC1=0)] |
| <>0 | Result not equal to zero | [(CC0=0) & (CC1=1) v (CC0=1) & (CC1=0)] |
| ==0 | Result equal to zero | [(CC0=0) & (CC1=0)] |
| UO | Result invalid (unordered) | [(CC0=1) & (CC1=1)] |
| OV | Overflow | [OV=1] |
| OS | Stored overflow | [OS=1] |
| BR | Binary result | |

## Evaluation with jump functions

You can evaluate the RLO and BR status bits, all combinations of CC0 and CC1 and the OV and OS status bits with the relevant jump functions (Table 15.4). Chapter 16 "Jump Functions" contains a detailed description.

## Notes on evaluating a number range overflow

A calculation result outside the number range defined for the data type sets the OV (overflow) status bit and the OS (stored overflow) status bit in parallel.

If the result of a subsequent function (in the case of a chain calculation, for example) is within the permissible number range, the OV flag is reset. The OS flag, however, remains set, so that a result overflow within a chain calculation can also be detected at the end of the calculation. OS is not reset until the JOS jump function or a block change (call or block end).

**Table 15.4** Evaluating the Status Bits using Jump Functions

| RLO | BR | CC0 | CC1 | OV | OS | Executed jump functions |
|---|---|---|---|---|---|---|
| "1" | – | – | – | – | – | JC, JCB |
| "0" | – | – | – | – | – | JCN, JNB |
| – | "1" | – | – | – | – | JBI |
| – | "0" | – | – | – | – | JNBI |
| – | – | 0 | 0 | – | – | JZ, JMZ, JPZ |
| – | – | 0 | 1 | – | – | JN, JP, JPZ |
| – | – | 1 | 0 | – | – | JN, JM, JMZ |
| – | – | 1 | 1 | – | – | JUO |
| – | – | – | – | 1 | – | JO |
| – | – | – | – | – | 1 | JOS |

You can evaluate an overflow with:

- Binary checks

```
L    Value1;
L    Value2;
+I   ;
A    OV;      //Individual evaluation
=    Status1;
L    Value3;
+I   ;
A    OV;      //Individual evaluation
=    Status2;
L    Value4;
+I   ;
A    OS;      //Overall evaluation
=    Status_overall;
T    Result;
```

- Jump functions

```
L    Value1;
L    Value2;
+I   ;
JO   STA1;    //Individual evaluation
L    Value3;
+I   ;
JO   STA2;    //Individual evaluation
L    Value4;
+I   ;
JOS  STAG;    //Overall evaluation
T    Result;
```

You can evaluate a number overflow either after every calculation operation (check the OV status bit) or after the overall calculation (check the OS status bit).

## 15.4 Using the Binary Result

STEP 7 uses the binary result to represent the EN/ENO mechanism in the ladder diagram LAD and function block diagram FBD programming languages. You can ignore this if you program only in STL; you then have the binary result at your disposal as an additional RLO memory.

However, you can use BR as a group error flag, even with pure STL programming, in order to indicate errors in block processing (as used by the SFB and SFC system blocks and some standard blocks).

### EN/ENO mechanism

In the LAD and FBD programming languages, all boxes have an enable input EN and an enable output ENO. If the enable input has "1", the function in the box is executed. If the box is executed properly, the enable output then also has signal state "1". If an error occurs during execution of the box, (for example, overflow in the case of an arithmetic function), ENO is set to "0". If EN has signal state "0", ENO is also set to "0".

You can use these characteristics of EN and ENO to link several boxes in a chain, with the enable output ENO leading to the enable input EN of the next box (Figure 15.1). This makes it possible to "switch off" the entire chain (no box is processed if input Input1 in the example has signal state "0") or the remainder of the chain is no longer processed if one box signals an error.

The EN input and the ENO output are not block parameters but statement results that the LAD/FBD Editor generates itself prior to and following all boxes (even in the case of functions and function blocks). Here, the LAD/FBD Editor uses the binary result to store the signal state at EN during block processing or to check the error message from the box. You can find the statement sequence shown in Figure 15.1 in Network 6 of FB 115 in the "Program Flow Control" program. If you view this network FB 115 on screen, you can switch to Ladder diagram representation with VIEW → LAD. The Editor then displays the LAD graphics.

If you write your own functions or function blocks and you want to use these in, for example, ladder or function block diagram representation, you must influence the binary result in such a way that BR will be set to "0" if an error is detected (see below).

### Group error message in blocks

You can use the binary result as a group error message in blocks. If a block has been executed properly, set BR to "1". BR is set to "0" if a block signals an error.

Example: At the start of the block, BR is set to "1". If an error now occurs during execution of the block, for example, a result exceeds the defined range, so that further processing must be stopped, set the binary result to "0" with JNB and jump to the block end, for example (in the event of an error, the condition must supply signal state '0').

The "Clock entry" example in Chapter 26.4 "Brief Description of the Message Frame Example" also uses BR as a group error message.

```
SET  ;
SAVE ;           //BR = "1"
...  ;
L    10_000;
L    Result;     //if result > 10000
<=I  ;           //then BR = "0"
JNB  ERR;        //and jump to ERR
...
```

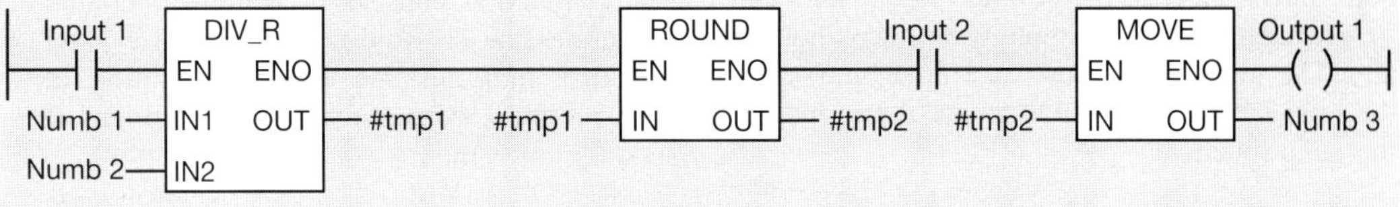

```
     A (
     A (
     A (
     A    Input1
     JNB  M001      //Evaluation of the EN input
     L    Numb1
     L    Numb2     //of the box
     /R
     T    tmp1
     AN   OV        //Error evaluation in the box
     SAVE           //and set BR
     CLR
M001:A    BR        //BR check for ENO output
     )
     JNB  M002      //EN input next box
     L    tmp1
     RND
     T    tmp2
     AN   OV
     SAVE
     CLR
M002:A    BR
     )
     A    Input2
     JNB  M003      //EN input last box
     L    tmp2
     T    Numb3
     SET
     SAVE
     CLR
M003:A    BR
     )
     =    Output1
```

**Figure 15.1** Example of the EN/ENO Mechanism

# 16 Jump Functions

With jump functions you can interrupt linear execution of the program and continue it at another point in the block. This program branching can be executed either conditionally or unconditionally. The jump distributor (case branching) and the loop jump are available as special forms of the jump functions.

**Overview**

| | | |
|---|---|---|
| JU | *label* | Unconditional jump |
| JC | *label* | Jump if RLO = "1" |
| JCN | *label* | Jump if RLO = "0" |
| JCB | *label* | Jump if RLO = "1" and save RLO |
| JNB | *label* | Jump if RLO = "0" and save RLO |
| JBI | *label* | Jump if BR = "1" |
| JNBI | *label* | Jump if BR = "0" |
| JZ | *label* | Jump if result is zero |
| JN | *label* | Jump if result is not zero |
| JP | *label* | Jump if result is greater than zero (positive) |
| JPZ | *label* | Jump if result is greater than or equal to zero |
| JM | *label* | Jump if result is less than zero (negative) |
| JMZ | *label* | Jump if result is less than or equal to zero |
| JUO | *label* | Jump if result is invalid |
| JO | *label* | Jump if overflow |
| JOS | *label* | Jump if stored overflow |
| JL | *label* | Jump distributor |
| LOOP | *label* | Loop jump |

The examples in this chapter are also presented on the diskette accompanying the book under the "Program Flow Control" program in function block FB 116 or source file Chap_16.

## 16.1 Programming a Jump Function

A jump function consists of a jump operation that defines the checked condition, and a jump label that indicates the program location at which program execution is to be continued if the condition is met.

A jump label consists of up to 4 characters which can include alphanumeric characters and the underscore. A jump label must not start with a numeric character. A jump label followed by a colon indicates the statement (line) that is to be processed after the executed jump statement.

Figure 16.1 gives an example. The condition for the jump here is a comparison operation; it supplies an RLO. This RLO is the jump condition for the JC jump. If the comparison is fulfilled, the RLO also equals "1" and the jump to jump label GR50 is executed. Program execution is then continued here. An unfulfilled comparison supplies RLO = "0", so that in this example the jump function is not executed. The program is continued at the next statement.

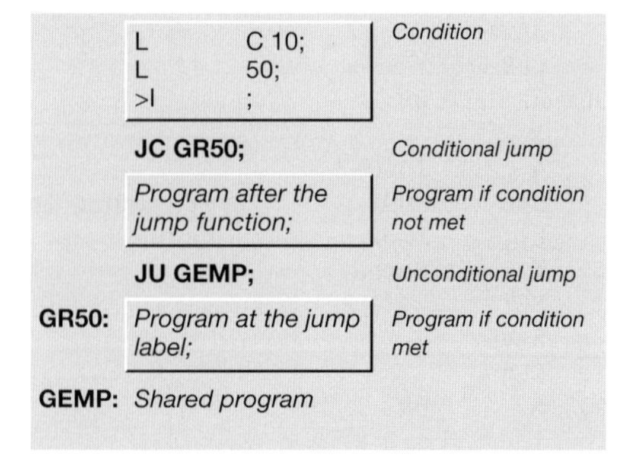


**Figure 16.1** Example of Program Branching

A jump can be made forward (in the direction of program processing; in the direction of higher line numbers) as well as backward. The jump can only take place within a block; that is, the jump destination must be in the same block as the jump statement. Subdivision into networks has no effect on the jump function.

If you use the Master Control Relay MCR, the jump destination must be in the same MCR zone or the same MCR area as the jump statement.

The jump destination must have a unique ID, that is, you can assign any given jump label only once in a block. The jump destination can be jumped to from several locations.

STL stores the the jump label designations in the non-execution-relevant section of the block on the data medium of the programming device. Only the jump widths are stored in the work memory of the CPU (in the compiled block). For this reason, program modifications made to blocks on-line at the CPU must also always be updated on the programming device data medium in order to retain the original designations. If this update is not made or if blocks are transferred from the CPU to the programming device, the non-execution-relevant block sections will be overwritten or deleted. The Editor then generates its own jump label designations (M001, M002 etc.) on screen or in the printout.

## 16.2 Unconditional Jump

The JU jump function is always executed, that is, regardless of any conditions. JU interrupts linear execution of the program and continues it at the location indicated by the jump label.

The JU jump function does not affect the status bits. If there are check statements, for example, AI, OI, etc., both immediately prior to the jump function and at the jump destination, these are treated as a single logic operation.

## 16.3 Jump Functions with RLO and BR

A program branch can be made dependent on the signal states of the RLO and BR status bits (Table 16.1). In addition, it is also possible to store the RLO in the BR status bit at the same time that it is checked.

### Setting the status bits

The jump functions conditional on the RLO set the STA and RLO status bits to “1” and OR and /FC to “0” whether the condition is fulfilled or not. This results in the following consequences for the use of these jump functions: The RLO is always set to “1”. If the statements contain operations conditional on the RLO immediately following these jump functions, they will be executed if the jump does not take place. If there are check statements, such as AI, OI etc., immediately following these jump functions, these checks are treated as first checks, that is, a new logic operation starts.

The jump functions conditional on the binary result set the STA status bit to “1” and the OR and /FC status bits to “0” whether the condition is fulfilled or not. The RLO and BR status remain unchanged. This has the following consequences for use: These jump functions terminate a logic operation; a new logic operation starts following this jump function or at the jump destination.

**Table 16.1** Jump Functions with RLO and BR

| RLO | BR | Executed Jumps | |
|---|---|---|---|
| “1” | – | JC | Jump if RLO = “1” |
| “1” | R “1” | JCB | Jump if RLO = “1” and save RLO |
| “0” | – | JCN | Jump if RLO = “0” |
| “0” | R “0” | JNB | Jump if RLO = “0” and save RLO |
| – | ”1” | JBI | Jump if BR = “1” |
| – | ”0” | JNBI | Jump if BR = “0” |

The RLO is retained and can be evaluated with a memory function following the jump function.

### Jump if RLO = "1"

The jump function JC is only executed if the RLO is "1" when the function is processed. If it is "0", the jump is not executed and program processing is continued with the next statement.

### Jump if RLO = "0"

The jump function JCN is only executed if the RLO is "0" when the function is processed. If it is "1", the jump is not executed and program processing is continued with the next statement.

### Jump if RLO = "1" and save the RLO

The jump function JCB is only executed if the RLO is "1" when the function is processed. Simultaneously, JCB sets the binary result to "1". If the RLO is "0", the jump is not executed and program processing is continued with the next statement. JCB sets the binary result in this case to "0" (the RLO is transferred in each case to the binary result).

### Jump if RLO = "0" and save the RLO

The jump function JNB is only executed if the RLO is "0" when this function is processed. Simultaneously, JNB sets the binary result to "0". If the RLO is "1", the jump is not executed and program processing is continued with the next statement. JNB sets the binary result in this case to "1" (the RLO is transferred in each case to the binary result).

### Jump if BR = "1"

The jump function JBI is only executed if the binary result is "1" when this function is processed. If the binary result is "0", the jump is not executed and program processing is continued with the next statement.

### Jump if BR = "0"

The jump function JBIN is only executed if the binary result is "0" when this function is processed. If the binary result is "1", the jump is not executed and program processing is continued with the next statement.

## 16.4 Jump Functions with CC0 and CC1

A program branch can be made conditional on the CC0 and CC1 status bits (Figure 16.2). This allows you to, for example, check to see if the result of a calculation is positive, zero or negative. See Chapter 15 "Status Bits" for details of when the CC0 and CC1 status bits are set.

### Setting the status bits

The jump functions conditional on the CC0 and CC1 status do not change any status bits. When the jump is made, the RLO is "taken along" and can be combined further (no change to /FC).

The binary checks are another method of checking the status bits (see Chapter 15 "Status Bits").

**Table 16.2** Jump Functions with CC0 and CC1

| CC0 | CC1 | Executed Jumps |
|---|---|---|
| 0 | 0 | JZ Jump if zero<br>JMZ Jump if zero or less than zero<br>JPZ Jump if zero or greater than zero |
| 1 | 0 | JM Jump if less than zero<br>JMZ Jump if zero or less than zero<br>JN Jump if not zero |
| 0 | 1 | JP Jump if greater than zero<br>JPZ Jump if zero or greater than zero<br>JN Jump if not zero |
| 1 | 1 | JUO Jump if invalid result |

### Jump if result is zero

The jump function JZ is only executed if CC0 = "0" and CC1 = "0". This is the case if

- accumulator 1 contains zero after an arithmetic or math function,
- accumulator 2 contains the same as accumulator 1 in a comparison function,
- accumulator 1 contains zero after a digital logic operation and
- the value of the last shifted bit is "0" after a shift function.

In all other cases, JZ continues program processing with the next statement.

### Jump if result is not zero

The jump function JN is only executed if status bits CC0 and CC1 have different signal states. This is the case if

- accumulator 1 does not contain zero after an arithmetic or math function,
- the contents of accumulator 2 are not the same as the contents of accumulator 1 in a comparison function
- accumulator 1 does not contain zero after a digital logic operation and
- the value of the last shifted bit is "1" after a shift function.

In all other cases, JN continues program processing with the next statement.

### Jump if result is greater than zero

The jump function JP is only executed if CC0 = "0" and CC1 = "1". This is the case if

- the contents of accumulator 1 are within the permissible positive number range following an arithmetic or math function (you check for a number range violation with JO or JOS),
- the contents of accumulator 2 are greater than the contents of accumulator 1 in a comparison function,
- accumulator 1 does not contain zero after a digital logic operation and
- the value of the last shifted bit is "1" after a shift function.

In all other cases, JP continues program processing with the next statement.

### Jump if result is greater than or equal to zero

The jump function JPZ is only executed if CC0 = "0". This is the case

- if the contents of accumulator 1 are within the permissible positive number range or are equal to zero following an arithmetic or math function (you check for a number range violation with JO or JOS),
- if the contents of accumulator 2 are greater than or equal to the contents of accumulator 1 in the case of a omparison function,
- after every digital logic operation and
- after every shift function.

In all other cases, JPZ continues program processing with the next statement.

### Jump if result is less than zero

The jump function JM is only executed if CC0 = "1" and CC1 = "0". This is the case if

- the contents of accumulator 1 are within the permissible negative number range following an arithmetic or math function (you check for a number range violation with JO or JOS) and
- the contents of accumulator 2 are less than the contents of accumulator 1 in a comparison operation.

In all other cases, JM continues program processing with the next statement.

### Jump if result is less than or equal to zero

The jump function JMZ is only executed if CC1 = "0". This is the case if

- the contents of accumulator 1 are within the permissible negative number range or are equal to zero following an arithmetic or math function (you check for a number range violation with JO or JOS), and
- the contents of accumulator 2 are less than or equal to the contents of accumulator 1 in the case of a comparison operation.

In all other cases, JMZ continues program processing with the next statement.

### Jump if invalid result

The jump function JUO is only executed if CC0 = "1" and CC1 = "1". This is the case if

- division by zero is made in an arithmetic function and
- an invalid REAL number is specified as the input value or is produced as the result.

In all other cases, JUO continues program processing with the next statement.

## 16.5 Jump Functions with OV and OS

A program branch can be executed dependent on the OV and OS status bits. This is a check to see if the result of a calculation is still within the permissible number range. See Chapter 15 "Status Bits" for details of when the OV and OS status bits are set.

### Jump if overflow

The jump function JO is only executed if the OV status bit has been set to "1". This is the case if the permissible number range has been exceeded following execution of an operation. The following functions can set the OV status bit:

- Arithmetic functions,
- Math functions,
- Two's complement,
- Comparison functions with REAL numbers and
- Conversion functions INT/DINT to BCD and REAL to DINT.

If OV = "0", JO continues program processing with the next statement.

In the case of a chain calculation with several calculations performed one after the other, the OV status bit must be evaluated after each calculation function since OV is reset again following the next calculation operation whose result is within the permissible number range. Check the OS status bit in order to evaluate a possible number range overflow at the end of the chain calculation.

### Jump if stored overflow

The jump function JOS is only executed if the OS status bit has been set to "1". This is always the case if a number range overflow sets the OV status bit (see above). In contrast to OV, OS remains set if a result is then in the permissible number range. The following functions reset OS:

- Block call and block end
- Jump if stored overflow JOS

If OS = "0", JOS continues program processing with the next statement.

## 16.6 Jump Distributor

The jump distributor JL allows specific (calculated) jumping to a program section in the block conditional on a number of shifts.

JL works in conjunction with a list of JU jump functions. This list immediately follows JL and can contain up to 255 entries. There is a jump label at JL that points to the end of the list (to the first statement after the list).

General example:

```
      L   Number_of_positions;
      JL  End;
      JU  M0;
      JU  M1;
      . . .
      JU  Mx;
End: . . .
```

In the example, the variable *Number_of_positions* loads a number into accumulator 1. This is followed by the jump distributor JL with the jump label to the end of the list of JU statements.

The number of the jump to be executed is in the right-hand byte of accumulator 1. If accumulator 1 contains 0, the first jump statement is executed, and if it contains 1, the second is executed, and so on. If the number is greater than the length of the list, JL branches to the end of the list (to the statement following the last jump).

JL is not subject to conditions and does not change the status bits.

Only JU statements are permissible in the list without gaps. You can assign the jump label designations as you please within the general rules.

## 16.7 Loop Jump

The loop jump LOOP allows simplified programming of program loops.

LOOP interprets the right-hand word of accumulator 1 as a signless 16-bit number in the range 0 to 65535. LOOP first decrements the accumulator contents by 1. If the value is then not zero, the jump is executed to the jump label specified. If the value is equal to zero after decrementing, the jump is not executed and the next statement is processed.

The value in accumulator 1 thus corresponds to the number of program loops to be passed. You must store this number in a loop counter. You can use any digital operand as the loop counter.

General example:

```
      L   Number;
Next: T   Counter;
      . . .
      . . .
      . . .
      L   Counter;
      LOOP      Next;
      . . .
```

The variable *Number* contains the number of loop passes. The variable *Counter* contains the loop passes still to be executed. At the first pass, *Counter* is preassigned with the number of loop passes. At the end of the program loop, the contents of *Counter* are loaded into the accumulator and decremented by the LOOP statement. If the accumulator does not contain zero following this, the jump to the specified jump label – here: Next – is executed and the variable *Counter* is updated.

The loop jump does not change the status bits.

# 17 Master Control Relay

With contact controls, a Master Control Relay activates or de-activates a section of the control that can consist of one or more rungs. A de-activated rung

- switches all non-retentive contactors off and
- retains the state of retentive contactors.

You can only change the state of the contactors again when the Master Control Relay (MCR) is active.

These characteristics mean that the Master Control Relay is used especially in the LAD (ladder logic).

This chapter describes the statements required for implementing the Master Control Relay. You can use these statements to emulate the properties of a Master Control Relay in the statement list. You can also find Master Control Relay examples on the diskette accompanying the book under the "Program Flow Control" program in function block FB 117 or source file Chap_17.

*Please note that switching off with the "software" Master Control Relay is no substitute for an EMERGENCY OFF or safety facility! Treat Master Control Relay switching in exactly the same way as switching with a memory function!*

STL provides the following statements for implementing the Master Control Relay (MCR):

- MCRA Activate MCR area
- MCR( Open MCR zone
- )MCR Close MCR zone
- MCRD Deactivate MCR area

The statements MCRA and MCRD identify an area in your program in which MCR dependency is to take effect. Within this area, you use the statements MCR(and)MCR to define one or more zones in which MCR dependency can be switched on and off. You can also nest the MCR zones. The result of logic operation (RLO) immediately prior to an MCR zone switches MCR dependency on or off within this zone.

## 17.1 MCR Dependency

MCR dependency affects the operations shown in Table 17.1. Independently of a preceding bit logic operation or digital logic operation, an assignment writes a "0" or a transfer operation writes zero to the specified operand if MCR dependency is switched on. Operands in set or reset operations do not change their signal states when MCR dependency is switched on.

You switch on MCR dependency in a zone if the RLO is "0" immediately prior to opening the zone (analogous to switching off the Master Control Relay). If you open an MCR zone with RLO "1" (Master Control Relay switched on), processing within this MCR zone takes place without MCR dependency.

**Table 17.1** MCR-Dependent Operations

| Operation | MCR is switched on | MCR is switched off |
|---|---|---|
| = Assignment | The associated operand is reset to "0" | Normal processing |
| S Set | No effect (status retained) | |
| R Reset | | |
| T Transfer | Zero is transferred | |

Example:

```
MCRA ;          //Activate MCR
A     Input0;
MCR( ;          //Open MCR zone
A     Input1;
A     Input2;
=     Output0;
)MCR ;          //Close MCR zone
MCRD ;          //Deactivate MCR
```

In the example, *Input0* = "0" also sets the operand *Output0* to "0". If *Input0* has signal state "1", you control the operand *Output1* with *Input1* and *Input2*.

## 17.2 MCR Area

To be able to use the characteristics of the Master Control Relay, define an MCR area with the statements MCRA and MCRD. MCR dependency is active within an MCR area (but not yet switched on).

```
MCRA;    //Activate MCR
. . .
. . .    //MCR area
. . .
MCRD;    //Deactivate MCR
```

MCRA defines the start of an MCR area and MCRD its end. If you call a block within an MCR area, MCR dependency is de-activated in the called block (Figure 17.1). An MCR area only starts again with the MCRA statement. When a block is exited, MCR dependency is set as it was before the block was called, regardless of the MCR dependency with which the block was exited.

## 17.3 MCR Zone

You define an MCR zone with the statements MCR( and )MCR. Within this zone, you can switch MCR dependency on with RLO = "0" and off with RLO = "1".

```
. . .            //Switch on MCR
. . .            //with "0"
A     Input3;
MCR( ;           //Start of dependency
. . .
. . .            //MCR zone
. . .
)MCR ;           //End of dependency
. . .
```

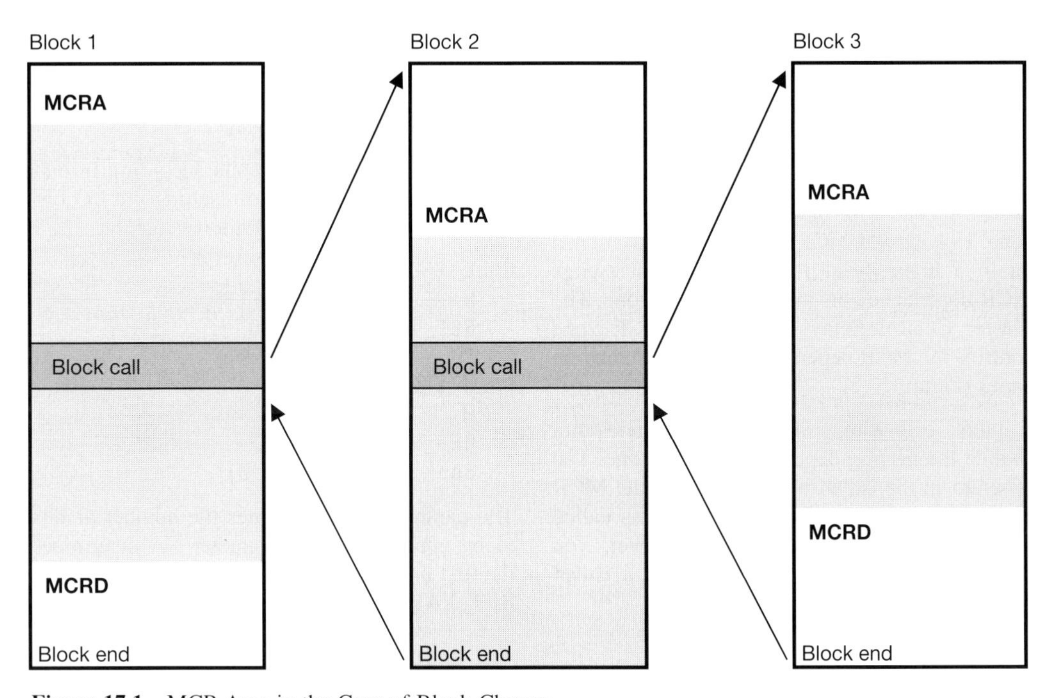


**Figure 17.1** MCR Area in the Case of Block Change

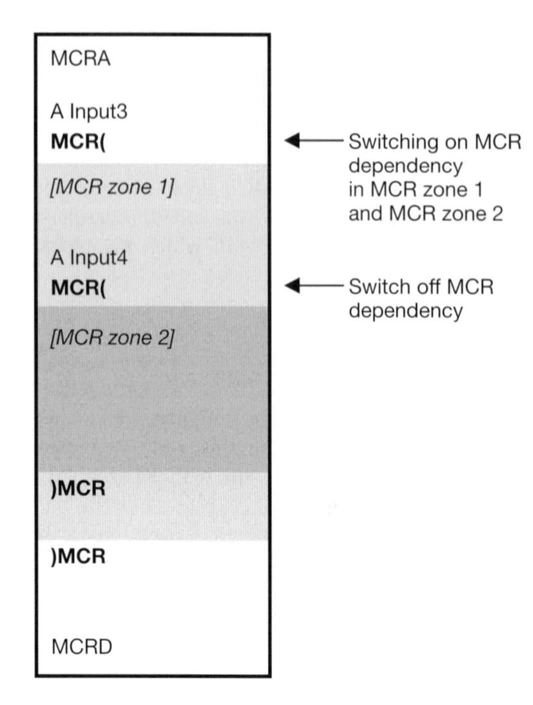


**Figure 17.2** MCR Dependency in the Case of Nested MCR Zones

The statements MCR( and )MCR end a bit logic combination.

You can open another MCR zone within an MCR zone. The nesting depth for MCR zones has the value 8; that is, you can open up to eight zones before having to close one.

You control the MCR dependency of a switched on MCR zone with the RLO when opening the zone. However, if MCR dependency is switched on in a higher-level zone, you cannot switch MCR dependency off in a lower-level zone. The Master Control Relay of the first MCR zone controls the MCR dependency in all switched on zones (Figure 17.2).

A block call within an MCR zone does not change the nesting depth of an MCR zone. The program in the called block is still in the MCR zone that was open when the block was called (and is controlled form here). However, you must re-activate MCR dependency in a called block by opening the MCR area.

In Figure 17.3 the operands *Input5* and *Input6* control the MCR dependencies. With *Input5* you can switch MCR dependency on in both zones (with "0"), regardless of the signal state of *Input6*. If the MCR dependency of Zone 1 is switched of with *Input5* = "1", you can control the MCR dependency of zone 2 with *Input6* Table 17.2).

## 17.4 Setting and Resetting I/O Bits

Despite MCR dependency being switched on, you can set or reset the bits of an I/O area with the system functions. A requirement for this is that the bits to be controlled are in the process-image output or a process-image output has been defined for the I/O area to be controlled.

The system function SFC 79 SET is available for setting the I/O bits, and SFC 80 RSET for resetting (Table 17.3). You call these system functions in an MCR zone. The system functions are only effective if MCR dependency is switched on; if MCR dependency is switched off, the calls of these SFCs remain without effect.

Setting and resetting the I/O bits also simultaneously updates the process-image output. The I/O are affected byte-by-byte. The bits not selected with the SFCs (in the first and in the last byte) retain the signal states as they are currently available in the process-image.

Example: Setting the I/O bits according to outputs Q 12.0 to Q 12.7 and resetting the I/O bits according to outputs Q 13.5 to Q 15.5.

```
CALL SFC 79 (
   N          := 8,
   RET_VAL    := MW 10,
   SA         := P#12.0);
CALL SFC 80 (
   N          := 16,
   RET_VAL    := MW 12,
   SA         := P#13.5);
```

The parameter N determines the number of bits to be controlled and parameter SA determines the first bit (Data type POINTER). The SFC uses RET_VAL to return any detected error.

**Table 17.2** MCR Dependency in the Case of Nested MCR Zones (Example)

| *Input5* | *Input6* | Zone 1 | Zone 2 |
|---|---|---|---|
| "1" | "1" | No MCR dependency | |
| "1" | "0" | No MCR dependency | MCR dependency switched on |
| "0" | "1" or "0" | MCR dependency switched on | |

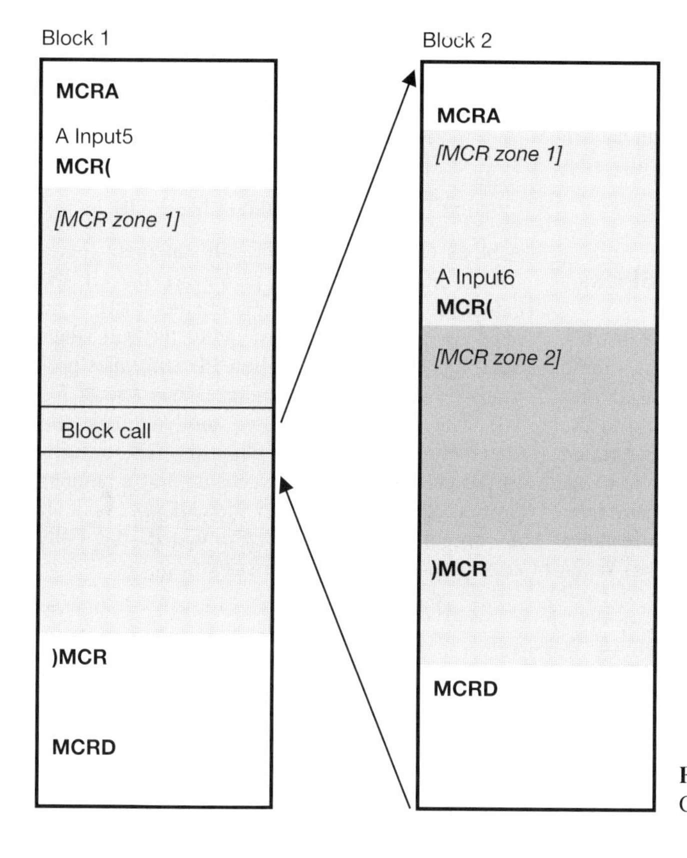


**Figure 17.3** MCR Zones at Block Change

**Table 17.3** Parameters of the SFCs for Controlling the I/O Bits

| SFC | Parameter | Declaration | Data Type | Assignment, Description |
|---|---|---|---|---|
| 79 | N | INPUT | INT | Number of bits to be set |
| | RET_VAL | OUTPUT | INT | Error information |
| | SA | OUTPUT | POINTER | Pointer to the first bit to be set |
| 80 | N | INPUT | INT | Number of bits to be reset |
| | RET_VAL | OUTPUT | INT | Error information |
| | SA | OUTPUT | POINTER | Pointer to the first bit to be reset |

# 18 Block Functions

In this chapter, you will learn how to call and terminate code blocks and how to work with operands from data blocks. The next chapter then deals with using block parameters.

You will find examples of the block functions on the diskette accompanying the book under the "Program Flow Control" program in function block FB 118 or source file Chap_18.

## 18.1 Block Functions for Code Blocks

Block functions for code blocks include instructions for calling and terminating blocks (Table 8.1). Code blocks are called and processed with CALL. You can pass data for processing to the called block and take over data from the called block. This data transfer is carried out via block parameters. CALL transfers the block parameter to the called block and also opens the instance data block in the case of function blocks. When code blocks have no block parameters, they can also be called with UC or CC. A block is terminated with a block end statement.

### 18.1.1 Block Calls: General

If a code block is to be processed, it must be "called". Figure 18.1 gives an example of calling function FC 10 in organization block OB 1.

A block call consists of the call statement (here: CALL FC 10) and the parameter list. If the called block has no block parameters, there is no need for the parameter list. After the call statement has been executed, the CPU continues program processing in the called block (here: FC 10). The block is processed until a block end statement is encountered. Then the CPU returns to the calling block (here: OB 1) and continues processing this block after the call statement. If an organization block is terminated, the CPU continues in the operating system.

**Table 18.1** Block Functions for Code Blocks

| *Calling a Function Block* | | |
|---|---|---|
| with data block and with block parameter<br>**CALL FB 10, DB 10 (**<br>In1 := Number1,<br>In2 := Number2,<br>Out := Number3); | as local instance and with block parameter<br>**CALL *name* (**<br>In1 := Number1,<br>In2 := Number2,<br>Out := Number3); | without block parameter, unconditionally and conditionally<br>**UC FB 11;**<br>**CC FB 11;** |
| *Calling a function* | | |
| with function value and with block parameter<br>**CALL FC 10 (**<br>In1 := Number1,<br>In2 := Number2,<br>**RET_VAL** := Number3); | without function value and with block parameter<br>**CALL FC 10 (**<br>In1 := Number1,<br>In2 := Number2,<br>Out := Number3); | without block parameter, unconditionally and conditionally<br>**UC FC 11;**<br>**CC FC 11;** |
| *Block end statements* | | |
| Conditional block end<br>**BEC** | Unconditional block end<br>**BEU** | Block end<br>**BE** |

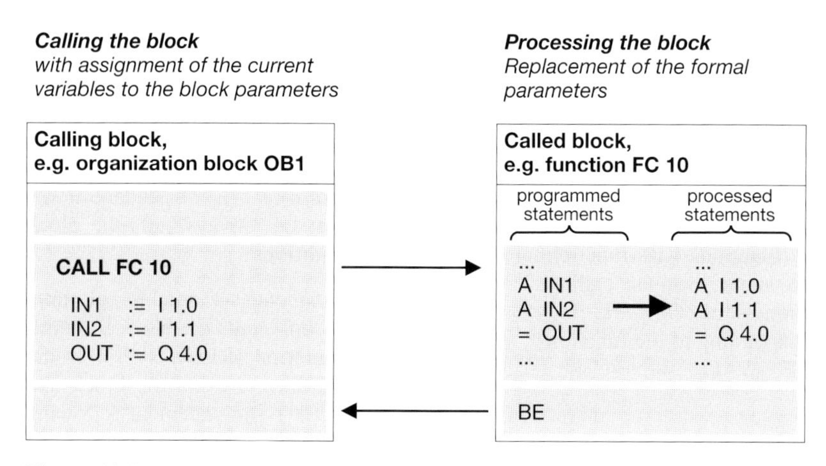


**Figure 18.1** Block Call Example

The information the CPU requires to make its return to the calling block is stored in the block stack (B stack). With each block call, a new stack element is created that includes the return address, the contents of the data register and the address of the local data stack of the calling block. If the CPU goes to the Stop state as a result of an error, you can use the programming device to see from the contents of the B stack which blocks were processed up to the triggering error.

The block parameters are the data interface to the called block. You are advised to avoid data transfer via internal registers (for example, accumulators, address registers, RLO) since the contents of these registers can be changed at a block change (as a result of "concealed" statements from the Editor).

### 18.1.2 CALL Call Statement

You call FBs, FCs, SFBs and SFCs with CALL. CALL is an unconditional call, that is, the specified block is always called and processed independently of any conditions. (You cannot call organization blocks; they are called by the operating system depending on events.)

#### Calling function blocks

You call a function block FB by specifying, after CALL, the function block and, separated with a comma, the instance data block associated with the call. You can address both blocks either absolutely or symbolically. The assignment of the absolute address to the symbol address is made in the symbol table, with an instance data block having the associated function block as data type. The CALL operation is followed by the list with the block parameters. In source-oriented input, the list of the block parameters is placed between round brackets; the block parameters are each separated by a comma.

With function blocks, you need not initialize all block parameters when the block is called. The uninitialized block parameters retain their current value. If you do not specify any parameters, the brackets are also dispensed with in source-oriented input.

If you have generated function blocks with the block attribute "multi-instance-capable", you can also call these as a local instance within other "multi-instance-capable" function blocks. Here, the called function block uses the instance data block of the calling function block to store its local data. You declare the local instance in the static local data of the calling function block and you can then call the function block in the program (without specifying an instance data block). The local instance is treated like a complex data type within the "higher-level" function block.

**Calling functions**

You call a function FC by specifying the function, after CALL, either with absolute addressing or with symbolic addressing. This is followed by the parameter list, in brackets in the case of source-oriented input. You must initialize all existing parameters; however, the parameters can be in any order. Calling functions with function value takes exactly the same form as calling functions without function value. Only the first output parameter – corresponding to the function value – has the name RET_VAL.

**Calling system blocks**

The operating system of the CPU contains system functions SFCs and system function blocks SFBs, that you can use. The number and type of system blocks is CPU-specific. All system blocks are called with CALL.

You call a system function block in the same way as a function block you have written yourself; you set up the associated instance data block in the user memory with the data type of the SFB.

You call a system function in the same way as a function you have written yourself.

System blocks are only available in the operating system of the CPU. When calling system blocks during off-line programming, the Editor requires a description of the call interface in order to be able to initialize the parameters. The interface description is located in library "Standard Library V3.x" supplied under "System Function Blocks". From here, the Editor copies the relevant interface description into the off-line container "Blocks" when you call a system block. The copied interface description then appears as a "normal" block object.

### 18.1.3 UC and CC Call Statements

You can call function blocks and functions with UC and CC. It is a requirement that the called function has no block parameters and the called function block has no instance data block - and therefore also no block parameters and no static local data. However, the Editor does not check this.

You can use the UC and CC operations if a block is too long or not clear enough for you, by simply "breaking down" the block into sections and then calling the sections in sequence. The UC and CC call operations do not distinguish between functions FCs and function blocks FBs. Both block types are handled in the same way.

- The *UC call statement* is an unconditional statement, that is, UC always calls the block regardless of conditions.
- The *CC call statement* is a conditional statement, that is, CC only calls the block if the result of logic operation (RLO) is "1". If the RLO is "0", CC does not call the block and sets the RLO to "1". The statement following CC is then processed.
- Effect on the indicator bits (condition code bits): The OS status bit is reset at block change; the CC0, CC1 and OV status bits are not affected, the /FC status bit is reset; that is, a new logic operation begins with the first check statement in the new block or following a block call.
- Binary nesting stack at a block change: You can also call a code block within a binary nesting expression. The current stack depth of the binary nesting stack does not change at a block change. The possible nesting stack depth in a block that can be called within a binary nest is therefore the difference between the maximum possible nesting depth and the current nesting depth at the block call.
- Master Control Relay at a block change: MCR dependency is deactivated at a block call. The MCR is switched off in the called block, regardless of whether the MCR was switched on or off prior to the block call. When a block is exited, MCR dependency is set as it was prior to the block being called.
- Accumulators and address registers at block change: The contents of accumulators and of the address registers are not changed at a block change with UC or CC.
- Data blocks at a block change: Calling a block saves the data block register in the B stack; the block end statement restores its contents when the called block is exited. The global data block current prior to the block call and the instance data block are also opened following the block call. If no data block was opened prior to the block call (for example, no in-

stance data block in OB 1), there will also be no data block open following the block call, regardless of the data blocks open in the called block.

Additional possibilities:

- Indirect addressing of FB and FC calls with UC and CC
- Calling via block parameters with UC
- Calling via block parameters with CC also in function blocks

### 18.1.4 Block End Functions

The BEC statement terminates program processing in a block conditional on the RLO, and the BEU and BE statements end a block unconditionally.

#### Conditional block end BEC

Execution of BEC depends on the RLO. If the RLO is "1" when BEC is processed, the statement is executed and the block currently being processed is terminated. A return jump is then made to the previously processed block containing the block call.

If the RLO is "0" when BEC is processed, the statement is not executed. The CPU sets the RLO to "1" and processes the statement following BEC. A subsequently programmed check statement is always a first check.

#### Unconditional block end BEU

When BEU is processed, the block currently being processed is exited. A return jump is then made to the previously processed block containing the block call.

In contrast to the BE statement, you can program BEU several times within one block. The program section following BEU is only processed if it is jumped to with a jump function.

#### Block end BE

When BE is processed, the block currently being processed is exited. A return jump is then made to the previously processed block containing the block call.

BE is always the last statement in a block.

Programming BE is a matter of choice. With incremental input, you terminate block programming by closing the block; with source-oriented input, the keyword replaces the block end, for example, END_FUNCTION_BLOCK instead of the BE statement.

### 18.1.5 Temporary Local Data

You use the temporary local data for intermediate storage of results occurring during block processing. Temporary local data are only available during block processing; after the block is terminated, its data are lost.

Temporary local data are operands that are located in the local data stack (L stack) in the CPU system memory. The operating system of the CPU provides the temporary local data for every code block when that code block is called. The values in the L stack are semi random when the block is called. In order to make meaningful use of the local data, you must first write them before reading. After a block is terminated, the L stack is assigned to the next called block.

The volume of temporary local data required by a block is shown in the block header. In this way, the operating system learns how many bytes are available in the L stack when the block is called. You can also see from the entry in the block header how many local data bytes the block requires (in the Editor when the block is open with FILE → PROPERTIES or in the SIMATIC Manager when the block is marked with EDIT → OBJECT PROPERTIES, on "General – Part 2" tab in each case).

#### Declaration of temporary local data

You declare the temporary local data in the declaration section of the code block:

- with incremental programming under "temp" or
- with source-oriented programming between VAR_TEMP and END_VAR.

Figure 18.2 gives an example of declaring temporary local data. The variable *temp1* is located in the temporary local data and is of data type INT, the variable *temp2* is of data type REAL.

The temporary local data are stored in the L stack in the order of their declaration accord-

*Incremental programming*

| Address | Declaration | Name | Type | Initial value |
|---|---|---|---|---|
| 0.0 | in | In | INT | 0 |
| | out | | | |
| | inout | | | |
| 2.0 | stat | Total | INT | 0 |
| 0.0 | temp | temp1 | INT | |
| 2.0 | temp | temp2 | REAL | |

*Source-oriented programming*

```
VAR_INPUT
  In     : INT := 0;
END_VAR
VAR_OUTPUT . . . END_VAR
VAR_IN_OUT  . . . END_VAR
VAR
  Total : INT := 0;
END_VAR
VAR_TEMP
  temp1 : INT;
  temp2 : REAL
END_VAR
```

**Figure 18.2** Example of Declaring Local Data in a Function Block

ing to data type. Chapter 26.2 "Data Storage of Variables" contains more detailed information on data storage in the L stack.

### Symbolic addressing of temporary local data

You address the temporary local data with their symbolic names. You assign the names in accordance with the rules for block-local symbols. All operations that are also valid for the memory bits are permissible for the temporary local data. However, please note that a temporary local data bit is not suitable as an edge memory bit since it does not retain its signal state beyond processing of the block.

You can only access the temporary local data of a block in the block itself. (Exception: The temporary local data of the calling block can be accessed via block parameters.)

### Size of the L stack

The size of the overall L stack is CPU-specific. The quantity of temporary local bytes available in a priority class, that is in the program of an organization block, is also fixed. On the S7-300, the quantity is fixed; for example, 256 bytes per priority class on the CPU 314. On the S7-400, you can adjust the quantity of local data bytes to your requirements when parameterizing the CPU. This quantity must be shared between the blocks called in the relevant organization block and the blocks called in turn from within these blocks.

Please note in this regard that the Editor also uses temporary local data, for example when transferring block parameters. You do not see these temporary local data at the programming interface.

### Start information

The operating system of the CPU transfers start information in the temporary local data when an organization block is called. This start information is 20 bytes long in each organization block and has an almost identical structure in each block. Chapters 20 "Main Program", 21 "Interrupt Handling", 22 "Start-Up Characteristics" and 23 "Error Handling" describe the start information assignments for the individual organization blocks.

These 20 bytes of start information must always be available in every priority class used. If you program evaluation of synchronous errors (programming and access errors), you must provide an additional 20 bytes for the start information of these error organization blocks since these error OBs are processed in the same priority class.

You declare the start information when programming an organization block. This is mandatory. The standard library "Standard Library V3.x" contains templates for declaration under "Organization Blocks". If you do not require the start information, it is sufficient to declare the first 20 bytes as, for example, a field (as shown in Figure 18.3).

### Absolute addressing of temporary local data

Normally, you access the temporary local data via symbolic addressing, with absolute addressing being the exception. If you are familiar with data storage in the L stack, you can work out for yourself the addresses at which the static local are located. You can also see the addresses in the variable declaration table of the compiled block.

The operand identifier for the temporary local data is L; a bit is addressed with L, a byte with LB, a word with LW and a doubleword with LD.

Example: For absolute addressing, you want to keep 16 bytes of temporary local data whose individual values you then want to access both as byte and as bit. Create this area as a field right at the start of the temporary local data so that the addressing starts at 0. In an organization block, you would locate this field declaration immediately following the declaration of the start information, so that in this case, addressing begins at 20 (Figure 18.3).

Note: Declare *all* the temporary local data you use, including those you address exclusively with absolute addresses. The Editor also uses temporary local data, for example when transferring block parameters, and only takes account of the temporary local data *declared* by you when assigning its own temporary local data. There is a danger of address overlaps occurring if you do not declare used temporary local data.

Chapter 26 "Direct Variable Access" describes how you learn the address of a variable in the temporary local data at runtime.

### ANY data type

A variable in the temporary local data can be declared – as an exception – with data type ANY. This is required in order to generate an ANY pointer that can be changed at runtime. See Section 26.3.3 "Variable ANY Pointers" for more details.

## 18.1.6 Static Local Data

Static local data are operands that a function block stores in its instance data block.

The static local data are the "memory" of a function block. They retain their value until this is changed by the program, just like data operands in global data blocks. The volume of static local data is limited by the data type of the variables and by the CPU-specific length of a data block.

### Declaration of static local data

You declare the static local data in the declaration section of the function block:

- with incremental programming under "stat" or
- with source-oriented programming between VAR and END_VAR.

Figure 18.2 in Section 18.1.5 gives an example of variable declaration in a function block. The block parameters are declared first, followed by the static local data and finally the temporary local data.

*Incremental programming*

| Address | Declaration | Name | Type |
|---|---|---|---|
| 0.0 | temp | SINFO | ARRAY [1.. 20] |
| | | | BYTE |
| 20.0 | temp | Lbyte | ARRAY [1..16] |
| | | | BYTE |

*Source-oriented programming*

```
VAR_TEMP
  SINFO : ARRAY[1..20] OF BYTE;

  Lbyte : ARRAY[1..16] OF BYTE;
END_VAR
```

**Figure 18.3** Example of Declaration of Temporary Local Data in an Organization Block

The static local are stored in the instance data block after the block parameters in order of declaration and according to data type. Chapter 26.2 "Data Storage of Variables" contains more detailed information on data storage in data blocks.

### Symbolic addressing of static local data

You access the static local data with their symbolic names. You assign the names in accordance with the rules for block-local symbols.

You can access static local data with all operations that can also be used in conjunction with data operands in global data blocks.

Example: The function block "Totalizer" adds an input value to a value stored in the static local data and then stores the total in the static local again. At the next call, the input value is added to this total again, and so on (Figure 18.4 top).

Total is a variable in the data block "Totalizer-Data", the instance data block of the function block "Totalizer" (you can define the names of all blocks yourself in the symbol table within the permissible framework). The instance data block has the data structure of the function block; in the example, it contains two INT variables with the names *In* and *Total*.

### Accessing static local data from outside the function block

The static local data are usually only processed in the function block itself. However, they are stored in a data block, you can access the static local data at any time in the same way as you ac-

FB "Totalizer"

| Address | Declaration | Name | Type |
|---|---|---|---|
| + 0.0 | in | In | INT |
| + 2.0 | stat | Total | INT |

```
     L        #In;
     L        #Total;
     +I;
     T        #Total;
```

DB "TotalizerData"

| Address | Declaration | Name | Type |
|---|---|---|---|
| + 0.0 | in | In | INT |
| + 2.2 | stat | Total | INT |

FB "Evaluation"

| Address | Declaration | Name | Type |
|---|---|---|---|
| 0.0 | in | Add | BOOL |
| 0.1 | in | Delete | BOOL |
| 2.0 | stat | EM_Add | BOOL |
| 2.1 | stat | EM_Del | BOOL |
| 4.0 | stat | Memory | Totalizer |

```
     A        #Add;
     FP       #EM_Add;
     JCN      M1;
     CALL     #Memory
          (In:="Value2");

M1:  A        #Delete;
     FP       #EM_Del;
     JCN      End;
     L        #Memory.Total;
     T        "Result";
     L        0;
     T        #Memory.Total;
```

DB "EvaluationData"

| Address | Declaration | Name | Type |
|---|---|---|---|
| 0.0 | in | Add | BOOL |
| 0.1 | in | Delete | BOOL |
| 2.0 | stat | EM_Add | BOOL |
| 2.1 | stat | EM_Del | BOOL |
| 4.0 | stat:in | Memory.In | INT |
| 6.0 | stat | Memory.Total | INT |

*In the **Data view**, the data block shows all individual variables so that the variables of a local instance appear with their full names*

*Simultaneously, you see the corresponding absolute addresses*

**Figure 18.4** Example of Static Local Data and Local Instances

cess variables in a global data block with *"DataBlockName".OperandName*.

In our little example, the data block is called *"TotalizerData"* and the data operand *Total*. An access could take the following form:

```
L     "TotalizerData".Total;
T     MW 20;
L     0;
T     "TotalizerData".Total;
```

## Local instances

When you call a function block, you normally specify the instance data block provided for the call. The function block then stores its block parameters and its static local data in the instance data block. From STEP 7 V2, you can generate "multi instances", that is, you can call a function block in another function block. The static local data (and the block parameters) of the called function block are then a subset of the static local data of the calling block. A requirement for this is that both the calling and the called function block have block version 2, that is, they have "multi-instance capability".

Example (Figure 18.4 bottom): In the static local data of the function block "Evaluate", you declare a variable *Memory* that corresponds to the function block "Totalizer" and has the same structure. Now you can call the function block "Totalizer" via the variable *Memory,* without, however, specifying a data block because the data for *Memory* are located 'block-local' in the static local data (*Memory* is the local instance of the FB "Totalizer").

You access the static local data of *Memory* in the program of the function block "Evaluate" in the same way as you access structure components by specifying the structure name (Memory) and the component name (Total).

The instance data block "EvaluateData" therefore contains the variables *Memory.In and Memory.Total,* that you can also access as global variables, for example as *"EvaluateData".Memory.Total*.

You can find this example of the use of a local instance in function blocks FB 6, 7 and 8 in the "Program Flow Control" program on the diskette accompanying the book. The example in Section 19.5.3 "Feed" contains further applications of local instances.

## Absolute addressing of static local data

Normally you access static local data using symbolic addresses with absolute addressing being the exception. Within a function block, the instance data block is opened via the DI register. Operands in this data block, static local data as well as block parameters, therefore have the operand identifier DI. You address a bit with DIX, a byte with DIB, a word with DIW and a doubleword with DID.

If you are familiar with storing data in a data block, you can work out yourself the addresses at which the static local data are stored. You can also see the addresses in the variable declaration table of the compiled block. But a word of caution! These addresses are relative to the start of the instance. They are only valid if you call the function block with a data block. If you call the function block as a local instance, the local data of the local instance are located in the middle of the instance data block of the calling function block. You can see the absolute addresses in, for example, the compiled instance data block which contains all local instances. Select VIEW → DATA VIEW, if you want to read the addresses of individual local data operands.

If we consider our example, we could access the variable Total in the function block "Totalizer" with DIW 2 if the FB "Totalizer" is called with a data block (cf. the operand assignment in the DB "TotalizerData"), and with DIW 6, if the FB "Totalizer" is called as a local instance in the FB "Evaluate" (cf. the operand assignment in the DB "EvaluateData").

However, if we program a function block without knowing whether it is called with a data block or as a local instance, that is, one that is to be "multi-instance-capable", how can we then assign absolute addresses to the static local data? Put briefly, we add to the address of the variable the offset of the local instance from address register AR2. See Chapter 25 "Indirect Addressing" and Chapter 26 "Direct Variable Access" for more detailed information.

## 18.2 Block Functions for Data Blocks

You store your program data in the data blocks. In principle, you can also use the bit memory area for storing data; however, with the data blocks, you have significantly more possibilities with regard to data volume, data structuring and data types. This chapter shows you

- how to work with data operands,
- how to call data blocks and
- how to create, delete and test data blocks at runtime.

You can use data blocks in two versions: as *global data blocks,* that are not assigned to any code block, and as *instance data blocks,* that are assigned to a function block. The data in the global data blocks are, in a manner of speaking, "free" data that every code block can make use of. You yourself determine their volume and structure direct through programming the global data block. An instance data block contains only the data with which the associated function block works; this function block then also determines the structure and storage location of the data in 'its' instance data block.

The number and length of data blocks are CPU-specific. The numbering of the data block begins at 1; there is no data block DB 0. You can use each data block either as a global data block or as an instance data block.

You must first create ("set up") the data blocks you use in your program, either by programming, such as code blocks, or at runtime using the system function SFC 22 CREAT_DB.

Data blocks must be stored in work memory so that they can be read and written to from the user program. You can also leave data blocks in load memory by using the block attribute "Unlinked" (keyword UNLINKED in source-oriented input). Such data blocks do not occupy space in work memory. However, you can only read data blocks in load memory with the system function SFC 20 BLKMOV. This procedure is suitable for data blocks with parameterization data or recipe data that are required relatively rarely for controlling the plant or the process.

If you set the attribute "The data block is write-protected in the programmable controller" in the block properties (corresponds to the keyword READ_ONLY in source-oriented input), you can then only read from this DB.

### 18.2.1 Two Data Block Registers

The CPU provides two data block registers for processing data operands. These registers contain the numbers of the current data blocks; these are the data blocks with whose operands processing is currently taking place. Before accessing a data block operand, you must open the data block containing the operand. If you use fully-addressed access to data operands (with specification of the data block, see below), you need not be concerned with opening the data blocks and with the assignments of the data block register. The Editor generates the necessary instructions from your specifications.

The Editor uses the first data block register preferably for accessing global data blocks and the second data block register for accessing instance data blocks. For this reason, these registers are given the names "Global data block register" (DB register) and "Instance Data Block Register" (DI register). The handling of the registers by the CPU is absolutely identical. Each data block can be opened via one of the two registers (or also via both simultaneously).

When you load a data word, you must specify which of the two possible open data blocks contains the data word. If the data block has been opened via the DB register, the data word is called DBW; if the data word is in the data block opened via the DI register, it is called DIW. The other data operands are named accordingly (Table 18.2).

**Table 18.2** Data Operands

| Data operand | located in a data block opened via the DB register | located in a data block opened via the DI register |
|---|---|---|
| Data bit | DBX y.x | DIX y.x |
| Data byte | DBB y | DIB y |
| Data word | DBW y | DIW y |
| Data doubleword | DBD y | DID y |

*x = Bit address, y = Byte address*

### 18.2.2 Accessing Data Operands

You can use the following methods for accessing data operands:

- Symbolic addressing with full addressing,
- Absolute addressing with full addressing and
- Absolute addressing with part addressing.

See Chapter 25 "Indirect Addressing" for further addressing methods.

Symbolic access to the data operands in global data blocks requires the minimum system knowledge. For absolute access or for using both data block registers, you must observe the notes described below.

#### Symbolic addressing of data operands

I recommend you use symbolic addressing of data operands as far as possible. Symbolic addressing

- makes it easier to read and understand the program (if meaningful terms are used as symbols),
- reduces write errors in programming (the Editor compares the terms used in the symbol table and in the program; "number switching errors" such as DBB 156 and DBB 165 that can occur when using absolute addresses, cannot occur here) and
- does not require programming knowledge at the machine code level (which data block has the CPU opened currently?).

Symbolic addressing uses fully-addressed access (data block together with data operand), so that the data operand always has a unique address.

You determine the symbolic address of a data operand in two steps:

- Assignment of the data block in the symbol table
  Data blocks are global data that have unique addresses within a program. In the symbol table, you assign a symbol (e.g. Motor1) to the absolute address of the data block (e.g. DB 51).
- Assignment of the data operands in the data block
  You define the names of the data operands (and the data type) during programming of the data block. The name applies only in the associated block (it is "block-local"). You can also use the same name in another block for another variable.

#### Fully-addressed access to data operands

In the case of fully-addressed access, you specify the data block together with the data operand. This method of addressing can be symbolic or absolute:

```
L      MOTOR1.ACTVAL;
L      DB 51.DBW 20;
```

MOTOR1 is the symbolic address that you have assigned to a data block in the symbol table. ACTVAL is the data operand you defined when programming the data block. The symbolic name MOTOR1.ACTVAL is just as unique a specification of the data operand as the specification DB 51.DBW 20.

Fully-addressed data access is only possible in conjunction with the global data block register (DB register). With fully-addressed data operands, the Editor executes two statements: First, the data block is opened via the DB register and this is then followed by access to the data operands.

You can use fully-addressed access with all operations permissible for the data type of the addressed data operand. These are the bit logic operations, the memory functions for binary operands and the load and transfer functions for digital operands. You can also specify fully-addressed data operands at the block parameters, for example (strongly recommended, see Chapter 19 "Block Parameters").

#### Absolute addressing of data operands

For absolute addressing of data operands, you must know the addresses at which the Editor places the data operands when setting up. You can find out the addresses by outputting them after programming and compiling the data block. You will then see from the address column the absolute address at which the relevant variable begins. This procedure is suitable for all data blocks, both those you use as global data blocks as well as those you use as instance data blocks (for local instances see Section 18.2.4 below). In this way, you can also see where the Editor stores the block parameters and the static local data in the case of function blocks.

If you want to calculate the address, Section 26.2 "Data Storage of Variables" provides valuable information.

Data operands are addressed bytewise like the bit memory, for example; they are also used in conjunction with the same operations (Table 18.3) and are executed in exactly the same way.

If you intend to assign exclusively absolute addresses to the operands of a data block, reserve the required quantity of bytes via a field declaration.

### 18.2.3 Open Data Block

| | |
|---|---|
| OPN DB *x* | Open a data block via the DB register with absolute address |
| OPN DB *name* | Open a data block via the DB register with symbolic address |
| OPN DI *x* | Open a data block via the DI register with absolute address |
| OPN DI *name* | Open a data block via the DI register with symbolic address |

Data blocks are opened regardless of any conditions. Opening does not affect the RLO and the contents of the accumulators; the nesting depth of the block calls does not change. The opened data block must be in work memory.

Example: The value of data word DBW 10 from data block DB 12 is to be transferred to data word DBW 12 of data block DB 13 (Figure 18.5). The values in data words DBW 14 from data blocks DB 12 and DB 13 are to be added; the result is to be stored in data word DBW 14 of data block DB 14.

When a data block is opened it remains "valid" until another data block is opened. Under certain circumstances – not visible to you – this can be done by the Editor (see "Special Points in Data

**Table 18.3** Operations with Data Blocks

| Statement | | Meaning |
|---|---|---|
| A | – | Check for signal state "1" and combine according to logic AND of a |
| O | – | Check for signal state "1" and combine according to logic OR of a |
| X | – | Check for signal state "1" and combine according to logic exclusive OR of a |
| AN | – | Check for signal state "0" and combine according to logic AND of a |
| ON | – | Check for signal state "0" and combine according to logic OR of a |
| XN | – | Check for signal state "0" and combine according to logic exclusive OR of a |
| = | – | Assignment to a |
| S | – | Set a |
| R | – | Reset a |
| FP | – | Edge evaluation for positive edge with a |
| FN | – | Edge evaluation for negative edge with a |
| | DBXy.x | Data bit via the DB register |
| | DIXy.x | Data bit via the DI register |
| | DBz.DBXy.x | Fully-addressed data bit |
| L | – | Load a |
| T | – | Transfer a |
| | DBBy | Data byte via the DB register |
| | DBWy | Data word via the DB register |
| | DBDy | Data double word via the DB register |
| | DIBy | Data byte via the DI register |
| | DIWy | Data word via the DI register |
| | DIDy | Data doubleword via the DI register |
| | DBz.DBBy | Fully-addressed data byte |
| | DBz.DBWy | Fully-addressed data word |
| | DBz.DBDy | Fully-addressed data doubleword |

*x = Bit address, y = Byte address, z = Number of the data block*

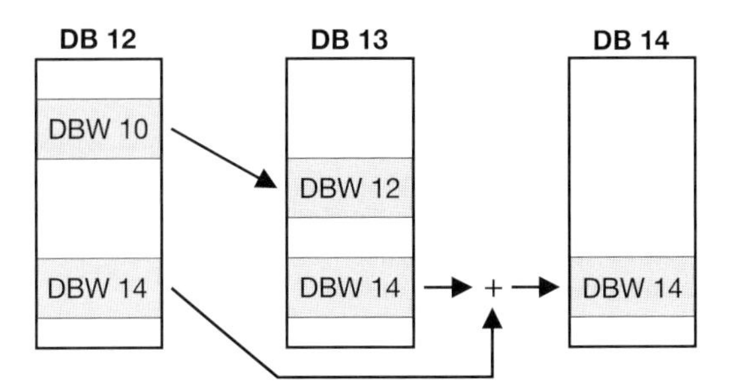


**Figure 18.5** Opening Data Blocks (Figure for Programming Example)

You can program this example in two ways:

| Programming with part addressing | Programming with full addressing |
|---|---|
| `OPN DB 12;` | |
| `L DBW 10;` | `L DB 12.DBW 10;` |
| `OPN DB 13;` | |
| `T DBW 12;` | `T DB 13.DBW 12;` |
| `OPN DB 12;` | |
| `L DBW 14;` | `L DB 12.DBW 14;` |
| `OPN DB 13;` | |
| `L DBW 14;` | `L DB 13.DBW 14;` |
| `+I ;` | `+I ;` |
| `OPN DB 14;` | |
| `T DBW 14;` | `T DB 14.DBW 14;` |

Addressing" below). For example, a block call with CALL in conjunction with parameter transfer can change the contents of the data block registers.

With a block change using UC or CC, the contents of the data block registers are retained. On returning to the calling block, the block end statement restores the old contents of the registers.

### 18.2.4 Exchanging the Data Block Registers

CDB Exchange data block registers

The statement CDB exchanges the contents of the data block registers. It is executed regardless of conditions and does not affect either the status bits or the other registers.

Example: With the statement CDB, you can take a "detour" via the DB register to open, via the DI register, a data block transferred as a block parameter (not possible direct).

```
CDB    ;
OPN    #Data2;
CDB    ;
```

With CDB, you transfer the contents of the DB register temporarily to the DI register. Then you open, via the block parameter #Data2, the data block transferred as an actual parameter; that is, you write its number into the DB register. After renewed exchange, the old value is again in the DB register and the DI register contains the number of the parameterized data block.

### 18.2.5 Data Block Length and Number

| | | |
|---|---|---|
| L | DBLG | Load the length of the data block opened via the DB register |
| L | DBNO | Load the number of the data block opened via the DB register |
| L | DILG | Load the length of the data block opened via the DI register |
| L | DINO | Load the number of the data block opened via the DI register |

The statement L DBLG loads the length of the data block that was opened via the DB register into accumulator 1. The length is the same as the quantity of data bytes. The statement L DILG is the same for the DI register.

The statement L DBNO loads the number of the data block that was opened via the DB register into accumulator 1. L DINO shows you the number of the current data block that was opened via the DI register. It is not possible to write the number back; you can only influence the data block registers via OPN DB or OPN DI and CDB (exchange data block registers).

These statements transfer the previous contents of accumulator 1 into accumulator 2 in accordance with a “normal” load function. If no data block has been opened via the relevant register, zero is loaded both as the length and as the number.

### 18.2.6 Special Points in Data Addressing

#### Changing the assignments in the DB register

With the following functions, the Editor generates additional statements that can affect the contents of one of the two data block registers:

- Full addressing of data operands
  Each time data operands are addressed fully, the Editor first opens the data block with the statement OPN DB, and then accesses the data operands. The DB register is overwritten each time here. This applies also when initializing block parameters with fully-addressed data operands.
- Access to block parameters
  Access to the following block parameters changes the contents of the DB register: With functions, all block parameters of complex data type and with function blocks, in/out parameters of complex data type.
- Prior to the actual block call, CALL FB stores the number of the current instance data block in the DB register (by exchanging the data block registers) and opens the instance data block for the called function block. In this way, the associated instance data block is always open in a called function block. Following the actual block call, CALL FB exchanges the data block registers again, so that the current instance data block is once again available in the calling function block. In this way, CALL FB changes the contents of the DB register.
- In function blocks the DI register is permanently assigned the number of the current instance data block. All accesses to block parameters or static local data are made via the DI register and, incidentally, also via the address register AR2 in the case of “multi-instance-capable” function blocks. Please note this permanent assignment if you change the contents of the DI register with CDB or OPN DI.

If, for example, you want to use both data block registers simultaneously for data exchange, you must first save the register contents and then restore them. The example shown in Figure 18.6 describes a relevant method.

#### Making changes to data block assignments at a later stage

In programmable controllers, the global operands are permanently assigned to an address. This applies to inputs, outputs, peripheral I/O, bit memory, timers, counters and also blocks. When plugged into the mounting rack, a module has a fixed address that maps to the operand areas peripheral I/O as well as inputs and outputs. Inserting or deleting an operand, for example, an input byte, followed by shifting of the subsequent operands does not make sense here; you would have to change not only the program but also the wiring. This characteristic has been transferred to the bit memory, the timers and the counters. No-one would think, for example, of deleting say function block FB 35 in the number band and then shifting all the subsequent block numbers so that all these blocks would be assigned a new number.

However, when programming data blocks, it makes sense to insert or delete data operands. It might even be desirable here to shift the subsequent operands (or more precisely: the values of the operands). In the case of absolute addressing, it is obvious that you must then correct (update) the addresses of the (shifted) data operands in the program.

An example: Data block DB 21 ‘Motor1’ is assigned the INT variables *Actval* (data word DBW 10) and *Setpoint* (DBW 12). If you insert the INT variable *MaxCurrent* prior to the variable *Actval,* the values shift and consequently the variable *Actval* shifts to data word DBW 12 and *Setpoint* shifts to data word DBW 14.

Assignments prior to insertion:

| | *Actval* | *Setpoint* | |
|---|---|---|---|
| DBW 8 | DBW 10 | DBW 12 | DBW 14 |

```
VAR_TEMP
  ZW_DB  :  WORD;                     //Intermediate buffer for global data block
  ZW_DI  :  WORD;                     //Intermediate buffer for instance data block
END_VAR

//Save data block registers
  L    DBNO;                          //Buffer global data block number
  T    ZW_DB;
  L    DINO;                          //Buffer instance data block number
  T    ZW_DI;

//Working with part-addressed data operands
//when using both data block registers

  OPN  DB 12;                         //Open data block DB 12 via the DB register
  OPN  DI 13;                         //Open data block DB 13 via the DI register
  L    DBW 12;                        //##########
  T    DIW 14;                        //# Be careful with symbolic addressing in this
  L    DID  0;                        //# program section, e.g. when using block
  L    DBD  4;                        //# parameters, block-local variables and
  +R   ;                              //# fully-addressed data operands
  T    DID  8;                        //##########

//Restore data block registers
  OPN  DB[ZW_DB];                     //Open original global data block
  OPN  DI[ZW_DI];                     //Open original instance data block
```

**Figure 18.6** Example of the Direct Use of Both Data Block Registers

Assignments following insertion:

| | *MaxCurrent* | *Actval* | *Setpoint* |
|---|---|---|---|
| DBW 8 | DBW 10 | DBW 12 | DBW 14 |

If you have previously addressed data word DBW 10, in order, for example, to compare the actual value, you must specify data word DBW 12 following the change.

*The same behavior also exists in symbolic addressing!* If you create a program in the LAD programming language, the Editor works in incremental input mode. With symbolically addressed variables, it stores the absolute address of the associated operand in the program (if you program "Input1", for example, the compiled block contains the operand I 1.1). When outputting the program in symbolic representation, the Editor searches the symbol table and the declaration section of the blocks for the stored addresses, and when it finds them, uses the symbolic names.

To return to the previous example, if you have addressed the variable *"Motor1".Actval* in the program, the Editor writes DB21.DBW10 in the program. If you then output the program with symbolic addressing following the change in the data block, *"Motor1".MaxCurrent* now stands at this position, because the address DBW 10 now contains the variable *MaxCurrent*.

A change in the assignments in the data block therefore brings with it the consequence that you must update the addresses in all program sections that access the relevant data operands either with absolute *or symbolic* addressing.

This also applies to changes in instance data blocks or to inserting or deleting static local data in function blocks. While it is the case that in code block programming, all changes to all local variables are correctly updated "block-locally" prior to the Editor storing the blocks in the program, this does not apply to accesses "from outside", such as when you access 'foreign' instance data from other blocks. This access then takes place like an access to global data operands and it must be updated with regard to addresses following an operand shift.

## 18.3 System Functions for Data Blocks

There are three system functions for handling data blocks. Their parameters are described in Table 18.4.

- SFC 22 CREAT_DB
  Create data block
- SFC 23 DEL_DB
  Delete data block
- SFC 24 TEST_DB
  Test data block

### 18.3.1 Creating a Data Block

System function SFC 22 creates a data block in the work memory. As the data block number, the system function takes the lowest free number in the number band given by the input parameters LOW_LIMIT and UP_LIMIT. The numbers specified at these parameters are included in the number band. If both values are the same, the data block is generated with this number. The output parameter DB_NUMBER supplies the number of the actually created data block. With the input parameter COUNT, you specify the length of the data block to be created. The length corresponds to the number of data bytes and must be an even number.

Creating the data block is not the same as calling it. The current data block is still valid. A data block created with the system function contains random data. For meaningful use, data must first be written to a data block created in this way before the data can be read.

In the event of an error, the data block is not created, the output parameter is assigned as undefined and an error number is signaled via the function value.

### 18.3.2 Deleting a Data Block

System function SFC 23 deletes the data block in RAM (work and load memory) whose number is specified at the input parameter DB_NUMBER. In doing so, the data block can be currently open, even in a lower-level block or on a lower priority level.

In the event of an error, the data block is not deleted and an error number is signaled in the function value.

### 18.3.3 Testing a Data Block

System function SFC 24 supplies the number of available data bytes for a data block in the work memory (at output parameter DB_LENGTH) and the write-protection ID (at output parameter WRITE_PROT, where signal state "1" signifies write-protected). You specify the number of the selected data block at input parameter DB_NUMBER.

**Table 18.4** SFCs for Handling Data Blocks

| SFC | Name | Declaration | Data Type | Assignment, Description |
|---|---|---|---|---|
| 22 | LOW_LIMIT | INPUT | WORD | Lowest number of the data block to be created |
| | UP_LIMIT | INPUT | WORD | Highest number of the data block to be created |
| | COUNT | INPUT | WORD | Length of the data block in bytes (even number) |
| | RET_VAL | OUTPUT | INT | Error information |
| | DB_NUMBER | OUTPUT | WORD | Number of the created data block |
| 23 | DB_NUMBER | INPUT | WORD | Number of the data block to be deleted |
| | RET_VAL | OUTPUT | INT | Error information |
| 24 | DB_NUMBER | INPUT | WORD | Number of the data block to be tested |
| | RET_VAL | OUTPUT | INT | Error information |
| | DB_LENGTH | OUTPUT | WORD | Length of the data block (in bytes) |
| | WRITE_PROT | OUTPUT | BOOL | "1" = write-protected |

In the event of an error, the output parameter is assigned as undefined and an error number is signaled in the function value.

## 18.4 Null Operations

Null operations have no effect when processed by the CPU. STL has NOP 0, NOP 1 and BLD statements as null operations.

### 18.4.1 NOP Statements

You can use the statements NOP 0 (bit pattern 16x "0") and NOP 1 (bit pattern 16x "1") to enter a statement that has no effect. Please note that null operations occupy memory space (2 bytes) and have an instruction execution time.

Example: There must always be a statement at a jump label. If you want to have a jump in your program but do not want anything further to be executed, use NOP 0.

```
        A    I 1.0
        JC   MXX1
        ...
MXX1:   NOP  0
        ...
```

You can enter an empty line for clarity more effectively by simply entering an (empty) line comment (this does not require user memory space and involves no loss of execution time since no code is entered.

### 18.4.2 Program Display Statements

The Editor uses the program display instruction BLD nnn to incorporate decompiling information into the program. The BLD statements are normally not displayed. If the Editor comments out the block calls in the event of a time stamp conflict, it also displays the BLD statements used (for example, BLD1 and BLD2 in the case of an FC).

# 19 Block Parameters

This chapter describes how to use block parameters. You will learn

- how to declare block parameters,
- how to work with block parameters,
- how to initialize block parameters
- how to "pass on" block parameters.

Block parameters represent the transfer interface between the calling and the called block. All functions and function blocks can be provided with block parameters.

## 19.1 Block Parameters in General

### 19.1.1 Defining the Block Parameters

Block parameters make it possible to parameterize the processing instruction in a block, the block function. Example: You want to write a block as an adder that you can use in your program several times with different variables. You transfer the variables as block parameters; in our example, three input parameters and one output parameter (Figure 19.1). Since the adder need not store any values internally, a function is suitable as the block type.

You define a block parameter as an *input parameter* if you only check or load its value in the block program. If you only write a block parameter (assign, set, reset, transfer), you use an *output parameter*. You must always use an *in/out parameter* if a block parameter is to be both checked and written. The Editor does not check the use of the block parameters.

### 19.1.2 Processing the Block Parameters

In the adder program, the names of the block parameters stand as place holders for the later current variables. You use the block parameters in

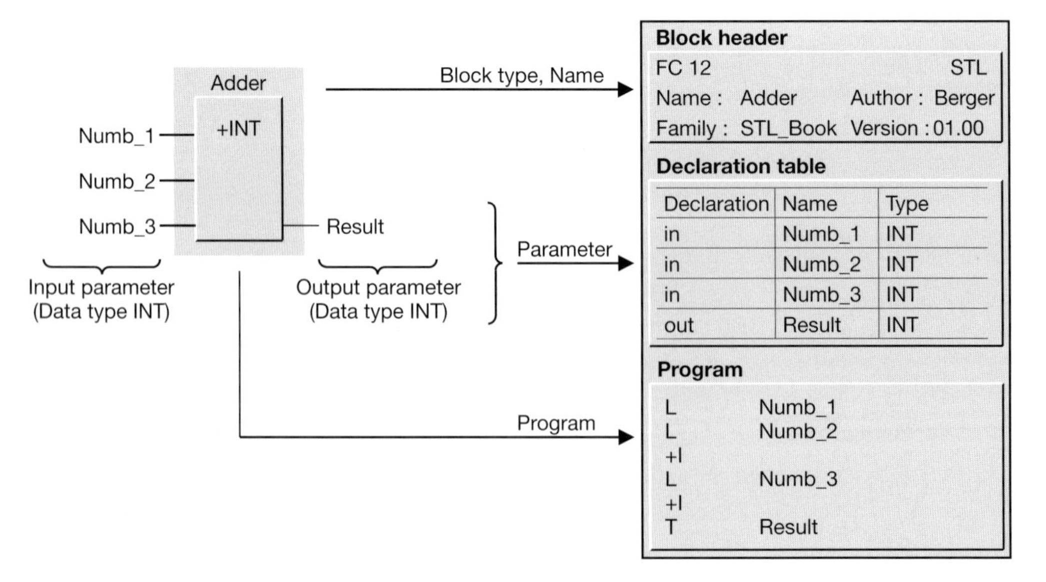


**Figure 19.1** Example of Block Parameters

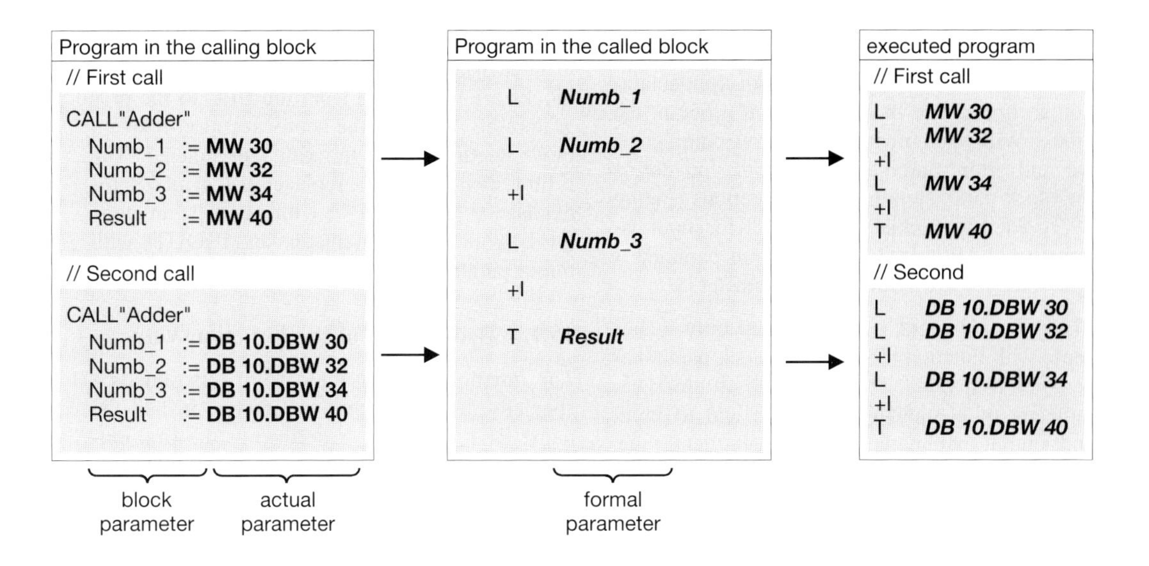


the same way as symbolically addressed variables; in the program, they are called *formal parameters*.

You can call the Adder function several times in your program. With each call, you transfer other values to the adder at the block parameters (Figure 19.2). The values can be constants, operands or variables; they are called *Actual parameters*.

At runtime, the CPU replaces the formal parameters with the actual parameters. The first call in the example adds the contents of memory words MW 30, MW 32 and MW 34 and stores the result in memory word MW 40. The same block with the actual parameters of the second call adds data words DBW 30, DBW 32 and DBW 34 of data block DB 10 and stores the result in data word DBW 40 of data block DB 10.

*Incremental programming*

| Address | Declaration | Name | Typ | Initial value | Comment |
|---|---|---|---|---|---|
| 0.0 | in | *Name* | *data_type* | *pre_assignment* | *Parameter comment* |
| | out | *Name* | *data_type* | *pre_assignment* | *Parameter comment* |
| | inout | *Name* | *data_type* | *pre_assignment* | *Parameter comment* |

*Source-oriented programming*

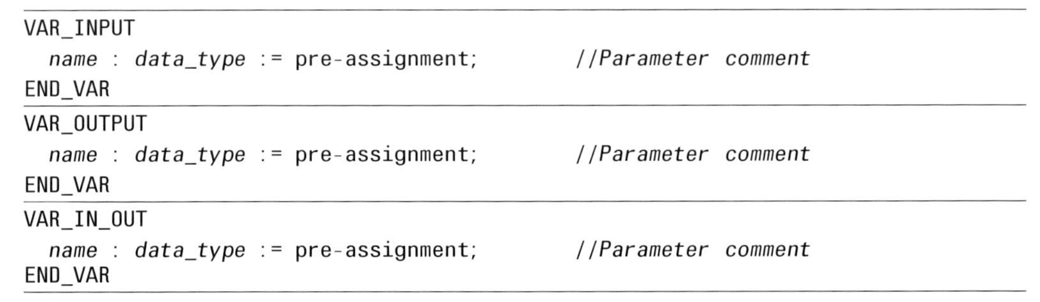

```
VAR_INPUT
  name : data_type := pre-assignment;        //Parameter comment
END_VAR
VAR_OUTPUT
  name : data_type := pre-assignment;        //Parameter comment
END_VAR
VAR_IN_OUT
  name : data_type := pre-assignment;        //Parameter comment
END_VAR
```

**Figure 19.3** Declaration of the Block Parameters

### 19.1.3 Definition of the Block Parameters

You define the block parameters in the declaration section of the block when you program the block. With incremental input, you complete a list and with source-oriented input you define the block parameters in specific sections (Figure 19.3). The keyword is VAR_INPUT for input parameters, VAR_OUTPUT for output parameters and VAR_IN_OUT for in/out parameters.

The pre-assignment is optional and only makes sense with function blocks if the block parameter is stored as a value. This applies to all block parameters of elementary data type and to input and output parameters of complex data type. Specification of a parameter comment is optional and always possible.

The *Block parameter name* can be up to 24 characters in length. It must consist only of alphanumeric characters (without national characters) and the underscore. The name must not be a keyword.

For the *Data type of a block parameter* all elementary, complex and user-defined data types are permissible. In addition, you can use the data types (parameter types) listed in Table 19.1 with block parameters.

STEP 7 stores the names of the block parameters in the non-execution-relevant section of the blocks on the data medium of the programming device. The work memory of the CPU (in the compiled block) contains only the declaration types and the data types. For this reason, program changes made to blocks on-line at the CPU must always be updated on the data medium of the programming device, in order to retain the original names. If the update is not made, or if blocks are transferred from the CPU to the programming device, the non-execution-relevant block sections are overwritten or deleted. The Editor then generates replacement symbols for display or printout (IN*n* with input parameters, OUT*n* with output parameters and INOUT*n* with in/out parameters, with *n* beginning at 1).

### 19.1.4 Declaration of the Function Value

The function value in the case of functions is a specially treated output parameter. It has the name RET_VAL and is defined as the first output parameter. You declare the function value by specifying the data type of the function value following the block type and separated by a colon.

The above-named example of the adder can also be programmed with the total as the function value:

```
FUNCTION FC 12 : INT
VAR_INPUT
  Numb_1 : INT;
  Numb_2 : INT;
  Numb_3 : INT;
END_VAR
BEGIN
  L     Numb_1;
  L     Numb_2;
  +I    ;
  L     Numb_3;
  +I    ;
  T     RET_VAL;
END_FUNCTION
```

**Table 19.1** Parameter Types

| Parameter Type | | Permissible as | Remark |
|---|---|---|---|
| TIMER | Timer function | Input parameter | Data types TIMER and COUNTER also in the Symbol table |
| COUNTER | Counter function | | |
| BLOCK_DB | Data block | | – |
| BLOCK_SDB | System data block | | – |
| BLOCK_FC | Function | | – |
| BLOCK_FB | Function block | | – |
| POINTER | Simple variable and area pointer or DB pointer | Input and in/out parameter | – |
| ANY | Any variable and ANY pointer | (with functions also output parameter) | Also permissible with temporary local variables |

In the example, the function value is of data type INT. With T RET_VAL, the function value is assigned the total from Numb_1, Numb_2 and Numb_3.

All elementary data types are permissible as the data type of the function value and, in addition, the data types DATE_AND_TIME, STRING, POINTER, ANY and user-defined data types UDT are also permissible. ARRAY and STRUCT are not permissible.

### 19.1.5 Initializing Block Parameters

When calling a block with block parameters, you initialize the block parameters with actual parameters. These can be constants, absolute-addressed operands, fully-addressed data operands or symbolically addressed variables. The actual parameter must be of the same data type as the block parameter.

The call sequence of the block parameters (in source-oriented input) is not important; the Editor "sorts" the block parameters according to the specifications in the declaration when outputting the compiled block.

You must initialize all block parameters of a function at every call. In the case of function blocks, initialization of individual or all block parameters is optional.

## 19.2 Formal Parameters

In this chapter, you will learn how to access the block parameters within a block. Table 19.2 shows that it is possible to access block parameters of elementary data types, components of a field or a structure, and timer and counter functions without restriction. Section 19.4 shows you how you can "pass on" block parameters to called blocks.

Access to complex data types and with parameter types POINTER and ANY is currently not supported by STL. However, you can initialize acquired blocks or system blocks that have such parameters with the relevant variables. Chapter 26 "Direct Variable Access" shows you how

**Table 19.2** Access to Block Parameters

| Data Types | Permissible with | | | Access in the block | |
|---|---|---|---|---|---|
| | IN | I_O | OUT | possible | with |
| Elementary data types | | | | | |
| BOOL | x | x | x | x | Binary checks, memory operations |
| BYTE, WORD, DWORD, CHAR, INT, DINT, REAL, S5TIME, TIME, TOD, DATE | x | x | x | x | Load and transfer operations |
| Complex data types | | | | | |
| DT, STRING | x | x | x | – | Not possible direct |
| ARRAY, STRUCT | | | | | |
| Individual binary components | x | x | x | x | Binary checks, memory operations |
| Individual digital components | x | x | x | x | Load and transfer operations |
| Complete variables | x | x | x | – | Not possible direct |
| Parameter types | | | | | |
| TIMER | x | – | – | x | All operations for timer functions |
| COUNTER | x | – | – | x | All operations for counter functions |
| BLOCK_FC, BLOCK_FB | x | – | – | x | Calling with UC and CC[2] |
| BLOCK_DB | x | – | – | x | Opening with OPN DB |
| BLOCK_SDB | x | – | – | – | Not possible[3] |
| POINTER, ANY | x | x | x[1] | – | Not possible direct |

[1] Only with functions
[2] CC not with functions
[3] Only meaningful with system blocks

you can nevertheless use parameters with these data types in blocks you have written yourself.

### Block parameters of data type BOOL

Block parameters of data type BOOL can be individual binary variables or binary components of fields and structures. You can check input parameters and in/out parameters with contacts or with binary box inputs, and you can influence output parameters and in/out parameters with memory functions (Table 19.3). With functions FCs you *must* assign a value to a binary output parameter in the block or else you must set or reset this value. You must not, for example, exit the block first.

When programming, you use the formal parameter in place of the block parameter xxxx.

After the CPU has used the actual parameter specified as the block parameter, it processes the statement as described in the chapters 4 "Binary Logic Operations" and 5 "Memory Functions".

### Block parameters of digital data type

Block parameters of digital data type occupy 8, 16 or 32 bits (all elementary data types except BOOL). They can be individual digital variables or digital components of fields and structures. You read input parameters and in/out parameters with the load function, and you write output parameters and input parameters with the transfer function. With functions FCs, you *must* transfer a value to a digital output parameter. You must not, for example, exit the block first.

| | | |
|---|---|---|
| L | xxxx | Load an input or in/out parameter |
| T | xxxx | Transfer to an output or input parameter |

When programming, you use the formal parameter in place of the block parameter xxxx.

After the CPU has used the actual parameter, it processes the statements as described in the Chapter 6 "Transfer Functions".

### Block parameters of data type DT and STRING

Direct access to block parameters of data type DT and STRING is not currently possible. In function blocks, you can "pass on" parameters of data types DT and STRING to parameters of called blocks.

Chapter 26 "Direct Variable Access" shows you how to program access to parameters of a higher data type yourself.

### Block parameters of data type ARRAY and STRUCT

Direct access to block parameters of data type ARRAY and STRUCT is possible on a compo-

**Table 19.3** Accessing Block Parameters of Data Type BOOL

| | | |
|---|---|---|
| A | – | AND logic operation with check for signal state "1" |
| AN | – | AND logic operation with check for signal state "0" |
| O | – | OR logic operation with check for signal state "1" |
| ON | – | OR logic operation with check for signal state "0" |
| X | – | Exclusive OR logic operation with check for signal state "1" |
| XN | – | Exclusive OR logic operation with check for signal state "0" |
| – | xxxx | of an input or in/out parameter of data type BOOL |
| – | xxxx | of an input parameter of data type TIMER |
| – | xxxx | of an input parameter of data type COUNTER |
| S | – | Set |
| R | – | Reset |
| = | – | Assignment |
| – | xxxx | of an output or in/out parameter of data type BOOL |
| FP | – | Edge evaluation positive |
| FN | – | Edge evaluation negative |
| – | xxxx | of an in/out parameter of data type BOOL |

nent-wise basis, that is, you can access individual binary or digital components with the relevant operations (binary logic operations, memory functions, load and transfer functions).

Access to the complete variable (entire field or entire structure) is not currently possible and neither is access to individual components of combined or user-defined data type. In function blocks, you can 'pass on' parameters of data type ARRAY and STRUCT to parameters of called blocks.

Chapter 26 "Direct Variable Access" shows you how to program access to parameters of a higher data type yourself.

### Block parameters of user-defined data type

You handle block parameters of user-defined data type in the same way as block parameters of data type STRUCT.

Direct access to block parameters of data type UDT is possible on a component-wise basis, that is, you can access individual binary or digital components with the relevant operations (binary logic operations, memory functions, load and transfer functions).

Access to the complete variable is not currently possible and neither is access to individual components of combined or user-defined data type. In function blocks, you can "pass on" parameters of data type UDT to parameters of called blocks.

Chapter 26 "Direct Variable Access" shows you how to program access to parameters of a higher data type yourself.

### Block parameters of data type TIMER

In addition to the check statements listed in Table 19.3, you can program a block parameter of data type TIMER with the following statements:

SP – Start as pulse
SD – Start as ON delay
SE – Start as extended pulse
SS – Start as retentive ON delay
SF – Start as OFF delay

R – Reset
FR – Enable

– xxxx input parameter of data type TIMER

When programming, you use the formal parameter in place of the block parameter xxxx.

After using the formal parameter, the CPU processes this STL statement in exactly the same way as described in Chapter 7 "Timer Functions". When a timer is started, the time value can also be a block parameter of data type S5TIME.

### Block parameters of data type COUNTER

In addition to the check statements listed in Figure 19.3, you can program a block parameter of data type COUNTER with the following statements:

S – Set counter
CU – Count up
CD – Count down

R – Reset
FR – Enable

– xxxx of an input parameter of data type COUNTER

When programming, you use the formal parameter in place of the block parameter xxxx.

After using the formal parameter, the CPU processes this STL statement exactly as described in Chapter 8 "Counter Functions". When setting a counter, the count value can also be a block parameter of, for example, data type WORD.

### Block parameters of data type BLOCK_xx

OPN – Open a data block (parameter type BLOCK_DB)
UC – Call a function (parameter type BLOCK_FC)
UC – Call a function block (parameter type BLOCK_FB)
CC – Conditional call of a function (parameter type BLOCK_FC)
CC – Conditional call of a function block (parameter type BLOCK_FB) (see text)
– xxxx via an input parameter

When programming, you use the formal parameter in place of the block parameter xxxx.

When opening a data block via a block parameter, the CPU always uses the global data block register (DB register).

The functions and function blocks transferred with block parameters must themselves not contain block parameters. A conditional block call via a block parameter is only possible if it is the block parameter of a function block.

As the instance data block in a function block call, you can also use a data block that you have transferred as a block parameter. Since the Editor has no means of checking the data type of the data block used at runtime, you must yourself ensure that the transferred data block is also suitable as an instance data block for the called function block.

Example: You can specify a block parameter of type BLOCK_DB with the name *Data* as the instance data block in a function block call:

```
CALL FB 10, Data
```

### Block parameters of data type POINTER and ANY

Direct access to block parameters of data type POINTER and ANY is currently not possible.

Chapter 26 "Direct Variable Access" shows you how to program access to parameters of data types POINTER and ANY yourself.

## 19.3 Actual Parameters

When you call a block, you initialize its block parameters with constants, operands or variables with which it is to operate. These are the actual parameters. If you call the block often in your program, you usually use different actual parameters each time it is called.

The actual parameter must agree in data type with the block parameter: You can only apply a binary actual parameter (for example, a memory bit) to a block parameter of data type BOOL; you can only initialize a block parameter of data type ARRAY with an identically dimensioned field variable. Table 19.4 gives an overview of which operands you can use as actual parameters with which data type.

When calling functions, you must initialize all block parameters with actual parameters.

When calling function blocks, it is not necessary to initialize the block parameters. STEP 7 stores all block parameters of elementary data type, input and output parameters of complex data type and input parameters of data types TIMER, COUNTER and BLOCK_xx as a value or as a number. In/out parameters of complex data types and block parameters of data types POINTER and ANY are stored as pointers to the actual parameters. So that a meaningful value is entered here, you should initialize at least the last named block parameters – at least, at the first call.

You can also access the block parameters of the function block direct. Since they are located in a data block, you can handle the block parameters like data operands. Example: A function block with the instance data block "Lift_stat_1" controls a binary output parameter with the name *Up*. Following processing in the function block

**Table 19.4** Initialization with Actual Parameters

| Data Type of the Block Parameter | Permissible Actual Parameters |
|---|---|
| Elementary data type | • Simple operands, fully-addressed data operands, constants<br>• Components of fields or structures of elementary data type<br>• Block parameter of the calling block<br>• Components of block parameters of the calling block of elementary data type |
| Complex data type | • Variables or block parameters of the calling block |
| TIMER, COUNTER and BLOCK_xx | • Timers, counters and blocks |
| POINTER | • Simple operands, fully-addressed data operands<br>• Range pointer or DB pointer |
| ANY | • Variables of any data type<br>• ANY pointer |

(after its call), you can check the parameter as follows, without having initialized the output parameter:

```
A   'Lift_stat_1'.Up;
```

You program this check instead of initializing the parameter.

### Initializing block parameters of elementary data types

The actual parameters listed in Table 19.5 are permissible as actual parameters of elementary data types.

You can assign either absolute or symbolic addresses to input, output and memory bit operands. Input operands should be placed only at input parameters and output operands at output parameters (however, this is not mandatory). Memory bit operands are suitable for all declaration types. You must apply peripheral inputs only to input parameters and peripheral outputs only to output parameters.

When you use part-addressed data operands, you must ensure that when you access the block parameter (in the *called* block), the currently open data block is also the "correct" one. Since the Editor may in certain circumstances change the data block when the block is called, part addressing is not recommended for data operands. Use only fully-addressed data operands for this reason.

Temporary local data are usually symbolically addressed. They are located in the L stack of the calling block (and are declared in the calling block).

If the calling block is a function block, you can also use its static local data as actual parameters (see "Passing On Block Parameters" below). Static data are usually symbolically addressed. If you use absolute addressing via the DI register (DI operands), you must ensure that when accessing the block parameter (in the *called* block) the data block currently opened via the DI register is also the "correct" one. Please note in this regard that when using the called block as a local instance, the absolute address of the block-local variable depends on the declaration of the local instance in the called block.

With a block parameter of data type BOOL, you can apply the constant TRUE (signal state "1") or FALSE (signal state "0"), and with block parameters of digital data type, you can apply all constants corresponding to the data type. Initialization with constants is only meaningful with input parameters.

You can also initialize a block parameter of elementary data type with components of fields and structures, provided such a component is of the same data type as the block parameter.

**Table 19.5** Actual Parameters of Elementary Data Types

| Operands | Permissible with IN | I_O | OUT | Binary operand or symbolic name | Digital operand or symbolic name |
|---|---|---|---|---|---|
| Inputs (process image) | x | x | x | I y.x | IB y, IW y, ID y |
| Outputs (process image) | x | x | x | Q y.x | QB y, QW y, QD y |
| Memory bits | x | x | x | M y.x | MB y, MW y, MD y |
| Peripheral inputs | x | – | – | – | PIB y, PIW y, PID y |
| Peripheral outputs | – | – | x | – | PQB y, PQW y, PQD y |
| Global data | | | | | |
| Part addressing | x | x | x | DBX y.x | DBB y, DBW y, DBD y |
| Full addressing | x | x | x | DB z.DBX y.x | DB z.DBB y, etc. |
| Temporary local data | x | x | x | L y.x | LB y, LW y, LD y |
| Static local data | x | x | x | DIX y.x | DIB y, DIW y, DID y |
| Constants | x | – | – | TRUE, FALSE | all digital constants |
| Components of ARRAY or STRUCT | x | x | x | Complete component name | Complete component name |

### Initializing block parameters of complex data types

Every block parameter can be of the complex data type or of the user-defined data type. Variables of the same data type are permissible as actual operands.

For initializing block parameters of data type DT or STRING, individual variables or components of fields or structures of the same data type are permissible. If you initialize a function block with a STRING variable, this variable must have the same maximum length as the STRING block parameter.

For initializing block parameters of data type ARRAY or STRUCT, variables with exactly the same structure as the block parameters are permissible.

Parameter assignment with complex data types is described in Section 26.4 "Brief Description Message Frame Example" in the examples "Composing the Message Frame" and "Read Time of Day".

### Initializing block parameters of user-defined data type

With complex or extensive data structures, the use of user-defined data types (UDTs) is recommended. First, you define the UDT and then you use it, for example, to apply the variable in the data block or to declare the block parameter. Following this, you can use the variable when initializing the block parameter. It is also the case here, that the actual parameter (the variable) must be of the same data type (the same UDT) as the block parameter.

Parameter assignment with user-defined data types is shown in Section 26.4 "Brief Description Message Frame Example" in the example "Message Frame Data".

### Initializing block parameters of type TIMER, COUNTER and BLOCK_xx

You initialize a block parameter of type TIMER with a timer function, and a block parameter of type COUNTER with a counter function. To block parameters of parameter types BLOCK_FC and BLOCK_FB, you can apply only blocks without their own parameters. These blocks are then called in the case of access with UC (and also CC in the case of function blocks). You initialize BLOCK_DBs with a data block that is opened in the called block via the DB register.

Block parameters of types TIMER, COUNTER and BLOCK_xx must only be input parameters.

### Initializing block parameters of type POINTER

Pointers (constants) are permissible for block parameters of parameter type POINTER. These pointers are either range pointers (32-bit pointers) or DB pointers (48-bit pointers). The operands are of elementary data type and can also be fully-addressed data operands.

Output parameters of type POINTER are not permissible with function blocks.

### Initializing block parameters of type ANY

Variables of all data types are permissible for block parameters of parameter type ANY. The programming within the called block determines which variables (operands or data types) must be applied to the block parameters, or which variables are feasible. You can also specify a constant in the format of the ANY pointer "P#[data block.]Operand Data type Number" and so define an absolute-addressed area.

An exception is the initialization of an ANY parameter with a temporary local data item of data type ANY. In this case, rather than generating a pointer to the variable, the Editor assumes that a pointer of data type ANY already exists in the temporary local data. This gives you the ability to apply to an ANY parameter an ANY pointer that you can change at runtime. The "variable ANY pointer" can be particularly useful in conjunction with the system function SFC 20 BLKMOV (see the "Buffer Entry" example in Section 26.4 "Brief Description Message Frame Example").

Output parameters of type ANY are not permissible with function blocks.

## 19.4 "Passing On" Block Parameters

"Passing on" block parameters is a special form of access and of initializing block parameters. The block parameters of the calling block are "passed on" to the parameters of the called block. Here, the formal parameter of the calling block then becomes the actual parameter of the called block.

In general, it is also the case here that the actual parameter must be of the same type as the formal parameter (that is, the relevant block parameters must agree in their data types). In addition, you can apply an input parameter of the calling blocks only at an input parameter of the called block, and similarly, an output parameter at an output parameter. You can apply an in/out parameter of the calling block to all declaration types of the called block.

There are restrictions with regard to data types caused by the variations in block parameter storage between functions and function blocks. Block parameters of elementary data type can be passed on without restriction in accordance with the information in the previous paragraph. Complex data types at inputs and output parameters can only be passed on if the calling block is a function block. Block parameters of parameter types TIMER, COUNTER and BLOCK_xx can only be passed on from one input parameter to another if the calling block is a function block. These statements are represented in Table 19.6.

You can "pass on" parameter types TIMER, COUNTER and BLOCK_DB in the case of functions with the help of indirect addressing. The relevant parameter is first assigned data type WORD or INT; you initialize it with a constant or a variable that contains the number of the timer, the counter or the block to be transferred. You can pass on this parameter to other blocks since it is of elementary data type. In the "last" block, you use a load function to transfer the contents of the parameter to a temporary local data word and you process the timer function, the counter or the block memory-indirect.

## 19.5 Examples

### 19.5.1 Conveyor Belt Example

The example shows the transfer of signal states via block parameters. For this purpose, we use the function of a conveyor belt control explained in Chapter 5 "Memory Functions". The conveyor belt control is to be located in a function block and all inputs and outputs are to be routed via block parameters, so that the function block can be used repeatedly (for several conveyor belts). Figure 19.4 shows the input and output parameters for the function block as well as the static local data used.

Distributing the parameters is quite simple in this case: All binary operands that were inputs have become input parameters, all outputs have become output parameters and all memory bits have become static local data.

The function block "Conveyor_belt" is to control two conveyor belts. For this purpose, it will be called twice; the first time with the inputs and outputs of conveyor belt 1 and the second time with those of conveyor belt 2. For each call, the function block requires an instance data block where it stores the data for the conveyor belt in each case. The data block for conveyor belt 1 is to be called "Belt_data1" and the data block for conveyor belt 2 is to be called "Belt_data2".

**Table 19.6** Permissible Combinations when Passing On Block Parameters

| Calling → called | FC calls FC | | | FB calls FC | | | FC calls FB | | | FB calls FB | | |
|---|---|---|---|---|---|---|---|---|---|---|---|---|
| declaration type | E | C | P | E | C | P | E | C | P | E | C | P |
| Input → Input | x | – | – | x | x | – | x | – | x | x | x | x |
| Output → Output | x | – | – | x | x | – | x | – | – | x | x | – |
| In/out → Input | x | – | – | x | – | – | x | – | – | x | – | – |
| In/out → Output | x | – | – | x | – | – | x | – | – | x | – | – |
| In/out → In/out | x | – | – | x | – | – | x | – | – | x | – | – |

E = Elementary data types
C = Complex data types
P = Parameter types TIMER, COUNTER and BLOCK_xx

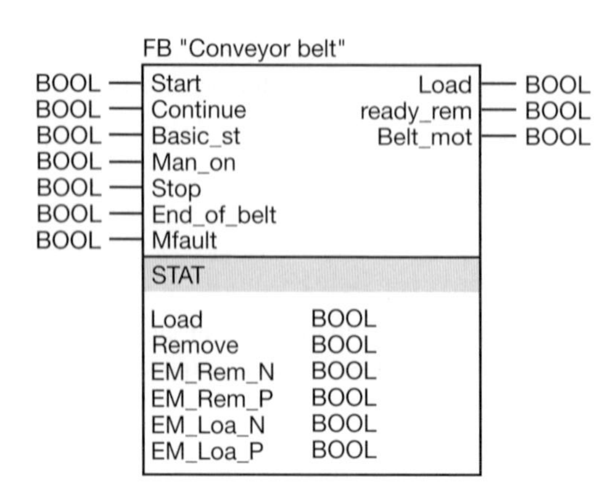


**Figure 19.4** Function Block for the Conveyor Belt Example

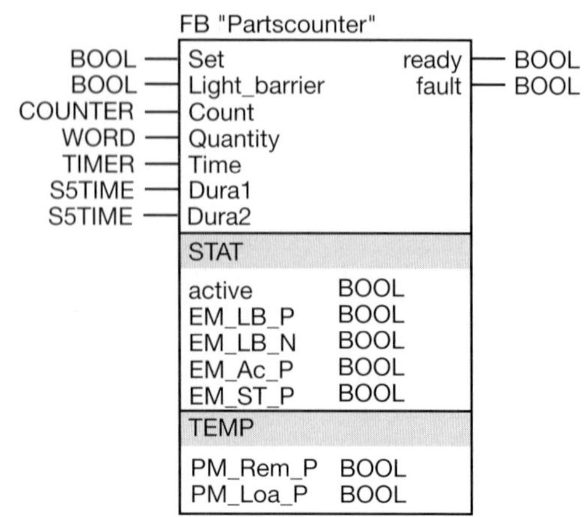


**Figure 19.5** Function Block for the Parts Counter Example

You can find the executed programming example in the "Conveyor Example" program on the diskette accompanying the book. The source program contains the programming of the function block with the input parameters, the output parameters and the static local data. This is followed by the programming of the instance data blocks; here, it is sufficient to specify the function block as the declaration section. You can use any data blocks as the instance block, for example, DB 21 for "Belt_data1" and DB 22 for "Belt_data2". In the symbol table, these data blocks have the data type of the function block.

At the end of the source program, you see another two complete calls of the function block, such as they might be found in OB1, for example. The inputs and output from the symbol table are used as the actual parameters. In those cases where these global symbols contain special characters, you must place these symbols between quotation marks in the program.

## 19.5.2 Parts Counter Example

The example demonstrates the handling of block parameters of elementary data types. The "Parts Counter" example from Chapter 8 "Counter Functions" is the basis of the function. The same function is implemented here as a function block, with all global variables declared either as block parameters or as static local data.

Timer and counter functions are transferred via block parameters of parameter types TIMER and COUNTER. These block parameters must be input parameters. The initial values of the counter (Quantity) and the timer function (Dura1 and Dura2) can also be transferred as block parameters; the data type of the block parameters corresponds here to the actual parameters.

The edge memory bits are stored in the static local data and the pulse memory bits are stored in the temporary local data.

You can find the executed programming example in the "Conveyor Example" program on the diskette accompanying the book. The source program contains the function block "Parts_counter", the associated instance data block "CountDat" and the call of the function block with instance data block.

## 19.5.3 Feed Example

The same functions as described in the two previous examples can also be called as local instances. In our example, this means that we program a function block "Feed" that is to control four conveyor belts and count the conveyed parts. In this function block, the FB "Conveyor Belt" is called four times and the FB "Parts_counter" is called once. The call does not take place in each case with its own instance data block, but the called FBs are to store their data in the instance data block of the function block "Feed".

Figure 19.6 shows how the individual conveyor belt controls are connected together (the FB "Parts_counter" is not represented here). The start signal is connected to the *Start* input of the controller of belt 1, the *ready_rem* output is connected to the *Start* input of belt 2, etc. Finally, the *ready_rem* output of belt 4 is connected to the *Remove* output of "Feed". The same signal sequence leads in the reverse direction from *Removed* via *Continue* and *Readyload* to *Load*.

Belt_motor, Light_barrier and /Mfault (motor fault) are individual signals of the conveyor belts; Reset, Startup and Stop control all conveyor belts via Basic_st, Man_on and Stop.

The following program for the function block "Feed" is designed in the same way. The input and output parameters of the function block can be seen from the figure. In addition, the digital values for the parts counter *Quantity, Dura1* and *Dura2* are designed as input parameters here. We declare the data of the individual conveyor belt controls and the data of the parts counter in the static local data in exactly the same way as for a user-defined data type, i.e. with name and data type. The variable "Belt1" is to receive the data structure of the function block "Conveyor_belt", also the variable "Belt2", etc.; the variable "Check" receives the data structure of the function block "Parts_counter".

The program in the function block starts with the initialization of the signals common to all conveyor belts. Here, we make use of the fact that the block parameters of the function blocks called as local instances are static local data in the current block and can be handled as such. The block parameter *Man_start* of the current function block controls the input parameter *Man_on* of all four conveyor belt controls with a simple assignment. We proceed in the same way with the signals *Stop* and *Reset*. And now the conveyor belt controls are initialized with the common signals. (You can, of course, also initialize these input parameters when the function block is called.)

The subsequent calls of the function blocks for conveyor belt control contain only the block parameters for the individual signals for each conveyor belt and the connection to the block parameters of "Feed". The individual signals are the light barriers, the commands for the belt motor and the motor faults. (We make use here of the fact that when a function block is called, not all block parameters have to be initialized.)

We program the connections between the individual belt controllers using assignments.

The FB "Parts_counter" is called as a local - instance even if it has no closer connection with the signals of the conveyor belt controls. The instance data block of "Feed" takes the FB data.

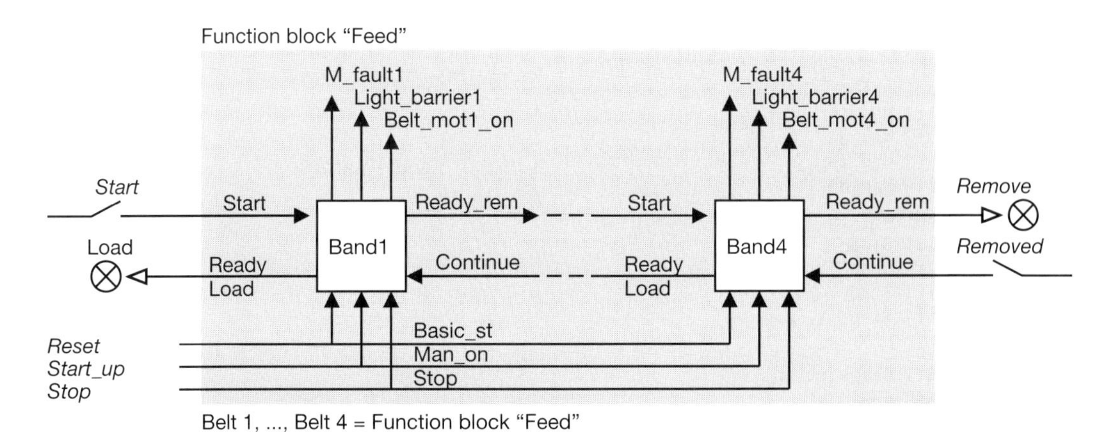


**Figure 19.6** Feed Programming Example

The input parameters *Quantity, Dura1* and *Dura2* of "Feed" need to be set only once. This can be done with the default (as in the example) or in the restart program in OB 100 (through direct assignment, for example, if these three parameters are treated as global data).

The source program in the "Conveyor Example" program contains the function block "Feed" and the associated instance data block "FeedData". At the end, the call of the function block 'Feed' is shown with the instance data block for the main program.

```
FUNCTION_BLOCK "Feed"
TITLE = Control of several conveyor belts
//Example of local instances (declaration, calls)
NAME    : Feed
AUTHOR  : Berger
FAMILY  : STL_Book
VERSION : 01.00

VAR_INPUT
  Start       : BOOL    := FALSE;      //Start conveyor belts
  Removed     : BOOL    := FALSE;      //Parts have been removed from belt
  Man_start   : BOOL    := FALSE;      //Startup conveyor belts manually
  Stop        : BOOL    := FALSE;      //Stop conveyor belts
  Reset       : BOOL    := FALSE;      //Set control to basic state
  Count       : COUNTER;               //Counter for the parts
  Quantity    : WORD    := W#16#0200;  //Number of parts
  Tim         : TIMER;                 //Timer function for the monitor
  Dura1       : S5TIME  := S5T#5s;     //Monitoring time for parts
  Dura2       : S5TIME  := S5T#10s;    //Monitoring time for gap
END_VAR
VAR_OUTPUT
  Load        : BOOL    := FALSE;      //Load new parts onto belt
  Remove      : BOOL    := FALSE;      //Remove parts from belt
END_VAR
VAR
  Belt1 : "Conveyor_belt";             //Control for belt 1
  Belt2 : "Conveyor_belt";             //Control for belt 2
  Belt3 : "Conveyor_belt";             //Control for belt 3
  Belt4 : "Conveyor_belt";             //Control for belt 4
  Check : "Parts_counter";             //Control for counting and monitoring
END_VAR

BEGIN
NETWORK
TITLE = Initializing the common signals

    A   Man_start;
    =   Belt1.Man_on;
    =   Belt2.Man_on;
    =   Belt3.Man_on;
    =   Belt4.Man_on;

    A   Stop;
    =   Belt1.Stop;
    =   Belt2.Stop;
    =   Belt3.Stop;
    =   Belt4.Stop;

    A   Reset;
    =   Belt1.Basic_st;
    =   Belt2.Basic_st;
    =   Belt3.Basic_st;
    =   Belt4.Basic_st;
```

```
NETWORK
TITLE = Calling the conveyor belt controls
 CALL Belt1 (
  Start               := Start,
  Readyload           := Load,
  End_of_belt         := Light_barrier1,
  Mfault              := "/Mfault1",
  Belt_motor_on       := Belt_mot1_on);
 A Belt2.Readyload;
 = Belt1.Continue;
 A Belt1.Ready_rem;
 = Belt2.Start;
 CALL Belt2 (
  End_of_belt         := Light_barrier2,
  Mfault              := "/Mfault2",
  Belt_motor_on       := Belt_mot2_on);
 A Belt3.Readyload;
 = Belt2.Continue;
 A Belt2.Ready_rem;
 = Belt3.Start;
 CALL Belt3 (
  End_of_belt         := Light_barrier3,
  Mfault              := "/Mfault3",
  Belt_motor_on       := Belt_mot3_on);
 A Belt4.Readyload;
 = Belt3.Continue;
 A Belt3.Ready_rem;
 = Belt4.Start;
 CALL Belt4 (
  Continue            := Removed,
  Ready_rem           := Remove,
  End_of_belt         := Light_barrier4,
  Mfault              := "/Mfault4",
  Belt_motor_on       := Belt_mot4_on);
NETWORK
TITLE = Call for counting and monitoring
 CALL Check (
  Set                 := Start,
  Light_barrier       := Light_barrier1,
  Count               := #Count,
  Quantity            := #Quantity,
  Tim                 := #Tim,
  Dura1               := #Dura1,
  Dura2               := #Dura2,
  Finished            := Finished,
  Fault               := "Fault");
NETWORK
TITLE = Block end
 BE
END_FUNCTION_BLOCK
```

**Figure 19.7** Program for the Feed Example

# Program Processing

This section of the book discusses the various methods of program processing.

- The **main program** executes cyclically. After each program pass, the CPU returns to the beginning of the program and executes it again. This is the "standard" method of processing PLC programs.
- Numerous system functions support the utilization of system services, such as controlling the real-time clock or communication via bus systems. In contrast to the static settings made when parameterizing the CPU, system functions can be used dynamically at program run time.
- The main program can be temporarily suspended to allow **interrupt servicing.** The various types of interrupts (hardware interrupts, watchdog interrupts, time-of-day interrupts, time-delay interrupts, multiprocessor interrupts) are divided into priority classes whose processing priority you may yourself, to a large degree, determine. Interrupt servicing allows you to react quickly to signals from the controlled process or implement periodic control procedures independently of the processing time of the main program.
- Before starting the main program, the CPU initiates a **start-up program** in which you can make specifications regarding program processing, define default values for variables, or parameterize modules.
- **Error handling** is also part of program processing. STEP 7 distinguishes between synchronous errors, which occur during processing of a statement, and asynchronous errors, which can be detected independently of program processing. In both cases you can adapt the error routine to suit your needs.

**20 Main program**
Program structure; scan cycle control; response time; program functions; multicomputing operation; data exchange with system functions; start information

**21 Interrupt handling**
Hardware interrupts; watchdog interrupts; time-of-day interrupts; time-delay interrupts; multiprocessor interrupt; handling interrupt events

**22 Start-up characteristics**
Power-up, memory reset, retentivity; complete restart; warm restart; ascertain module address; parameterize modules

**23 Error handling**
Synchronous errors (programming errors, access errors); handling synchronous error events; asynchronous errors; system diagnostics

# 20 Main Program

The main program is the cyclically scanned user program; cyclic scanning is the "normal" way in which programs execute in programmable logic controllers. The large majority of control systems use only this form of program execution. If event-driven program scanning is used, it is in most cases only in addition to the main program.

The main program is invoked in organization block OB 1. It executes at the lowest priority level, and can be interrupted by all other types of program processing. The mode selector on the CPU's front panel must be at RUN or RUN-P. When in the RUN-P position, the CPU can be programmed via a programming device. In the RUN position, you can remove the key so that no one can change the operating mode without proper authorization; when the mode selector is at RUN, programs can only be read.

## 20.1 General Remarks

### 20.1.1 Program Structure

To analyze a complex automation task means to subdivide that task into smaller tasks or functions in accordance with the structure of the process to be controlled. You then define the individual tasks resulting from this subdividing process by determining the functions and stipulating the interface signals to the process or to other tasks. This breakdown into individual tasks can be done in your program. In this way, the structure of your program corresponds to the subdivision of the automation task.

A subdivided user program can be more easily configured, and can be programmed in sections (even by several people in the case of very large user programs). And finally, but not lacking in importance, subdividing the program simplifies both debugging and service and maintenance.

The structuring of the user program depends on its size and its function. A distinction is made between three different "methods":

- In a *linear program,* the entire main program is in organization block OB 1. Each current path is in a separate network. STEP 7 numbers the networks in sequence. When editing and debugging, you can reference every network directly by its number.
- A *partitioned program* is basically a linear program which is subdivided into blocks. Reasons for subdividing the program might be because it is too long for organization block OB 1 or because you want to make it more readable. The blocks are then called in sequence. You can also subdivide the program in another block the same way you would the program in organization block OB 1. This method allows you to call associated process-related functions for processing from within one and the same block. The advantage of this program structure is that, even though the program is linear, you can still debug and run it in sections (simply by omitting or adding block calls).
- A *structured program* is used when the conceptual formulation is particularly extensive, when you want to reuse program functions, or when complex problems must be solved. Structuring means dividing the program into sections (blocks) which embody self-contained functions or serve a specific functional purpose and which exchange the fewest possible number of signals with other blocks. Assigning each program section a specific (process-related) function will produce easily readable blocks with simple interfaces to other blocks when programmed.

The STL programming language supports structured programming through functions with which you can create "blocks" (self-contained program sections). Chapter 3, "The STL Programming

Language", discusses under the header "Blocks" the different kinds of blocks and their uses. You will find a detailed description of the functions for calling and ending blocks in Chapter 18, "Block Functions". The blocks receive the signals and data to be processed via the call interface (the block parameters), and forward the results over this same interface. The options for passing parameters are described in detail in Chapter 19, "Block Parameters".

### 20.1.2 Program Organization

Program organization determines whether and in what order the CPU will process the blocks which you have generated. To organize your program, you program block calls in the desired sequence in the supraordinate blocks. You should chose the order in which the blocks are called so that it mirrors the process-related or function-related subdivision of the controlled plant.

#### Nesting depth

The maximum depth applies for a priority class (for the program in an organization block), and is CPU-dependent. On the CPU 314, for example, the nesting depth is eight, that is, beginning with one organization block (nesting depth 1), you can add seven more blocks in the "horizontal" direction (this is called "nesting"). If more blocks are called, the CPU goes to STOP with a "Block overflow" error. Do not forget to include system function block (SFB) calls and system function (SFC) calls when calculating the nesting depth.

A data block call, which is actually only the opening or selecting of a data area, has no effect on the nesting depth of blocks, nor is the nesting depth affected by calling several blocks in succession (linear block calls).

#### Practice-related program organization

In organization block OB 1, you should call the blocks in the main program in such a way as to roughly organize your program. A program can be organized on either a process-related or function-related basis.

The following points of discussion can give only a rough, very general view with the intention of giving the beginner some ideas on program structuring and on translating his control task into reality. Advanced programmers normally have sufficient experience to organize a program to suit the special control task at hand.

- A *process-related program structure* closely follows the structure of the plant to be controlled. The individual program sections correspond to the individual parts of the plant or of the process to be controlled. Subordinate to this rough structure are the scanning of the limit switches and operator panels and the control of the actuators and display devices (in different parts of the plant). Bit memory or global data are used for signal interchange between different parts of the plant.
- A *function-related program structure* is based on the control function to be executed. Initially, this method of program structuring does not take the controlled plant into account at all. The plant structure first becomes apparent in the subordinate blocks when the control function defined by the rough structure is subdivided further.
- *In practice*, a hybrid of these two concepts is normally used. Figure 20.1 shows an example:

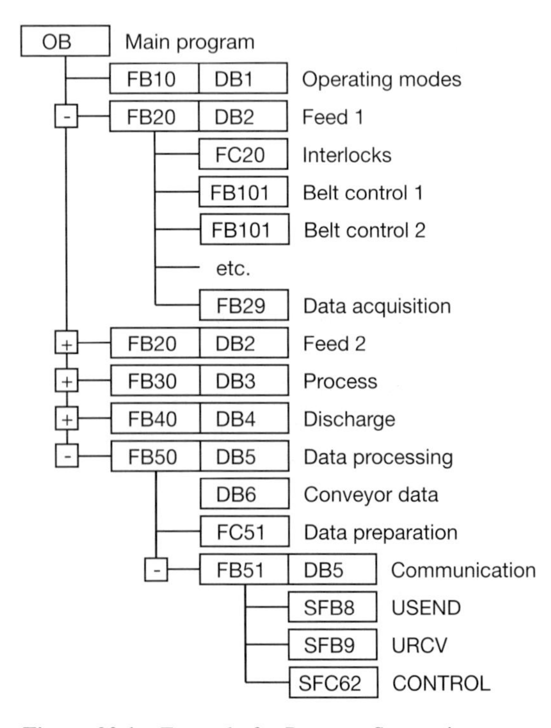


**Figure 20.1** Example for Program Structuring

A functional structure is mirrored in the operating mode program and in the data processing program which goes above and beyond the plant itself. Program sections Feeding Conveyor 1, Feeding Conveyor 2, Process and Discharging Conveyor are process-related.

The example also shows the use of different types of blocks. The main program is in OB 1; it is in this program that the blocks for the operating modes, the various pieces of plant equipment, and for data processing are called. These blocks are function blocks with an instance data block as data store. Feeding Conveyor 1 and Feeding Conveyor 2 are identically structured; FB 20, with DB 20 as instance data block for Feeding Conveyor 1 and with DB 21 as instance data block for Feeding Conveyor 2, is used for control.

In the conveyor control program, function FC 20 processes the interlocks; it scans inputs or memory bits and controls FB 20's local data. Function block FB 101 contains the control program for a conveyor belt, and is called once for each belt. The call is a local instance, so that its local data are in instance data block DB 20. The same applies for the data acquisition program in FB 29.

The data processing program in FB 50, which uses DB 50, processes the data acquired with FB 29 (and other blocks), which are located in global data block DB 60. Function FC 51 prepares these data for transfer. The transfer is controlled by FB 51 (with DB 51), in which system blocks SFB 8, SFB 9 and SFB 62 are called. Here, too, the SFBs save their instance data in "supraordinate" data block DB 51.

## 20.2 Scan Cycle Control

### 20.2.1 Process Image Updating

Before the operating system calls organization block OB 1, it first loads the process-image input table (the address area for inputs is loaded with the current signal states from the process). As a rule, the user program then works with these signal states and changes signal states in the process-image output table. When organization block OB 1 terminates, the operating system transfers the signal states from the process-image output table to the output modules (now, for the first time, the process is supplied with the processed data).

The process image is part of the CPU's internal system memory (Chapter 1.4, "CPU Memory Areas"). It begins at I/O address 0 and ends at the address stipulated by the relevant CPU. Normally, all digital modules lie in the process image address area, while all analog modules have addresses outside this area. If the CPU has free address allocation, you can use the configuration table to direct any module over the process image or address it outside the process image area.

#### Subprocess images

On the S7-400, the process image is divided into eight subprocess images whose sizes you can stipulate when you parameterize the CPU. System functions SFC 26 UPDAT_PI for the process-image input table and SFC 27 UPDAT_PO for the process-image output table are used to update these subprocess images (Table 20.1). You can update subprocess images by calling these SFCs at any time and at any location in the program. For instance, you can define a subprocess image for a priority class (a program scanning level) and then have it updated at the beginning and end of the relevant organization block when that priority class is processed.

The automatic updating of the entire process image by the operating system is programmable. You can completely disable automatic updating and allow only subprocess image updating or, by specifying the number 0 in the SFC call, you can

**Table 20.1** Parameters for the SFCs for Process Image Updating

| Parameter Name | SFC | | Declaration | Data Type | Contents, Description |
|---|---|---|---|---|---|
| PART | 26 | 27 | INPUT | BYTE | Number of the subprocess image (0 to 8) |
| RET_VAL | 26 | 27 | OUTPUT | INT | Error information |
| FLADDR | 26 | 27 | OUTPUT | WORD | On an access error: the address of the first byte to cause the error |

specify the time at which you want the entire process image updated.

The updating of a process image can be interrupted by calling a higher priority class.

If an error occurs during updating of a process image, it is reported via the SFC's function value. If an error occurs while the operating system is executing an automatic process image update, organization block OB 85, "Program run errors", is invoked.

### 20.2.2 Scan Cycle Monitoring Time

Program scanning in organization block OB 1 is monitored by the so-called "scan cycle monitor" or "scan cycle watchdog". The default value for the scan cycle monitoring time is 150 ms. You can change this value in the range from 1 ms to 6 s by parameterizing the CPU accordingly.

If the main program takes longer to scan than the specified scan cycle monitoring time, the CPU calls OB 80 ("Timeout"). If OB 80 has not been programmed, the CPU goes to STOP.

The scan cycle monitoring time includes the full scan time for OB 1. It also includes the scan times for higher priority classes which interrupt the main program (in the current cycle). Communication processes carried out by the operating system, such as GD communication or PG access to the CPU (program status!), also increase the runtime of the main program. The increase can be reduced in part by the way you parameterize the CPU ("cycle performance" parameter block). The CPU's cyclic memory test (S7-300) also increases the scan cycle time. You can limit or disable this test by parameterizing the CPU accordingly.

An SFC 43 RE_TRIGR system function call restarts the scan cycle monitoring time; the timer restarts with the new value. SFC 43 has no parameters.

#### Operating system run times

The scan cycle time also includes the operating system run times. These are composed of the following:

- System control of cyclic scanning ("no-load cycle"), fixed value
- Updating of the process image; dependent on the number of bytes to be updated
- Updating of the timers; dependent on the number of timers to be updated
- Communication load; dependent on the communication just completed (PG-CPU- or CPU-CPU-link); can be restricted to a value between 5% and 50% (default is 20%) of the scan cycle time by parameterizing the CPU accordingly.

All values at operating system runtime are properties of the relevant CPU.

### 20.2.3 Minimum Scan Cycle Time, Background Scanning

On an S7-400, you may specify a minimum scan cycle time. If the main program (including interrupts) takes less time, the CPU waits until the specified minimum scan cycle time has elapsed before beginning the next cycle by recalling OB 1.

The default value for the minimum scan cycle time is 0 ms, that is to say, the function is disabled. You can set a minimum scan cycle time of

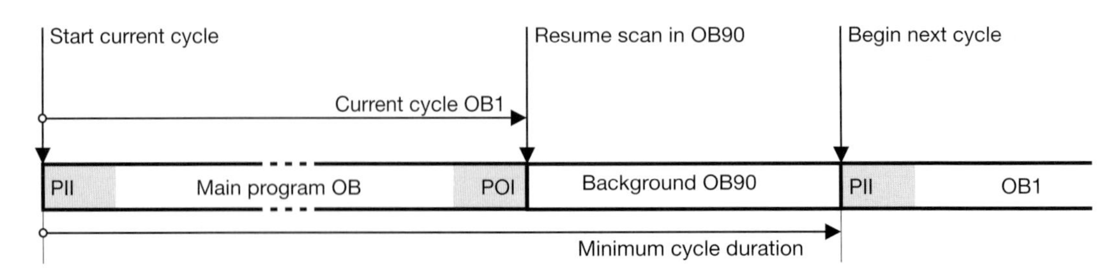


**Figure 20.2** Background Scanning with OB 90

from 1 ms to 6 s in index card "Cycle performance" when you parameterize the CPU.

### Background scanning OB 90

In the interval between the actual end of the cycle and expiration of the minimum cycle time, the CPU executes organization block OB 90 "Background scanning" (Figure 20.2). OB 90 is executed "in slices". When the operating system calls OB 1, execution of OB 90 is interrupted; it is then resumed at the point of interruption when OB 1 has terminated. OB 90 can be interrupted after each statement, any system block called in OB 90, however, is first scanned in its entirety. The length of a "slice" depends on the current scan cycle time of OB 1. The closer OB 1's scan time is to the minimum scan cycle time, the less time remains for executing OB 90. The program scan time is not monitored in OB 90.

OB 90 is scanned only in RUN mode. It can be interrupted by interrupt and error events, just like OB 1. You can activate and deactivate background scanning when parameterizing the CPU by entering a value of at least 20, or 0, in the "Local data" index card for priority class 29. The start information in the temporary local data (Figure 20.2) also tells which events cause OB 90 to execute from the beginning.

- B#16#91
  After a restart (complete restart or warm restart);
- B#16#92
  After a block processed in OB 90 was deleted or replaced;
- B#16#93
  After (re)loading of OB 90 in RUN mode;
- B#16#95
  After the program in OB 90 was scanned and a new background cycle begins.

## 20.2.4 Response Time

Before the main program is processed, the CPU's operating system loads the signal states of the digital input modules into the process-image input table. During program execution, the signal states of the inputs are then scanned and logically linked as per the functions in the user program; the outputs are set. After the program has terminated, the operating system transfers the signal states in the process-image output table to the output modules. A program cycle is then complete. Once the operating system has finished what it has to do, a new (user) program cycle begins with the updating of the process-image input table.

The introduction of a process image has given the programmable controller a response time which is dependent on the program execution time (scan cycle time). The response time lies between one and two scan cycles, as the following example explains.

When a limit switch is activated, for instance, it changes its signal state from "0" to "1". The pro-

**Table 20.2** Start Information for Background Scanning with OB 90

| Name | Data Type | Description | Contents |
|---|---|---|---|
| OB90_EV_CLASS | BYTE | Event class | B#16#11 = Call default OB |
| OB90_STRT_INF | BYTE | Start information | See text |
| OB90_PRIORITY | BYTE | Priority | B#16#1D = 29 |
| OB90_OB_NUMBR | BYTE | OB number | B#16#5A = 90 |
| OB90_RESERVED_1 | BYTE | Reserved | – |
| OB90_RESERVED_2 | BYTE | Reserved | – |
| OB90_RESERVED_3 | INT | Reserved | – |
| OB90_RESERVED_4 | INT | Reserved | – |
| OB90_RESERVED_5 | INT | Reserved | – |
| OB90_DATE_TIME | DT | Event occurrence | Instant of OB request |

grammable controller detects this change during the subsequent updating of the process image, and sets the inputs allocated to the limit switch to "1". The program evaluates this change by resetting an output, for example, in order to switch off the corresponding motor. The new signal state of the output that was reset is transferred at the end of the program scan; only then is the corresponding bit reset on the digital output module.

In a best-case situation, the process image is updated immediately following the change in the limit switch's signal. It would then take only one cycle for the relevant output to respond (Figure 20.3). In the worst-case situation, updating of the process image was just completed when the limit switch signal changed. It would then be necessary to wait approximately one cycle for the programmable controller to detect the signal change and set the input. After yet another cycle, the program can respond.

When so considered, the user program's execution time contains all procedures in one program cycle (including, for instance, the servicing of interrupts, the functions carried out by the operating system, such as updating timers, controlling the MPI interface and updating the process images).

The response time to a change in an input signal can thus be between one and two cycles. Added to the response time are the delays for the input modules, the switching times of contactors, and so on.

In some instances, you can reduce the response times by addressing the I/Os directly or calling program sections on an event-driven basis.

### 20.2.5 Scan Cycle Statistics

If your programmer is connected on-line with an active CPU, you can invoke the menu command PLC → MODULE INFORMATION to open a dialog box containing several index cards. The index card "Scan cycle time" shows the current scan cycle time as well as the shortest and longest scan cycle times. Also included in the display are the specified minimum scan cycle time and the scan cycle monitoring time.

The OB 1 start information (see below) also shows you the shortest and longest cycle times and the scan time of the preceding cycle. You can process this information directly in your program.

## 20.3 Program Functions

In addition to parameterizing the CPU via STEP 7, you can also select a number of program functions dynamically at runtime via the integrated system functions.

### 20.3.1 Real-Time Clock

The following system functions can be used to control the CPU's real-time clock:

- SFC 0 SET_CLK
  Set date and time

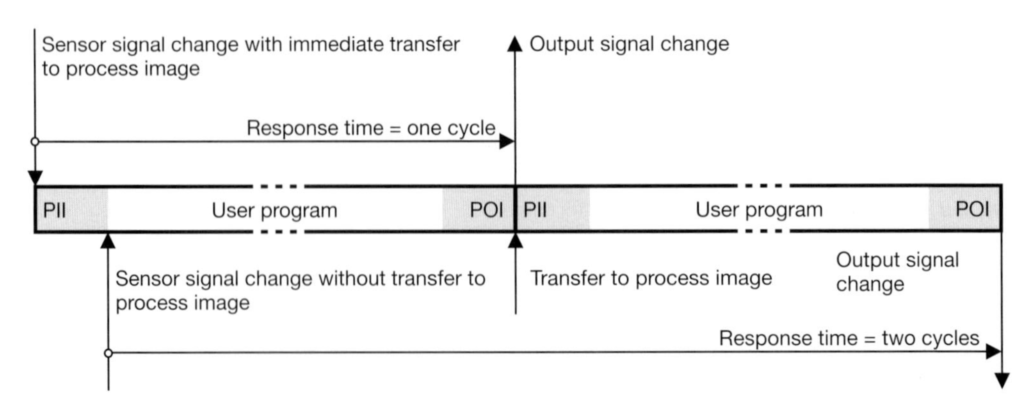


**Figure 20.3** Response Times of Programmable Controllers

**Table 20.3** SFC Parameters for the Real-Time Clock

| SFC | Parameter Name | Declaration | Data Type | Contents, Description |
|---|---|---|---|---|
| 0 | PDT | INPUT | DT | Date and time (new) |
| | RET_VAL | OUTPUT | INT | Error information |
| 1 | RET_VAL | OUTPUT | INT | Error information |
| | CDT | OUTPUT | DT | Date and time (current) |
| 48 | RET_VAL | OUTPUT | INT | Error information |

- SFC 1 READ_CLK
  Read date and time
- SFC 48 SNC_RTCB
  Synchronize CPU clocks

You will find a list of system function parameters in Table 20.3.

When several CPUs are connected to one another in a subnetwork, initialize one of the CPU's clocks as "master clock". When parameterizing the CPU, also enter the synchronization interval after which all clocks in the subnetwork are to be automatically synchronized to the master clock.

Call SFC 48 SNC_RTCB in the CPU with the master clock. This call synchronizes all clocks in the subnetwork independently of automatic synchronization. When you set a master clock with SFC 0 SET_CLK, all other clocks in the subnetwork are automatically synchronized to this value.

### 20.3.2 Read System Clock

A CPU's system clock starts running on power-up or on a complete restart. As long as the CPU is executing the restart routine or is in RUN mode, the system clock keeps running. When the CPU goes to STOP, the current system time is "frozen". If you initiate a warm restart on an S7-400 CPU, the system clock starts running again using the saved value as its starting time.

The system time has data format TIME, whereby it can assume only positive values (TIME#0ms to TIME#24d20h31m23s647ms). In the event of an overflow, the clock starts again at 0. An S7-300 CPU updates the system clock every 10 milliseconds, an S7-400 CPU every millisecond.

You can read the current system time with system function SFC 64 TIME_TCK. The RET_VAL parameter contains the system time in the TIME data format.

You can use the system clock, for example, to read out the current CPU runtime or, by computing the difference, to calculate the time between two SFC 64 calls. The difference between two values in TIME format is computed using DINT subtraction.

### 20.3.3 Run-Time Meter

A run-time meter in a CPU counts the hours. You can use the run-time meter for such tasks as determining the CPU runtime or ascertaining the runtime of devices connected to that CPU.

The number of run-time meters per CPU depends on the CPU. When the CPU is at STOP, the run-time meter also stops running; when the CPU is restarted (complete or warm restart), the run-time meter begins again with the previous value. When a run-time meter reaches 32767 hours, it stops and reports an overflow. A run-time meter can be set to a new value or reset to zero only via an SFC call.

The following system functions are available to control a run-time meter:

- SFC 2 SET_RTM
  Set run-time meter
- SFC 3 CTRL_RTM
  Start or stop run-time meter
- SFC 4 READ_RTM
  Read run-time meter

Table 20.4 shows the parameter for these system functions.

**Table 20.4** Parameters of the SFCs for the Run-Time Meter

| SFC | Parameter | Declaration | Data Type | Contents, Description |
|---|---|---|---|---|
| 2 | NR | INPUT | BYTE | Number of the run-time meter (B#16#01 to B#16#08) |
| | PV | INPUT | INT | New value for the run-time meter |
| | RET_VAL | OUTPUT | INT | Error information |
| 3 | NR | INPUT | BYTE | Number of the run-time meter (B#16#01 to B#16#08) |
| | S | INPUT | BOOL | Start (with "1") or stop (with "0") run-time meter |
| | RET_VAL | OUTPUT | INT | Error information |
| 4 | NR | INPUT | BYTE | Number of the run-time meter (B#16#01 to B#16#08) |
| | RET_VAL | OUTPUT | INT | Error information |
| | CQ | OUTPUT | BOOL | Run-time meter running ("1") or stopped ("0") |
| | CV | OUTPUT | INT | Current value of the run-time meter |

The NR parameter stands for the number of the run-time meter, and has the data type BYTE. It can be initialized using a constant or a variable (as can all input parameters of elementary data type). The PV parameter (data type INT) is used to set the run-time meter to an initial value. SFC 3's-S-parameter starts (with signal state "1") or stops (with signal state "0") the selected run-time meter. CQ indicates whether the run-time meter was running (signal state "1") or stopped (signal state "0") when scanned. The CV parameter records the hours in INT format.

### 20.3.4 Compressing CPU Memory

Multiple deletion and reloading of blocks, which often occur during on-line block modification, can result in gaps in the CPU's work memory and in the RAM load memory which decrease the amount of usable space in memory. When you call the "Compress" function, you start a CPU program which fills these gaps by pushing the blocks together. You can initiate the "Compress" function via a programming device connected to the CPU or by calling system function SFC 25 COMPRESS. The parameters for SFC 25 are listed in Figure 20.5.

The compression procedure is distributed over several program cycles. The SFC returns BUSY = "1" to indicate that it is still in progress, and DONE = "1" to indicate that it has completed the compression operation. The SFC cannot compress when an externally initiated compression is in progress, when the "Delete Block" function is active, or when PG functions are accessing the block to be shifted (for instance the program status).

Note that blocks of a particular CPU-specific maximum length cannot be compressed, so that gaps would still remain in CPU memory. Only the Compress function initiated via the PG while the CPU is at STOP closes all gaps.

### 20.3.5 Waiting and Stopping

The system function SFC 47 WAIT halts the program scan for a specified period of time; the system function SFC 48 STP terminates the program scan, and the CPU goes to STOP.

SFC 47 WAIT has input parameter WT of data type INT in which you can specify the waiting time in microseconds (µs). The maximum waiting time is 32767 microseconds; the minimum

**Table 20.5** Parameters for SFC 25

| SFC | Parameter | Declaration | Data Type | Contents, Description |
|---|---|---|---|---|
| 25 | RET_VAL | OUTPUT | INT | Error information |
| | BUSY | OUTPUT | BOOL | Compression still in progress (with "1") |
| | DONE | OUTPUT | BOOL | Compression completed (with "1") |

waiting time corresponds to the execution time of the system function, which is CPU-specific.

SFC 47 can be interrupted by higher-priority events. On an S7-300, this increases the waiting time by the scan time of the higher-priority interrupt routine.

SFC 46 STP has no parameters.

### 20.3.6 Multiprocessing Mode

The S7-400 enables multiprocessing. As many as four appropriately designed CPUs can be operated in one rack on the same P bus and K bus. Each CPU has a CPU number by which it is referenced in multiprocessing mode. You can assign this number when you configure the hardware (index card "Multicomputing"). The configuration data for all the CPUs must be loaded into the PLC, even when you make changes to only one CPU.

After assigning parameters to the CPUs, you must assign each module in the station to a CPU. This is done by parameterizing the module in the "Addresses" tab under "CPU Assignment". At the same time that you assign the module's address area, you also assign the module's interrupts to this CPU. With VIEW → FILTER → CPU NO.X-MODULES, you can emphasize the modules assigned to a CPU in the configuration tables.

The CPUs in a multiprocessing network all have the same operating mode. This means

- They must all be parameterized with the same restart mode;
- They all go to RUN simultaneously;
- They all go to HALT when you debug in single-step mode in one of the CPUs;
- They all go to STOP as soon as one of the CPUs goes to STOP.

When one rack in the station fails, organization block OB 86 is called in each CPU.

The user programs in these CPUs execute independently of one another; they are not synchronized.

An SFC 35 MP_ALM call starts organization block OB 60 "Multiprocessor interrupt" in all CPUs simultaneously (see section 21.6, "Multiprocessor Interrupt").

## 20.4 Data Interchange Using System Functions

This section deals with system blocks for data interchange

- With the distributed I/O;
- Between CPUs via global data communication;
- Between two partners in the same station;
- Between two partners in different stations;
- Via system function blocks (SFB communication).

The interface description for the system blocks described in this section can be found in library Standard Library V3.x under System Function Blocks.

Section 22.5 "Parameterizing Modules" describes how to parameterize modules or transfer data records in RUN mode using system functions.

### 20.4.1 System Functions for Distributed I/O

With PROFIBUS-DP, information is transferred in packets. It is sometimes necessary to transfer a particular area in one packet as associated (consistent) data. Variables, which can comprise as much as a doubleword, always contain consistent data. The data consistency of longer variables (arrays, structures) or data areas is 32 bytes on the S7-300 and 122 bytes on the S7-400. This data consistency is ensured when you use system functions SFC 14 DPRD_DAT and SFC 15 DPWR_DAT. Each station of distributed I/O can send so-called diagnostic data, which provide information on the status of the station.

The following SFCs can be used to address a standard distributed I/O slave:

- SFC 11 DPSYN_FR
  Send SYNC/FREEZE commands to DP standard slaves
- SFC 13 DPNRM_DG
  Read diagnostic data from a standard DP slave
- SFC 14 DPRD_DAT
  Read user data from a standard DP slave
- SFC 15 DPWR_DAT
  Write user data to a standard DP slave

**Table 20.6** Parameters for SFCs Used to Reference the Distributed I/O

| SFC | Parameter | Declaration | Data Type | Contents, Description |
|---|---|---|---|---|
| 11 | REQ | INPUT | BOOL | Initiating for sending with REQ = “1” |
| | LADDR | INPUT | WORD | Configured starting address (from the E area) |
| | GROUP | INPUT | BYTE | DP slave group (from the Hardware Configuration) |
| | MODE | INPUT | BYTE | Command (see text) |
| | RET_VAL | OUTPUT | INT | Error information |
| | BUSY | OUTPUT | BOOL | Write operation still running while BUSY = “1” |
| 13 | REQ | INPUT | BOOL | Read request with REQ = “1” |
| | LADDR | INPUT | WORD | Configured diagnostic address (from the I area) |
| | RET_VAL | OUTPUT | INT | Error information |
| | RECORD | OUTPUT | ANY | Destination area for the diagnostic data read |
| | BUSY | OUTPUT | BOOL | Read still in progress when BUSY = “1” |
| 14 | LADDR | INPUT | WORD | Configured start address (from the I area) |
| | RET_VAL | OUTPUT | INT | Error information |
| | RECORD | OUTPUT | ANY | Destination area for user data read |
| 15 | LADDR | INPUT | WORD | Configured start address (from the Q area) |
| | RECORD | INPUT | ANY | Source area for user data to be written |
| | RET_VAL | OUTPUT | INT | Error information |

The parameters for these SFCs are listed in Table 20.6.

The RECORD parameter describes the area in which the data that are read are to be stored or from which data are to be read. Actual parameters may be variables of data type ARRAY or STRUCT or an ANY pointer of data type BYTE (for example: P#DB*z*DBX*y.x* BYTE *nnn*).

- Send SYNC/FREEZE commands
  You use the SFC 11 DPSYN_FR to send the commands SYNC, UNSYNC, FREEZE and UNFREEZE to one or more DP slave groups. The send operation is initiated with REQ = “1” and is completed when BUSY signals “0”.

  In the GROUP parameter, one group occupies one bit (from bit 0 = Group 1 to bit 7 = Group 8). The commands in the MODE parameter are also organized on a bitwise basis:

| | |
|---|---|
| Bit 2 = “1” | UNFREEZE |
| Bit 3 = “1” | FREEZE |
| Bit 4 = “1” | UNSYNC |
| Bit 5 = “1” | SYNC |

  In this way, you can send several commands with one signal call even to several groups.

- Read diagnostic data
  SFC 13 DPNRM_DG reads diagnostic data from a DP slave. The read procedure is initiated with REQ = “1”, and is terminated when BUSY = “0” is returned. Function value RET_VAL then contains the number of bytes read. Depending on the slave, diagnostic data may comprise from 6 to 240 bytes. If there are more than 240 bytes, the first 240 bytes are transferred and the relevant overflow bit is then set in the data.

- Read user data
  SFC 14 DPRD_DAT reads consistent user data (data which are associated through their contents) from a standard DP slave. You specify the length of the consistent data when you parameterize the DP slave.

- Write user data
  SC 15 DPWR_DAT writes consistent user data (data which are associated through their contents) to a standard DP slave. You specify the length of the consistent data when you parameterize the DP slave.

### 20.4.2 System Functions for GD Communication

In S7-400 systems, you can also control GD communication in your program. Additionally or alternatively to the cyclic transfer of global data, you can send or receive a GD packet with the following SFCs:

- SFC 60 GD_SND
  Send GD packet
- SFC 61 GD_RCV
  Receive GD packet

The parameters for these SFCs are listed in Table 20.7. The prerequisite for the use of these SFCs is a configured global data table. After compiling this table, STEP 7 shows you, in the "GD Identifier" column, the numbers of the GD circles and GD packets which you need for parameter assignments (for instance, GD 1..2..3 corresponds to circle 1, packet 2, element 3).

The size of a GD packet may not exceed 22 bytes of net data on an S7-300 or 54 bytes on an S7-400, in both instances in the case of a single element of that length. An element is either an address or an address area, such as MB 0:15, which corresponds to memory bytes MB 0 to MB 15. The data consistency applies to an element, whereby the size of a consistent address area is CPU-specific (8 bytes on an S7-300, 16 bytes on an S7-412/413, 32 bytes on an S7-414/416). If an address area exceeds the length applicable for data consistency, blocks with consistent data in the applicable length are formed, beginning with the first byte.

If you want to ensure data consistency for the entire GD packet when transferring with SFCs 60 and 61, you must disable or delay higher-priority interrupts and asynchronous errors on both the Send and Receive side during processing of the SFC 60 or SFC 61 (see section 21.7, "Handling Interrupt Events").

Do not forget about the system resources when using the SFCs for sending GD packets; under some circumstances, you will have to simulate the scan rates per program.

### 20.4.3 System Functions for Data Interchange Within a Station

This section deals with system functions for internal station connections not configured in the connection table ("Communication via Non-configured Connections"). These functions address nodes directly allocated to the CPU, for instance via PROFIBUS-DP. The node identification is derived from the I/O address. Use the LADDR parameter to specify the module start address and the IOID parameter to indicate whether this address is located in the input or output area.

These system functions establish the required communication links dynamically and – if specified – break the connections when the job has been executed. If a connection cannot be set up due to lack of resources in either the transmitter or the receiver, "temporary lack of resources" is reported. The transfer must then be retried. Between two communication partners there can be no more than one connection in each direction.

By modifying the block parameters at run time, you can utilize a system function for different communication connections. An SFC may not interrupt itself. You may modify a program section in which one of these SFCs is used only in STOP mode; afterwards, you must execute a complete restart.

These SFCs transfer a maximum of 76 bytes of user data. A CPU's operating system combines the user data into blocks consistent within them-

**Table 20.7** SFC Parameters for GD Communication

| SFC | Parameter | Declaration | Data Type | Contents, Description |
|---|---|---|---|---|
| 60 | CIRCLE_ID | INPUT | BYTE | Number of the GD circle |
| | BLOCK_ID | INPUT | BYTE | Number of the GD packet to be sent |
| | RET_VAL | OUTPUT | INT | Error information |
| 61 | CIRCLE_ID | INPUT | BYTE | Number of the GD circle |
| | BLOCK_ID | INPUT | BYTE | Number of the GD packet to be received |
| | RET_VAL | OUTPUT | INT | Error information |

selves, without regard to the direction of transfer. In S7-300 systems, these blocks are 8 bytes long, in systems with a CPU 412/413 they are 16 bytes long, and in systems with a CPU 414/416 they are 32 bytes long. When two CPUs exchange data, the block size on the "passive" CPU is decisive for data consistency.

The following system functions handle data transfers between two CPUs in the same station:

- SFC 72 I_GET
  Read data
- SFC 73 I_PUT
  Write data
- SFC 74 I_ABORT
  Disconnect

The parameters for these SFCs are listed in Table 20.8.

The SD, RD and VAR_ADDR parameters describe the area from which the data to be transferred are to be read or to which the receive data are to be written. Actual parameters may be addresses, variables or data areas addressed with an ANY pointer. The Send and Receive data are not checked for identical data types.

A job is initiated with REQ = "1" and BUSY = "0" ("first call"). While the job is in progress, BUSY is set to "1". Changes to the REQ parameter no longer have any effect. When the job is completed, BUSY is reset to "0". If REQ is still "1", the job is immediately restarted.

- Read data
  When the read procedure has been initiated, the operating system in the partner CPU assembles and sends the requested data. An SFC call transfers the Receive data to the destination area. RET_VAL then shows the number of bytes transferred. If CONT is = "0", the communication link is broken. If CONT is = "1", the link is maintained. The data are also read when the communication partner is in STOP mode.
- Write data
  When the write procedure has been initiated, the operating system transfers all data from the source area to an internal buffer on the first call, and sends them to the partner in the link. There, the receiver writes the data into data area VAR_ADDR. BUSY is then set to "0". The data are also written when the receiving partner is at STOP.
- Disconnect
  REQ = "1" breaks a connection to the specified communication partner. With I_ABORT, you can break only those connections established in the same station with I_GET or I_PUT.

### 20.4.4 System Functions for Data Interchange Between Two Stations

This section covers system functions for external station connections not configured in the connection table ("Communication via Non-config-

**Table 20.8** SFC Parameters for Internal Station Communication

| Parameter | For SFC | | | Declaration | Data Type | Contents, Description |
|---|---|---|---|---|---|---|
| REQ | 72 | 73 | 74 | INPUT | BOOL | Initiate job with REQ = "1" |
| CONT | 72 | 73 | – | INPUT | BOOL | CONT = "1": Connection remains intact after job terminates |
| IOID | 72 | 73 | 74 | INPUT | BYTE | B#16#54 = Input area<br>B#16#55 = Output area |
| LADDR | 72 | 73 | 74 | INPUT | WORD | Module start address |
| VAR_ADDR | 72 | 73 | – | INPUT | ANY | Data area in partner CPU |
| SD | – | 73 | – | INPUT | ANY | Data area in own CPU which contains the Send data |
| RET_VAL | 72 | 73 | 74 | OUTPUT | INT | Error information |
| BUSY | 72 | 73 | 74 | OUTPUT | BOOL | Job in progress when BUSY = "1" |
| RD | 72 | – | – | OUTPUT | ANY | Data area in own CPU which will take the Receive data |

ured Connections"). These functions address nodes that are on the same MPI subnet. The node identification is derived from the MPI address (DEST_ID parameter).

These system functions set up the required communication links dynamically and – if specified – break them when the job has been executed. If a connection cannot be established because of a lack of resources in either the sender or the receiver, "temporary lack of resources" is reported. The transfer must then be retried. Between two communication partners, there can be only one connection in each direction.

By modifying the block parameters at run time, you can utilize a system function for different communication links. An SFC may not interrupt itself. You may modify a program section in which one of these SFCs is used only in STOP mode; a complete restart must then be executed.

These SFCs transfer a maximum of 76 bytes of user data. A CPU's operating system combines the user data into blocks consistent within themselves, without regard to the direction of transfer. In S7-300 systems, these blocks have a length of 8 bytes, in systems with a CPU 412/413 a length of 16 bytes, and in systems with a CPU 414/416 a length of 32 bytes. If two CPUs exchange data via X_GET or X_PUT, the block size of the "passive" CPU is decisive to data consistency. In the case of a SEND/RECEIVE connection, all data are consistent.

On a transition from RUN to STOP, all active connections (all SFCs except X_RECV) are cleared.

The following system functions handle data transfers between partners in different stations:

- SFC 65 X_SEND
  Send data
- SFC 66 X_RCV
  Receive data
- SFC 67 X_GET
  Read data
- SFC 68 X_PUT
  Write data
- SFC 69 X_ABORT
  Disconnect

The parameters for these SFCs are listed in Table 20.9.

The SD, RD and VAR_ADDR parameters describe the area from which the data to be sent are to be read or to which the Receive data are to be written. Actual parameters may be addresses, variables, or data areas addressed with an

**Table 20.9** SFC Parameters for External Station Communication

| Parameter | For SFC | | | | | Declaration | Data Type | Contents, Description |
|---|---|---|---|---|---|---|---|---|
| REQ | 65 | – | 67 | 68 | 69 | INPUT | BOOL | Job initiation with REQ = "1" |
| CONT | 65 | – | 67 | 68 | – | INPUT | BOOL | CONT = "1": Connection is maintained when job is completed |
| DEST_ID | 65 | – | 67 | 68 | 69 | INPUT | WORD | Partner's node identification (MPI address) |
| REQ_ID | 65 | – | – | – | – | INPUT | DWORD | Job identification |
| VAR_ADDR | – | – | 67 | 68 | – | INPUT | ANY | Data area in partner CPU |
| SD | 65 | – | – | 68 | – | INPUT | ANY | Data area in own CPU which contains the Send data |
| EN_DT | – | 66 | – | – | – | INPUT | BOOL | If "1": Accept Receive data |
| RET_VAL | 65 | 66 | 67 | 68 | 69 | OUTPUT | INT | Error information |
| BUSY | 65 | – | 67 | 68 | 69 | OUTPUT | BOOL | Job in progress when BUSY = "1" |
| REQ_ID | – | 66 | – | – | – | OUTPUT | DWORD | Job identification |
| NDA | – | 66 | – | – | – | OUTPUT | BOOL | When "1": Data received |
| RD | – | 66 | 67 | – | – | OUTPUT | ANY | Data area in own CPU which will accept the Receive data |

ANY pointer. Send and Receive data are not checked for matching data types. When the Receive data are irrelevant, a "blank" ANY pointer (NIL pointer) as RD parameter in X_RCV is permissible.

A job is initiated with REQ = "1" and BUSY = "0" ("first call"). While the job is in progress, BUSY is set to "1"; changes to the REQ parameter now no longer have any effect. When the job terminates, BUSY is set back to "0". If REQ is still "1", the job is immediately restarted.

- Send data
  When the Send request has been submitted, the operating system transfers all data from the source area to an internal buffer on the first call, then transfers the data to the partner CPU. BUSY is "1" for the duration of the send procedure. When the partner has signaled that it has fetched the data, BUSY is set to "0" and the send job terminated. If CONT is = "0", the connection is broken and the respective CPU resources are available to other communication links. If CONT is = "1", the connection is maintained. The REQ_ID parameter makes it possible for you to assign an ID to the Send data which you can evaluate with SFC X_RCV.

- Receive data
  The Receive data are placed in an internal buffer. Multiple packets can be put in a queue in the chronological order of their arrival. Use EN_DT = "0" to check whether or not data were received; if so, NDA is "1", RET_VAL shows the number of bytes of Receive data, and REQ_ID is the same as the corresponding parameter in SFC X_SEND. When EN_DT is = "1", the SFC transfers the first (oldest) packet to the destination area; NDA is then "1" and RET_VAL shows the number of bytes transferred. If EN_DT is "1' but there are no data in the internal queue, NDA is "0". On a complete restart, all data packets in the queue are rejected. In the event of a broken connection or a restart, the oldest entry in the queue, if already "queried" with EN_DT = "0", is retained; otherwise, it is rejected like the other queue entries.

- Read data
  When the read procedure has been initiated, the operating system in the partner CPU assembles and sends the data required under VAR_ADDR. On an SFC call, the Receive data are entered in the destination area. RET_VAL then shows the number of bytes transferred. If CONT is "0", the communication link is broken. If CONT is "1", the connection is maintained. The data are then read even when the communication partner is in STOP mode.

- Write data
  When the write procedure has been initiated, the operating system transfers all data from the source area to an internal buffer on the first call, then sends the data to the partner CPU. There, the partner CPU's operating system writes the Receive data to the VAR_ADDR data area. BUSY is then set to "0". The data are written even if the communication partner is in STOP mode.

- Disconnect
  REQ = "1" breaks an existing connection to the specified communication partner. X_ABORT can be used to break only those connections established in the CPU's own station with X_SEND, X_GET or X_PUT.

### 20.4.5 SFB Communication

The prerequisite for communication via system function blocks is a configured connection table ("Communication via Configured Connections"). The connection table is used to define and describe communication links.

A communication link is specified by a connection ID for each communication partner. STEP 7 assigns the connection IDs when it compiles the connection table. Use the "local ID" to initialize the SFB in the local or "own" module and the "remote ID" to initialize the SFB in the partner module.

The same logical connection can be used for different Send/Receive requests. A distinction must be made by adding a job ID to the connection ID in order to stipulate the relationship between Send and Receive block.

#### Two-way data interchange

The following SFBs are available for two-way data interchange:

**Table 20.10** SFB Parameters for Sending and Receiving Data

| Parameter | For SFB | | | | Declaration | Data Type | Contents, Description |
|---|---|---|---|---|---|---|---|
| REQ | 8 | – | 12 | – | INPUT | BOOL | Start data exchange |
| EN_R | – | 9 | – | 13 | INPUT | BOOL | Receive ready |
| R | – | – | 12 | – | INPUT | BOOL | Abort data exchange |
| ID | 8 | 9 | 12 | 13 | INPUT | WORD | Connection ID |
| R_ID | 8 | 9 | 12 | 13 | INPUT | DWORD | Job ID |
| DONE | 8 | – | 12 | – | OUTPUT | BOOL | Job terminated |
| NDR | – | 9 | – | 13 | OUTPUT | BOOL | New data fetched |
| ERROR | 8 | 9 | 12 | 13 | OUTPUT | BOOL | Error occurred |
| STATUS | 8 | 9 | 12 | 13 | OUTPUT | WORD | Job status |
| SD_1 | 8 | – | 12 | – | IN_OUT | ANY | First Send area |
| SD_2 | 8 | – | – | – | IN_OUT | ANY | Second Send area |
| SD_3 | 8 | – | – | – | IN_OUT | ANY | Third Send area |
| SD_4 | 8 | – | – | – | IN_OUT | ANY | Fourth Send area |
| RD_1 | – | 9 | – | 13 | IN_OUT | ANY | First Receive area |
| RD_2 | – | 9 | – | – | IN_OUT | ANY | Second Receive area |
| RD_3 | – | 9 | – | – | IN_OUT | ANY | Third Receive area |
| RD_4 | – | 9 | – | – | IN_OUT | ANY | Fourth Receive area |
| LEN | – | – | 12 | 13 | IN_OUT | WORD | Length of data block in bytes |

- SFB 8 USEND
  Uncoordinated sending of a data packet of CPU-specific length
- SFB 9 URCV
  Uncoordinated receiving of a data packet of CPU-specific length
- SFB 12 BSEND
  Sending of a data block of up to 64 Kbytes in length
- SFB 13 BRCV
  Receiving of a data block of up to 64 Kbytes in length

SFB 8 and SFB 9 or SFB 12 and SFB 13 must always be used as a pair. The parameters for these SFBs are listed in Table 20.10.

A positive edge at the REQ (request) parameter starts the data exchange, a positive edge at the R (reset) parameter aborts it. A "1" in the EN_R (enable receive) parameter signals that the partner is ready to receive data. Initialize the ID parameter with the connection ID, which STEP 7 enters in the connection table for both the local and the remote partner (the two IDs may differ). R_ID allows you to choose a specifiable but unique job ID which must be identical for the Send and Receive block. This allows several pairs of Send and Receive blocks to share a single logical connection (as each has a unique ID). The block transfers the actual values of the ID and R_ID parameters to its instance data block on the *first call*. The first call establishes the communication relationship (for this instance) until the next complete restart.

With a "1" in the DONE or NDR parameter, the block signals that the job terminated without error. An error, if any, is flagged in the ERROR parameter. A value other than zero in the STATUS parameter indicates either a warning (ERROR = "0") or an error (ERROR = "1"). You must evaluate the DONE, NDR, ERROR and STATUS parameters after *every* block call.

- SFB 8 USEND and SFB 9 URCV: The SD_n and RD_n parameters are used to specify the variable or the area you want to transfer. Send area SD_n must correspond to the respective Receive area RD_n. Use the parameters without gaps, beginning with 1. No values need be specified for unneeded parameters (like an FB, not all SFB parameters need be assigned values). The first time SFB 9 is called, a Receive mailbox is generated; on all subsequent calls, the Receive must fit into this mailbox.

**Table 20.11** SFB Parameters for Reading and Writing Data

| Parameter | For SFB | | Declaration | Data Type | Contents, Description |
|---|---|---|---|---|---|
| REQ | 14 | 15 | INPUT | BOOL | Start data exchange |
| ID | 14 | 15 | INPUT | WORD | Connection ID |
| NDR | 14 | – | OUTPUT | BOOL | New data fetched |
| DONE | – | 15 | OUTPUT | BOOL | Job terminated |
| ERROR | 14 | 15 | OUTPUT | BOOL | Error occurred |
| STATUS | 14 | 15 | OUTPUT | WORD | Job status |
| ADDR_1 | 14 | 15 | IN_OUT | ANY | First data area in partner CPU |
| ADDR_2 | 14 | 15 | IN_OUT | ANY | Second data area in partner CPU |
| ADDR_3 | 14 | 15 | IN_OUT | ANY | Third data area in partner CPU |
| ADDR_4 | 14 | 15 | IN_OUT | ANY | Fourth data area in partner CPU |
| RD_1 | 14 | – | IN_OUT | ANY | First Receive area |
| RD_2 | 14 | – | IN_OUT | ANY | Second Receive area |
| RD_3 | 14 | – | IN_OUT | ANY | Third Receive area |
| RD_4 | 14 | – | IN_OUT | ANY | Fourth Receive area |
| SD_1 | – | 15 | IN_OUT | ANY | First Send area |
| SD_2 | – | 15 | IN_OUT | ANY | Second Send area |
| SD_3 | – | 15 | IN_OUT | ANY | Third Send area |
| SD_4 | – | 15 | IN_OUT | ANY | Fourth Send area |

- SFB 12 BSEND and SFB 13 BRCV: Enter a pointer to the first byte of the data area in parameter SD_1 or RD_1 (the length is not evaluated); the number of bytes of Send or Receive data is in the LEN parameter. Up to 64 Kbytes may be transferred; the data are transferred in blocks (sometimes called frames), and the transfer itself is asynchronous to the user program scan.

### One-way data interchange

The following SFBs are available for one-way data interchange:

- SFB 14 GET
  Read data up to a CPU-specific maximum length
- SFB 15 PUT
  Write data up to a CPU-specific maximum length

The operating system in the partner CPU collects the data read with SFB 14; the operating system in the partner CPU distributes the data written with SFB 15. A Send or Receive (user) program in the partner CPU is not required. Table 20.11 lists the parameters for these SFBs.

A positive edge at parameter REQ (request) starts the data interchange. Set the ID parameter to the connection ID entered by STEP 7 in the connection table.

With a "1" in the DONE or NDR parameter, the block signals that the job terminated without error. An error, if any, is flagged with a "1" in the ERROR parameter. A value other than zero in the STATUS parameter is indicative of either a warning (ERROR = "0") or an error (ERROR = "1"). You must evaluate the DONE, NDR, ERROR and STATUS parameters after *every* block call.

Use the ADDR_n parameter to specify the variable or the area in the partner CPU from which you want to fetch or to which you want to send the data. The areas in ADDR_n must coincide with the areas specified in SD_n or RD_n. Use the parameters without gaps, beginning with 1. Unneeded parameters need not be specified (as in an FB, an SFB does not have to have values for all parameters).

### Transferring print data

SFB 16 PRINT allows you to transfer a format description and data to a printer via a CP 441

**Table 20.12** Parameters for SFB 16 PRINT

| Parameter | Declaration | Data Type | Contents, Description |
|---|---|---|---|
| REQ | INPUT | BOOL | Start data exchange |
| ID | INPUT | WORD | Connection ID |
| DONE | OUTPUT | BOOL | Job terminated |
| ERROR | OUTPUT | BOOL | Error occurred |
| STATUS | OUTPUT | WORD | Job status |
| PRN_NR | IN_OUT | BYTE | Printer number |
| FORMAT | IN_OUT | STRING | Format description |
| SD_1 | IN_OUT | ANY | First variable |
| SD_2 | IN_OUT | ANY | Second variable |
| SD_3 | IN_OUT | ANY | Third variable |
| SD_4 | IN_OUT | ANY | Fourth variable |

communications processor. Table 20.12 lists the parameters for this SFB.

A positive edge at the REQ parameter starts the data exchange with the printer specified by the ID and PRN_NR parameters. The block signals an error-free transfer by setting DONE to "1". An error, if any, is flagged by a "1" in the ERROR parameter. A value other than zero in the STATUS parameter is indicative of either a warning (ERROR = "0") or an error (ERROR = "1"). You must evaluate the DONE, ERROR and STATUS parameters after *every* block call.

Enter the characters to be printed in STRING format in the FORMAT parameter. You can integrate as many as four format descriptions for variables in this string, defined in parameters SD_1 to SD_4. Use the parameters without gaps, beginning with 1; do not specify values for unneeded parameters. You can transfer up to 420 bytes (the sum of FORMAT and all variables) per print request.

## Control functions

The following SFBs are available for controlling the communication partner:

- SFB 19 START
  Execute a complete restart in the partner controller
- SFB 20 STOP
  Switch the partner controller to STOP
- SFB 21 RESUME
  Execute a warm restart in the partner controller

These SFBs are for one-way data exchange. The parameters for them are listed in Table 20.13.

A positive edge at the REQ parameter starts the data exchange. Enter as ID parameter the connection ID which STEP 7 entered in the connection table.

With a "1" in the DONE parameter, the block signals that the job terminated without error. An

**Table 20.13** SFB Parameters for Partner Controller

| Parameter | For SFB | | | Declaration | Data Type | Contents, Description |
|---|---|---|---|---|---|---|
| REQ | 19 | 20 | 21 | INPUT | BOOL | Start data exchange |
| ID | 19 | 20 | 21 | INPUT | WORD | Connection ID |
| DONE | 19 | 20 | 21 | OUTPUT | BOOL | Job terminated |
| ERROR | 19 | 20 | 21 | OUTPUT | BOOL | Error occurred |
| STATUS | 19 | 20 | 21 | OUTPUT | WORD | Job status |
| PI_NAME | 19 | 20 | 21 | IN_OUT | ANY | Program name (P_PROGRAM) |
| ARG | 19 | – | 21 | IN_OUT | ANY | Irrelevant |
| IO_STATE | 19 | 20 | 21 | IN_OUT | BYTE | Irrelevant |

error, if any, is flagged by a "1" in the ERROR parameter. A value other than zero in the STATUS parameter is indicative of either a warning (ERROR = "0") or an error (ERROR = "1"). You must evaluate the DONE, ERROR and STATUS parameters after *every* block call.

Specify as PI_NAME an array variable with the contents "P_PROGRAM" (ARRAY [1..9] OF CHAR). The ARG and IO_STATE parameters are currently irrelevant, and need not be assigned a value.

- SFB 19 START executes a complete restart of the partner CPU. Prerequisite is that the partner CPU is at STOP and that the mode selector is positioned to either RUN or RUN-P.
- SFB 20 STOP sets the partner CPU to STOP. Prerequisite for error-free execution of this job request is that the partner CPU is not at STOP when the request is submitted.
- SFB 21 RESUME executes a warm restart of the partner CPU. Prerequisite is that the partner CPU is at STOP, that the mode selector is set to either RUN or RUN-P, and that a warm restart is permissible at this time.

**Monitoring functions**

The following system blocks are available for monitoring functions:

- SFB 22 STATUS
  Check partner status
- SFB 23 USTATUS
  Receive partner status
- SFC 62 CONTROL
  Check status of an SFB instance

Table 20.14 lists the parameters for the SFBs, Table 20.15 those for SFC 62.

**Table 20.14** SFB Parameters for Querying Status

| Parameter | For SFB | | Declaration | Data Type | Contents, Description |
|---|---|---|---|---|---|
| REQ | 22 | – | INPUT | BOOL | Start data exchange |
| EN_R | – | 23 | INPUT | BOOL | Ready to receive |
| ID | 22 | 23 | INPUT | WORD | Connection ID |
| NDR | 22 | 23 | OUTPUT | BOOL | New data fetched |
| ERROR | 22 | 23 | OUTPUT | BOOL | Error occurred |
| STATUS | 22 | 23 | OUTPUT | WORD | Job status |
| PHYS | 22 | 23 | IN_OUT | ANY | Physical status |
| LOG | 22 | 23 | IN_OUT | ANY | Logical status |
| LOCAL | 22 | 23 | IN_OUT | ANY | Status of an S7 CPU as partner |

**Table 20.15** Parameters for SFC 62 CONTROL

| Parameter | Declaration | Data Type | Contents, Description |
|---|---|---|---|
| EN_R | INPUT | BOOL | Ready to receive |
| I_DB | INPUT | BLOCK_DB | Instance data block |
| OFFSET | INPUT | WORD | Number of the local instance |
| RET_VAL | OUTPUT | INT | Error information |
| ERROR | OUTPUT | BOOL | Error detected |
| STATUS | OUTPUT | WORD | Status word |
| I_TYP | OUTPUT | BYTE | Block type identifier |
| I_STATE | OUTPUT | BYTE | Current status identifier |
| I_CONN | OUTPUT | BOOL | Connection status ("1" = connection exists) |
| I_STATUS | OUTPUT | WORD | STATUS parameter for SFB instance |

**Table 20.16** Start Information for OB 1

| Name | Data Type | Description | Contents |
|---|---|---|---|
| OB1_EV_CLASS | BYTE | Event class | B#16#11 = Call standard OB |
| OB1_STRT_INFO | BYTE | Start information | B#16#01 = 1st cycle after complete restart<br>B#16#02 = 1st cycle after warm restart<br>B#16#03 = Every other cycle |
| OB1_PRIORITY | BYTE | Priority | B#16#01 |
| OB1_OB_NUMBR | BYTE | OB Number | B#16#01 |
| OB1_RESERVED_1 | BYTE | Reserved | – |
| OB1_RESERVED_2 | BYTE | Reserved | – |
| OB1_PREV_CYCLE | INT | Previous scan cycle time | in ms |
| OB1_MIN_CYCLE | INT | Minimum scan cycle time | in ms |
| OB1_MAX_CYCLE | INT | Maximum scan cycle time | in ms |
| OB1_DATE_TIME | DT | Event occurrence | cyclic |

A positive edge at the REQ (request) parameter starts the data exchange. A "1" in the EN_R (enable receive) parameter signals readiness to receive. Enter in the ID parameter the connection ID which STEP 7 entered in the connection table.

With a "1" in the NDR parameter, the block signals that the job terminated without error. An error, if any, is flagged with a "1" in the ERROR parameter. A value other than zero in the STATUS parameter is indicative of either a warning (ERROR = "0") or an error (ERROR = "1"). You must evaluate the NDR, ERROR and STATUS parameters after *every* block call.

- SFB 22 STATUS fetches the status of the partner CPU and displays it in the PHYS (physical status), LOG (logical status) and LOCAL (operating status if the partner is an S7 CPU) parameters.
- SFB 23 USTATUS receives the status of the partner, which it sends, unbidden, in the event of a change (can be configured in the connection table; select connection, EDIT → OBJECT PROPERTIES, "Send operating mode messages"). The device status is displayed in the PHYS (physical status), LOG (logical status) and LOCAL (operating status if the partner is an S7 CPU) parameters.
- SFC 62 CONTROL determines the status of an SFB instance and the associated connection in the local controller. Enter the SFB's instance data block in the I_DB parameter. If the SFB is called as local instance, specify the number of the local instance in the OFFSET parameter (zero when no local instance, 1 for the first local instance, 2 for the second, and so on).

## 20.5 Start Information

The CPU's operating system forwards start information to organization block OB 1, as it does to every organization block, in the first 20 bytes of temporary local data. You can generate the declaration for the start information yourself or you can use information from standard library "Standard Library V3.x" under "Organization Blocks".

### 20.5.1 Start Information for OB 1

Table 20.16 shows this start information, the default symbolic designation, and the data types. You can change the designation at any time and choose names more acceptable to you. Even if you don't use the start information, you must reserve the first 20 bytes of temporary local data for this purpose (for instance in the form of a 20-byte array).

In SIMATIC S7, all event messages have a fixed structure which is specified by the event class. The start information for OB 1, for instance, reports event B#16#11 as a standard OB call. From the contents of the next byte you can tell whether the main program is in the first cycle

**Table 20.17** Parameters for SFC 6 RD_SINFO

| Parameter | Declaration | Data Type | Contents, Description |
|---|---|---|---|
| RET_VAL | OUTPUT | INT | Error information |
| TOP_SI | OUTPUT | STRUCT | Start information for the current OB |
| START_UP_SI | OUTPUT | STRUCT | Start information for the last OB started |

after power-up and is therefore calling, for instance, initialization routines in the cyclic program.

The priority and OB number of the main program are fixed. With three INT values, the start information provides information on the cycle time of the last scan cycle and on the minimum and maximum cycle times since the last power-up. The last value, in DATE_AND_TIME format, indicates when the priority control program received the event for calling OB 1.

Note that direct reading of the start information for an organization block is possible only in that organization block because that information consists of temporary local data. If you require the start information in blocks which lie on deeper levels, call system function SFC RD_SINFO at the relevant location in the program.

### 20.5.2 Reading Out Start Information

System function SFC 6 RD_SINFO makes the start information on the current organization block (that is, the OB at the top of the call tree) and on the start-up OB last executed available to you even at a deeper call level (Table 20.17). SFC 6 RD_SINFO can not only be called at any location in the main program but in every priority class, even in an error organization block or in the start-up routine.

Output parameter TOP_SI contains the first 12 bytes of start information on the current OB, output parameter START_UP_SI the first 12 bytes of start information on the last start-up OB executed. There is no time stamp in either case. If SFC 6 is called in the start-up routine, both parameters contain the same start information.

# 21 Interrupt Handling

Interrupt handling is always *event-driven*. When such an event occurs, the operating system interrupts scanning of the main program and calls the routine allocated to this particular event. When this routine has executed, the operating system resumes scanning of the main program at the point of interruption. Such an interruption can take place after every operation (statement).

Applicable events may be interrupts and errors. The order in which virtually simultaneous interrupt events are handled is regulated by a priority scheduler. Each event has a particular servicing priority. In S7-300 systems, the priorities are fixed; in S7-400 systems, you yourself can determine the priority class for an interrupt. Several interrupt events can be combined into priority classes.

Every routine associated with an interrupt event is written in an organization block in which additional blocks can be called. A higher-priority event interrupts execution of the routine in an organization block with a lower priority. You can affect the interruption of a program by high-priority events using system functions.

## 21.1 General Remarks

SIMATIC S7 provides the following interrupt events (interrupts):

- Hardware interrupt
  Interrupt from a module, either via an input derived from a process signal or generated on the module itself
- Watchdog interrupt
  An interrupt generated by the operating system at periodic intervals
- Time-of-day interrupt
  An interrupt generated by the operating system at a specific time of day, either once only or periodically
- Time-delay interrupt
  An interrupt generated after a specific amount of time has passed; a system function call determines the instant at which this time period begins
- Multiprocessor interrupt
  An interrupt generated by another CPU in a multiprocessor network

Other interrupt events are the synchronous errors which may occur in conjunction with program scanning and the asynchronous errors, such as diagnostic interrupts. The handling of these events is discussed in Chapter 23, "Error Handling".

### Priorities

An event with a higher priority interrupts a program being processed with lower priority because of another event. The main program has the lowest priority (priority class 1), asynchronous errors the highest (priority class 26), apart from the start-up routine. All other events are in the intervening priority classes. In S7-300 systems, the priorities are fixed; in S7-400 systems, you can change the priorities by parameterizing the CPU accordingly.

An overview of all priority classes, together with the default organization blocks for each, is presented in Section 3.1.2, "Priority Classes".

### Disabling interrupts

The organization blocks for event-driven program scanning can be disabled and enabled with system functions SFC 39 DIS_IRT and SFC 40 EN_IRT and delayed and enabled with SFC 41 DIS_AIRT and SFC 42 EN_AIRT (see Section 21.7, "Interrupt Handling").

### Current signal states

When an event-driven program is called, the process images are not automatically updated.

If you want to work with the input signal states current at the time of the call (and not with the signal states from the start of the cyclic program), you must scan the input modules directly (with Load statements via peripheral input area PI) or, in S7-400 systems, you must define a subprocess image and have it updated at the beginning of the event-driven program.

If you want to set or reset outputs immediately in the event-driven program (instead of at the end of the main program), you must address the output modules directly (with Transfer statements via peripheral output area PQ) or, in S7-400 systems, you must define a subprocess image and have it updated at the end of the event-driven program. Direct addressing of the peripheral output area automatically updates the process-image output table.

Note that updating of the inputs and outputs in the event-driven program is asynchronous to scanning of the main program in both cases. For example, this may result in an input or output having a different signal state in the main program prior to interrupt servicing than after interrupt servicing, which in turn would mean that you might have different input or output signal states in a single main program scan cycle.

### Start Information, temporary local data

Table 21.1 provides an overview of the start information for the event-driven organization blocks. In S7-300 systems, the available temporary local data have a fixed length of 256 bytes. In S7-400 systems, you can specify the length per priority class by parameterizing the CPU accordingly (parameter block "Local Data"), whereby the total may not exceed a CPU-specific maximum. Note that the minimum number of bytes for temporary local data for the priority class used must be 20 bytes so as to be able to accommodate the start information. Specify zero for unused priority classes.

## 21.2 Hardware Interrupts

Hardware interrupts are used to enable the immediate detection in the user program of events in the controlled process, making it possible to respond with an appropriate interrupt handling routine. STEP 7 provides organization blocks OB 40 to OB 47 for servicing hardware interrupts; which of these eight organization blocks are actually available, however, depends on the CPU.

Hardware interrupt handling is programmed in the hardware configuration data. With system functions SFC 55 WR_PARM, SFC 56 WR_DPARM and SFC 57 PARM_MOD, you can

**Table 21.1** Start Information for Interrupt Organization Blocks

| Byte | Multiprocessor Interrupt OB 60 | Hardware Interrupts OB 40 to OB 47 | Watchdog Interrupts OB 30 to OB 38 | Time-Delay Interrupts OB 20 to OB 23 | Time-of-Day Interrupts OB 10 to OB 17 |
|---|---|---|---|---|---|
| 0 | Event class | Event class | Event class | Event class | Event class |
| 1 | Start event | Start event | Start event | Start event | Start event |
| 2 | Priority class | Priority class | Priority class | Priority class | Priority class |
| 3 | OB number | OB number | OB number | OB number | OB number |
| 4 | – | – | – | – | – |
| 5 | – | Address identifier | – | – | – |
| 6..7 | Job identifier (INT) | Module start address (WORD) | Phase offset in ms (WORD) | Job identifier (WORD) | Interval (WORD) |
| 8..9 | – | Hardware interrupt | – | Expired delay | – |
| 10..11 | – | information (DWORD) | Time cycle in ms (INT) | (TIME) | – |
| 12..19 | Event instant (DT) | Event instant (DT) | Event instant ) (DT | Event instant (DT) | Event instant (DT) |

(re)parameterize the modules with hardware interrupt capability even in RUN mode.

### 21.2.1 Generating a Hardware Interrupt

A hardware interrupt is generated on the modules with this capability. This could, for example, be a digital input module that detects a signal from the process or a function module that generates a hardware interrupt because of an activity taking place on the module.

By default, hardware interrupts are disabled. A parameter is used to enable servicing of a hardware interrupt (static parameter), and you can specify whether the hardware interrupt should be generated for a coming event, a leaving event, or both (dynamic parameter). Dynamic parameters are parameters which you can modify at runtime using SFCs.

The hardware interrupt is acknowledged on the module when the organization block containing the service routine for that interrupt has finished executing.

#### Resolution on the S7-300

If an event occurs during execution of a hardware interrupt OB which itself would trigger generation of the same hardware interrupt, that hardware interrupt will be lost when the event that triggered it is no longer present following acknowledgment. It makes no difference whether the event comes from the module whose hardware interrupt is currently being serviced or from another module.

A diagnostic interrupt can be generated while a hardware interrupt is being serviced. If another hardware interrupt occurs on the same channel between the time the first hardware interrupt was generated and the time that interrupt was acknowledged, the loss of the latter interrupt is reported via a diagnostic interrupt to system diagnostics.

#### Resolution on the S7-400

If during execution of a hardware interrupt OB an event occurs on the same channel on the same module which would trigger the same hardware interrupt, that interrupt is lost. If the event occurs on another channel on the same module or on another module, the operating system restarts the OB as soon as it has finished executing.

### 21.2.2 Servicing Hardware Interrupts

#### Querying interrupt information

The start address of the module that triggered the hardware interrupt is in bytes 6 and 7 of the hardware interrupt OB's start information. If this address is an input address, byte 5 of the start information contains B#16#54; otherwise it contains B#16#55. If the module in question is a digital input module, bytes 8 to 11 contain the status of the inputs; for any other type of module, these bytes contain the interrupt status of the module.

#### Interrupt handling in the start-up routine

In the start-up routine, the modules do not generate hardware interrupts. Interrupt handling begins with the transition to RUN mode. Any hardware interrupts pending at the time of the transition are lost.

#### Error handling

If a hardware interrupt is generated for which there is no hardware interrupt OB in the user program, the operating system calls OB 85 (program execution error). The hardware interrupt is acknowledged. If OB 85 has not been programmed, the CPU goes to STOP.

Hardware interrupts deselected when the CPU was parameterized cannot be serviced, even when the OBs for these interrupts have been programmed. The CPU goes to STOP.

#### Disabling, delaying and enabling

Calling of the hardware interrupt OBs can be disabled and enabled with system functions SFC 39 DIS_IRT and SFC 40 EN_IRT, and delayed and enabled with SFC 41 DIS_AIRT and SFC 42 EN_AIRT.

### 21.2.3 Configuring Hardware Interrupts with STEP 7

Hardware interrupts are programmed in the hardware configuration data. Open the selected CPU with EDIT → OBJECT PROPERTIES and choose the "Interrupts" tab in the dialog box.

In S7-300 systems, the default priority for OB 40 is 16, and cannot be changed. In S7-400 systems, you can choose a priority between 2 and 24 for every possible OB (on a CPU-specific basis); priority 0 deselects execution of an OB. You should never assign the same priority twice because interrupts can be lost when more than 12 interrupt events with the same priority occur simultaneously.

You must also enable the triggering of hardware interrupts on the respective modules. To this purpose, these modules are parameterized much the same as the CPU.

When it saves the hardware configuration, STEP 7 writes the compiled data to the *System Data* object in off-line user program *Blocks;* from here, you can load the parameterization data into the CPU while the CPU is in STOP mode. The parameterization data for the CPU go into force immediately following loading; the parameter assignment data for the modules take effect after the next start-up.

## 21.3 Watchdog Interrupts

A watchdog interrupt is an interrupt which is generated at periodic intervals and which initiates execution of a watchdog interrupt OB. A watchdog interrupt allows you to execute a particular program periodically, independently of the processing time of the cyclic program.

In STEP 7, organization blocks OB 30 to OB 38 have been set aside for watchdog interrupts; which of these nine organization blocks are actually available depends on the CPU used.

Watchdog interrupt handling is set in the hardware configuration data when the CPU is parameterized.

### 21.3.1 Handling Watchdog Interrupts

#### Triggering watchdog interrupts in an S7-300

In an S7-300, the organization block for servicing watchdog interrupts is OB 35, which has the priority 12. You can set the interval in the range from 1 millisecond to 1 minute, in 1-millisecond increments, by parameterizing the CPU accordingly.

#### Triggering watchdog interrupts in an S7-400

You define a watchdog interrupt when you parameterize the CPU. A watchdog interrupt has three parameters: the interval, the phase offset, and the priority. You can set all three. Specifiable values for interval and phase offset are from 1 millisecond to 1 minute, in 1-millisecond increments; the priority may be set to a value between 2 and 24 or to zero, depending on the CPU (zero means the watchdog interrupt is not active).

STEP 7 provides the organization blocks listed in Table 21.2, in their maximum configurations.

#### Phase offset

The phase offset can be used to stagger the execution of watchdog interrupt handling routines despite the fact that these routines are timed to a multiple of the same interval. Use of the phase offset would put less of a time load on the main program.

The start time of the time interval and the phase offset is the instant of transition from STARTUP to RUN. The call instant for a watchdog interrupt OB is thus the time interval plus the phase offset. Figure 21.1 shows an example of this. No phase offset is set for time interval 1, time interval 2 is twice as long as time interval 1. Because of time interval 2's phase offset, the OBs for time interval 2 and those for time interval 1 are not called simultaneously. The lower-priority OB thus need not wait, and can precisely maintain its time interval.

**Table 21.2** Defaults for Watchdog Interrupts

| OB | Time Interval | Phase | Priority |
|---|---|---|---|
| 30 | 5 s | 0 ms | 7 |
| 31 | 2 s | 0 ms | 8 |
| 32 | 1 s | 0 ms | 9 |
| 33 | 500 ms | 0 ms | 10 |
| 34 | 200 ms | 0 ms | 11 |
| 35 | 100 ms | 0 ms | 12 |
| 36 | 50 ms | 0 ms | 13 |
| 37 | 20 ms | 0 ms | 14 |
| 38 | 10 ms | 0 ms | 15 |

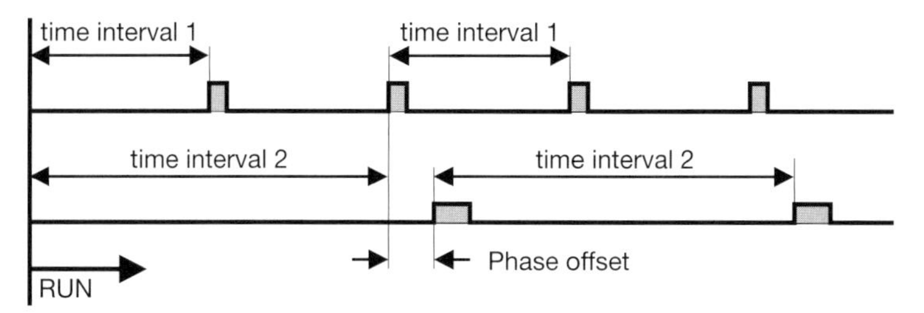


**Figure 21.1** Example of Phase Offset for Watchdog Interrupts

### Performance characteristics during startup

Watchdog interrupts cannot be serviced in the start-up OB. The time intervals do not begin until a transition is made to RUN mode.

### Performance characteristics on error

When the same watchdog interrupt is generated again while the associated watchdog interrupt handling OB is still executing, the operating system calls OB 80 (timing error). If OB 80 has not been programmed, the CPU goes to STOP.

The operating system saves the watchdog interrupt that was not serviced, servicing it at the next opportunity. Only one unserviced watchdog interrupt is saved per priority class, regardless of how many unserviced watchdog interrupts accumulate.

Watchdog interrupts that were deselected when the CPU was parameterized cannot be serviced, even when the corresponding OB is available. The CPU goes to STOP in this case.

### Disabling, delaying and enabling

Calling of the watchdog interrupt OBs can be disabled and enabled with system functions SFC 39 DIS_IRT and SFC 40 EN_IRT and delayed and enabled with SFC 41 DIS_AIRT and SFC 42 EN_AIRT.

### 21.3.2 Configuring Watchdog Interrupts with STEP 7

Watchdog interrupts are configured via the hardware configuration data. Simply open the selected CPU with EDIT → OBJECT PROPERTIES and choose the "Cyclic Interrupt" tab from the dialog box.

In S7-300 controllers, the processing priority is permanently set to 12. In S7-400 controllers, you may set a priority between 2 and 24 for each possible OB (CPU-specific); priority 0 deselects the OB to which it is assigned. You should not assign a priority more than once, as interrupts might be lost if more than 12 interrupt events with the same priority occur simultaneously.

The interval for each OB is selected under "Execution", the delayed call instant under "Phase Offset".

When it saves the hardware configuration, STEP 7 writes the compiled data to the *System Data* object in the off-line user program *Blocks*. From here, you can load the parameter assignment data into the CPU while the CPU is at STOP; the data take effect immediately.

## 21.4 Time-of-Day Interrupts

Time-of-day interrupts are used when you want to run a program at a particular time, either once only or periodically, for instance daily. In STEP 7, organization blocks OB 10 to OB 17 are provided for servicing time-of-day interrupts; which of these eight organization blocks are actually available depends on the CPU used.

You can configure the time-of-day interrupts in the hardware configuration data or control them at runtime via the program using system functions. The prerequisite for proper handling of the time-of-day interrupts is a correctly set real-time clock on the CPU.

### 21.4.1 Handling Time-of-Day Interrupts

#### General remarks

To start a time-of-day interrupt, you must first set the start time, then activate the interrupt. You can perform the two activities separately via the hardware configuration data or using SFCs. Note that when activated via the hardware configuration data, the time-of-day interrupt is started automatically following parameterization of the CPU.

You can start a time-of-day interrupt in two ways:

- Single-shot: the relevant OB is called once only at the specified time, or
- Periodically: depending on the parameter assignments, the relevant OB is started every minute, hourly, daily, weekly, monthly or yearly.

Following a single-shot time-of-day interrupt OB call, the time-of-day interrupt is canceled. You can also cancel a time-of-day interrupt with SFC 29 CAN_TINT.

If you want to once again use a canceled time-of-day interrupt, you must set the start time again, then reactivate the interrupt.

You can query the status of a time-of-day interrupt with SFC 31 QRY_TINT.

#### Performance characteristics during startup

During a complete restart, the operating system clears all settings made with SFCs. Settings made via the hardware configuration data are retained. On a warm restart, the CPU resumes servicing of the time-of-day interrupts in the first complete scan cycle of the main program.

You can query the status of the time-of-day interrupts in the start-up OB by calling SFC 31, and subsequently cancel or re-set and reactivate the interrupts. The time-of-day interrupts are serviced only in RUN mode.

#### Performance characteristics on error

If a time-of-day interrupt OB is called but was not programmed, the operating system calls OB 85 (program execution error). If OB 85 was not programmed, the CPU goes to STOP.

Time-of-day interrupts that were deselected when the CPU was parameterized cannot be serviced, even when the relevant OB is available. The CPU goes to STOP.

If you activate a time-of-day interrupt on a single-shot basis, and if the start time has already passed (from the real-time clock's point of view), the operating system calls OB 80 (timing error). If OB 80 is not available, the CPU goes to STOP.

If you activate a time-of-day interrupt on a periodic basis, and if the start time has already passed (from the real-time clock's point of view), the time-of-day interrupt OB is executed the next time that time period comes due.

If you set the real-time clock ahead, whether for the purpose of correction or synchronization, thus skipping over the start time for the time-of-day interrupt, the operating system calls OB 80 (timing error). The time-of-day interrupt OB is then executed precisely once.

If you set the real-time clock back, whether for the purpose of correction or synchronization, an activated time-of-day interrupt OB will no longer be executed at the instants which are already past.

If a time-of-day interrupt OB is still executing when the next (periodic) call occurs, the operating system invokes OB 80 (timing error). When OB 80 and the time-of-day interrupt OB have executed, the time-of-day interrupt OB is restarted.

Periodic time-of-day interrupts must correspond to an actual date. For example, monthly repetitions on the 31st day of each month will be carried out only in months having 31 days.

#### Disabling, delaying and enabling

Time-of-day interrupt OB calls can be disabled and enabled with SFC 39 DIS_IRT and SFC 40 EN_IRT, and delayed and enabled with SFC 41 DIS_AIRT and SFC 42 EN_AIRT.

### 21.4.2 Configuring Time-of-Day Interrupts with STEP 7

The time-of-day interrupts are configured via the hardware configuration data. Open the selected CPU with EDIT → OBJECT PROPERTIES and choose the "Time-Of-Day Interrupts" tab from the dialog box.

In S7-300 controllers, the processing priority is permanently set to 2. In S7-400 controllers, you can set a priority between 2 and 24, depending on the CPU, for each possible OB; priority 0 deselects an OB. You should not assign a priority more than once, as interrupts might be lost when more than 12 interrupt events with the same priority occur simultaneously.

The "Active" option activates automatic starting of the time-of-day interrupt. The "Execution" option screens a list which allows you to choose whether you want the OB to execute on a single-shot basis or at specific intervals. The final parameter is the start time (date and time).

When it saves the hardware configuration, STEP 7 writes the compiled data to the *System Data* object in the off-line user program *Blocks*. From here, you can load the parameter assignment data into the CPU while the CPU is at STOP; these data then go into force immediately.

### 21.4.3 System Functions for Time-of-Day Interrupts

The following system functions can be used for time-of-day interrupt control:

- SFC 28 SET_TINT
  Set time-of-day interrupt
- SFC 29 CAN_TINT
  Cancel time-of-day interrupt
- SFC 30 ACT_TINT
  Activate time-of-day interrupt
- SFC 31 QRY_TINT
  Query time-of-day interrupt

The parameters for these system functions are listed in Table 21.3.

- Set time-of-day interrupt
  You determine the start time for a time-of-day interrupt by calling system function SFC 28 SET_TINT. SFC 28 sets only the start time; to start the time-of-day interrupt OB, you must activate the time-of-day interrupt with SFC 30 ACT_TINT. Specify the start time in the SDT parameter in the format DATE_AND_TIME, for instance DT#1997-06-30-08:30. The operating system ignores seconds and milliseconds and sets these values to zero. Setting the start time will overwrite the old start time value, if any. An active time-of-day interrupt is canceled, that is, it must be reactivated.

**Table 21.3** SFC Parameters for Time-of-Day Interrupts

| SFC | Parameter | Declaration | Data Type | Contents, Description |
|---|---|---|---|---|
| 28 | OB_NR | INPUT | INT | Number of the OB to be called at the specified time on a single-shot basis or periodically |
| | SDT | INPUT | DT | Start date and start time in the format DATE_AND_TIME |
| | PERIOD | INPUT | WORD | Period on which start time is based:<br>W#16#0000 = Single-shot<br>W#16#0201 = Every minute<br>W#16#0401 = Hourly<br>W#16#1001 = Daily<br>W#16#1201 = Weekly<br>W#16#1401 = Monthly<br>W#16#1801 = Yearly |
| | RET_VAL | OUTPUT | INT | Error information |
| 29 | OB_NR | INPUT | INT | Number of the OB whose start time is to be deleted |
| | RET_VAL | OUTPUT | INT | Error information |
| 30 | OB_NR | INPUT | INT | Number of the OB to be activated |
| | RET_VAL | OUTPUT | INT | Error information |
| 31 | OB_NR | INPUT | INT | Number of the OB whose status is to be queried |
| | RET_VAL | OUTPUT | INT | Error information |
| | STATUS | OUTPUT | WORD | Status of the time-of-day interrupt |

- Activate time-of-day interrupt
  A time-of-day interrupt is activated by calling system function SFC 30 ACT_TINT. When a TOD interrupt is activated, it is assumed that a time has been set for the interrupt. If, in the case of a single-shot interrupt, the start time is already past, SFC 30 reports an error. In the case of a periodic start, the operating system calls the relevant OB at the next applicable time. Once a single-shot time-of-day interrupt has been serviced, it is, for all practical purposes, canceled. You can re-set and reactivate it (for a different start time) if desired.

- Cancel time-of-day interrupt
  You can delete a start time, thus deactivating the time-of-day interrupt, with system function SFC 29 CAN_TINT. The respective OB is no longer called. If you want to use this same time-of-day interrupt again, you must first set the start time, then activate the interrupt.

- Query time-of-day interrupt
  You can query the status of a time-of-day interrupt by calling system function SFC 31 QRY_TINT. The required information is returned in the STATUS parameter (see Table 21.4).

## 21.5 Time-Delay Interrupts

A time-delay interrupt allows you to implement a delay timer independently of the standard timers. In STEP 7, organization blocks OB 20 to OB 23 are set aside for time-delay interrupts; which of these four organization blocks are actually available depends on the CPU used.

**Table 21.4**
STATUS Parameter for SFC 31 QRY_TINT

| Bit | Meaning when bit = "1" |
|---|---|
| 0 | TOD interrupt disabled by operating system |
| 1 | New TOD interrupt rejected |
| 2 | TOD interrupt not activated and not expired |
| 3 | – |
| 4 | TOD interrupt OB loaded |
| 5 | No disable |
| 6 ... | – |

The priorities for time-delay interrupt OBs are programmed in the hardware configuration data; system functions are used for control purposes.

### 21.5.1 Handling Time-Delay Interrupts

#### General remarks

A time-delay interrupt is started by calling SFC 32 SRT_DINT; this system function also passes the delay interval and the number of the selected organization block to the operating system. When the delay interval has expired, the OB is called.

You can cancel servicing of a time-delay interrupt, in which case the associated OB will no longer be called.

You can query the status of a time-delay interrupt with SFC 34 QRY_DINT.

#### Performance characteristics during startup

On a complete restart, the operating system deletes all programmed settings for time-delay interrupts. On a warm restart, the settings are retained until processed in RUN mode, whereby the "residual cycle" is counted as part of the start-up routine.

You can start a time-delay interrupt in the start-up routine by calling SFC 32. When the delay interval has expired, the CPU must be in RUN mode in order to be able to execute the relevant organization block. If this is not the case, the CPU waits to call the organization block until the start-up routine has terminated, then calls the time-delay interrupt OB before the first network in the main program.

#### Performance characteristics on error

If no time-delay interrupt OB has been programmed, the operating system calls OB 85 (program execution error). If there is no OB 85 in the user program, the CPU goes to STOP.

If the delay interval has expired and the associated OB is still executing, the operating system calls OB 80 (timing error) or goes to STOP if there is no OB 80 in the user program.

Time-delay interrupts which were deselected during CPU parameterization cannot be serviced, even when the respective OB has been programmed. The CPU goes to STOP.

### Disabling, delaying and enabling

The time-delay interrupt OBs can be disabled and enabled with system functions SFC 39 DIS_IRT and SFC 40 EN_IRT, and delayed and enabled with SFC 41 DIS_AIRT and SFC 42 EN_AIRT.

## 21.5.2 Configuring Time-Delay Interrupts with STEP 7

Time-delay interrupts are configured in the hardware configuration data. Simply open the selected CPU with EDIT → OBJECT PROPERTIES and choose the "Interrupts" tab from the dialog box.

In S7-300 controllers, the priority is permanently preset to 3. In S7-400 controllers, you can choose a priority between 2 and 24, depending on the CPU, for each possible OB; choose priority 0 to deselect an OB. You should not assign a priority more than once, as interrupts could be lost if more than 12 interrupt events with the same priority occur simultaneously.

When it saves the hardware configuration, STEP 7 writes the compiled data to the *System Data* object in the off-line user program *Blocks*. From here, you can transfer the parameter assignment data while the CPU is at STOP; the data take effect immediately.

## 21.5.3 System Functions for Time-Delay Interrupts

A time-delay interrupt can be controlled with the following system functions:

- SFC 32 SRT_DINT
  Start time-delay interrupt
- SFC 33 CAN_DINT
  Cancel time-delay interrupt
- SFC 34 QRY_DINT
  Query time-delay interrupt

The parameters for these system functions are listed in Table 21.5.

- Start time-delay interrupt
  A time-delay interrupt is started by calling system function SFC 32 SRT_DINT. The SFC call is also the start time for the programmed delay interval. When the delay interval has expired, the CPU calls the programmed OB and passes the time delay value and a job identifier in the start information for this OB. The job identifier is specified in the SIGN parameter for SFC 32; you can read the same value in bytes 6 and 7 of the start information for the associated time-delay interrupt OB. The time delay is set in increments of 1 ms. The accuracy of the time delay is also 1 ms. Note that execution of the time-delay interrupt OB may itself be delayed when organization blocks with higher priorities are being processed when the time-delay interrupt OB is called.

**Table 21.5** SFC Parameters for Time-Delay Interrupts

| SFC | Parameter | Declaration | Data Type | Contents, Description |
|---|---|---|---|---|
| 32 | OB_NR | INPUT | INT | Number of the OB to be called when the delay interval has expired |
| | DTIME | INPUT | TIME | Delay interval; permissible: T#1ms to T#1m |
| | SIGN | INPUT | WORD | Job identification in the respective OB's start information when the OB is called (arbitrary characters) |
| | RET_VAL | OUTPUT | INT | Error information |
| 33 | OB_NR | INPUT | INT | Number of the OB to be canceled |
| | RET_VAL | OUTPUT | INT | Error information |
| 34 | OB_NR | INPUT | INT | Number of the OB whose status is to be queried |
| | RET_VAL | OUTPUT | INT | Error information |
| | STATUS | OUTPUT | WORD | Status of the time-delay interrupt |

You can overwrite a time delay with a new value by recalling SFC 32. The new time delay goes into force with the SFC call.

- Cancel time-delay interrupt
  You can call system function SFC 33 CAN_DINT to cancel a time-delay interrupt, in which case the programmed organization block is not called.

- Query time-delay interrupt
  System function SFC 34 QRY_DINT informs you about the status of a time-delay interrupt. You select the time-delay interrupt via the OB number, and the status information is returned in the STATUS parameter (Table 21.6).

## 21.6 Multiprocessor Interrupt

The multiprocessor interrupt allows a synchronous response to an event in all CPUs in multiprocessor mode. A multiprocessor interrupt is triggered using SFC 35 MP_ALM. Organization block OB 60, which has a fixed priority of 25, is the OB used to service a multiprocessor interrupt.

### General remarks

An SFC 35 MP_ALM call initiates execution of the multiprocessor interrupt OB. If the CPU is in single-processor mode, OB 60 is started immediately. In multiprocessor mode, OB 60 is started simultaneously on all participating CPUs, that is to say, even the CPU in which SFC 35 was called waits before calling OB 60 until all the other CPUs have indicated that they are ready.

**Table 21.6**
STATUS Parameter in SFC 34 QRY_DINT

| Bit | Meaning when bit = "1" |
|---|---|
| 0 | Time-delay interrupt disabled by operating system |
| 1 | New time-delay interrupt rejected |
| 2 | Time-delay interrupt activated and not expired |
| 3 | – |
| 4 | Time-delay OB loaded |
| 5 | No disable |
| 6 ... | – |

The multiprocessor interrupt is not programmed in the hardware configuration data; it is already present in every CPU with multicomputing capability. Despite this fact, however, a sufficient number of local data bytes (at least 20) must still be reserved in the CPU's "Local Data" tab under priority class 25.

### Performance characteristics during startup

The multiprocessor interrupt is triggered only in RUN mode. An SFC 35 call in the start-up routine terminates after returning error 32 929 (W#16#80A1) as function value.

### Performance characteristics on error

If OB 60 is still in progress when SFC 35 is recalled, the system function returns error code 32 928 (W#16#80A0) as function value. OB 60 is not started in any of the CPUs.

The unavailability of OB 60 in one of the CPUs at the time it is called or the disabling or delaying of its execution by system functions has no effect, nor does SFC 35 report an error.

### Disabling, delaying and enabling

The multiprocessor OB can be disabled and enabled with system functions SFC 39 DIS_IRT and SFC 40 EN_IRT, and delayed and enabled with SFC 41 DIS_AIRT and SFC 42 EN_AIRT.

### System functions for the multiprocessor interrupt

A multiprocessor interrupt is triggered with system function SFC 35 MP_ALM. Its parameters are listed in Table 21.7.

The JOB parameter allows you to forward a job identifier. The same value can be read in bytes 6 and 7 of OB 60's start information in all CPUs.

## 21.7 Handling Interrupts

The following system functions are available for handling interrupts and asynchronous errors:

- SFC 39 DIS_IRT
  Disable interrupts

- SFC 40 EN_IRT
  Enable disabled interrupts

**Table 21.7** Parameters for SFC 35 MP_ALM

| Parameter | Declaration | Data Type | Contents, Description |
|---|---|---|---|
| JOB | INPUT | BYTE | Job identification in the range B#16#00 to B#16#0F |
| RET_VAL | OUTPUT | INT | Error information |

- SFC 41 DIS_AIRT
  Delay interrupts
- SFC 42 EN_AIRT
  Enable delayed interrupts

These system functions affect all interrupts and all asynchronous errors. Table 21.8 lists the parameters for these system functions. System functions SFC 36 to SFC 38 are provided for handling synchronous errors.

### Disabling interrupts

System function SFC 39 DIS_IRT disables servicing of new interrupts and asynchronous errors. All new interrupts and asynchronous errors are rejected. If an interrupt or asynchronous error occurs following a Disable, the organization block is not executed; if the OB does not exist, the CPU does *not* go to STOP.

The Disable remains in force for all priority classes until it is revoked with SFC 40 EN_IRT. After a complete restart, all interrupts and asynchronous errors are enabled.

The MODE and OB_NR parameters are used to specify which interrupts and asynchronous errors are to be disabled. MODE = B#16#00 disables all interrupts and asynchronous errors. MODE = B#16#01 disables an interrupt class whose first OB number is specified in the OB_NR parameter. For example, MODE = B#16#01 and OB_NR = 40 disables all hardware interrupts; OB = 80 would disable all asynchronous errors. MODE = B#16#02 disables the interrupt or asynchronous error whose OB number you entered in the OB_NR parameter.

Regardless of a Disable, the operating system enters each new interrupt or asynchronous error in the diagnostic buffer.

### Enabling disabled interrupts

System function SFC 40 EN_IRT enables the interrupts and asynchronous errors disabled with SFC 39 DIS_IRT. An interrupt or asynchronous error occurring after the Enable will be serviced by the associated organization block; if that organization block is not in the user program, the CPU goes to STOP (except in the case of OB 81, the organization block for power supply errors).

The MODE and OB_NR parameters specify which interrupts and asynchronous errors are to be enabled. MODE = B#16#00 enables all interrupts and asynchronous errors. MODE = B#16#01 enables an interrupt class whose first OB number is specified in the OB_NR parame-

**Table 20.7** SFC Parameters for Interrupt Handling

| SFC | Parameter | Declaration | Data Type | Contents, Description |
|---|---|---|---|---|
| 39 | MODE | INPUT | BYTE | Disable mode (see text) |
| | OB_NR | INPUT | INT | OB number (see text) |
| | RET_VAL | OUTPUT | INT | Error information |
| 40 | MODE | INPUT | BYTE | Enable mode (see text) |
| | OB_NR | INPUT | INT | OB number (see text) |
| | RET_VAL | OUTPUT | INT | Error information |
| 41 | RET_VAL | OUTPUT | INT | (New) number of delays |
| 42 | RET_VAL | OUTPUT | INT | Number of delays remaining |

ter. MODE = B#16#02 enables the interrupt or asynchronous error whose OB number you entered in the OB_NR parameter.

### Delaying Interrupts

System function SFC 41 DIS_AIRT delays the servicing of higher-priority new interrupts and asynchronous errors. Delay means that the operating system saves the interrupts and asynchronous errors which occurred during the delay and services them when the delay interval has expired. Once SFC 41 has been called, the program in the current organization block (in the current priority class) will not be interrupted by a higher-priority interrupt; no interrupts or asynchronous errors are lost.

A delay remains in force until the current OB has terminated its execution or until SFC 42 EN_AIRT is called.

You can call SFC 41 several times in succession. The RET_VAL parameter shows the number of calls. You must call SFC 42 precisely the same number of times as SFC 41 in order to reenable the interrupts and asynchronous errors.

### Enabling delayed interrupts

System function SFC 42 EN_AIRT reenables the interrupts and asynchronous errors delayed with SFC 41. You must call SFC 42 precisely the same number of times as you called SFC 41 (in the current OB). The RET_VAL parameter shows the number of delays still in force; if RET_VAL is = 0, the interrupts and asynchronous errors have been reenabled.

If you call SFC 42 without having first called SFC 41, RET_VAL contains the value 32896 (W#16#8080).

# 22 Start-Up Characteristics

## 22.1 General Remarks

### 22.1.1 Operating Modes

Before the CPU begins processing the main program following power-up, it executes a start-up routine. STARTUP is one of the CPU's operating modes, as is STOP or RUN. This chapter describes the CPU's activities on a transition from and to STARTUP and in the start-up routine itself.

Following power-up ①, the CPU is in the STOP mode (Figure 22.1). If the keyswitch on the CPU's front panel is at RUN or RUN-P, the CPU switches to STARTUP mode ②, then to RUN mode ③. If an "unrecoverable" error occurs while the CPU is in STARTUP or RUN mode or if you position the keyswitch to STOP, the CPU returns to the STOP mode ④ ⑤.

The user program is tested in "single-step" operation in the HOLD mode (only S7-400 and then only STL). You can switch to this mode from both RUN and STARTUP, and return to the original mode when you abort the test ⑥ ⑦. You can also set the CPU to the STOP mode from the HOLD mode ⑧.

When you parameterize the CPU, you can define start-up characteristics with the "Startup" tab such as the maximum permissible amount of time for the Ready signals from the modules following power-up or whether the CPU is to start up when the configuration data do not coincide with the actual configuration or in what mode the CPU startup is to be in.

SIMATIC S7 has two start-up modes, namely the *complete restart* and the *warm restart*. On a complete restart, the main program is always processed from the beginning. A warm restart resumes the main program at the point of interruption, and "finishes" the cycle. S7-300 controllers support the complete restart; S7-400 controllers support both modes.

You can scan a program on a single-shot basis in STARTUP mode. STEP 7 provides organization blocks OB 100 (complete restart) and OB 101 (warm restart) expressly for this purpose. Sample applications are the parameterization of modules unless this was already taken care of by the CPU, and the programming of defaults for your main program.

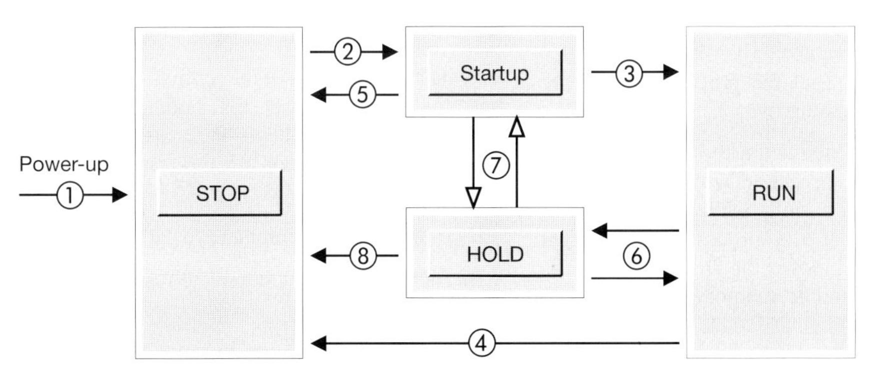


**Figure 22.1** CPU Operating Modes

### 22.1.2 HOLD Mode

In HOLD mode, the output modules are disabled. Writing to the modules affects the module memory, but does not switch the signal states "out" to the module outputs. The modules are not reenabled until you exit the HOLD mode.

In HOLD mode, everything having to do with timing is discontinued. This includes, for example, the processing of timers, clock memory and run-time meters, cycle time monitoring and minimum scan cycle time, and the servicing of time-of-day and time-delay interrupts. Exception: the real-time clock continues to function normally.

Every time the progression is made to the next statement in test mode, the timers for the duration of the single step run a little further, thus simulating a dynamic behavior similar to "normal" program scanning.

In HOLD mode, the CPU is capable of passive communication, that is, it can receive global data or take part in the unilateral exchange of data.

If the power fails while the CPU is in HOLD mode, battery-backed CPUs go to STOP on power recovery. CPUs without backup batteries execute an automatic complete restart.

### 22.1.3 Disabling the Output Modules

In the STOP and HOLD modes, modules are disabled (OD (output disable) signal). Disabled output modules output a zero signal or, if they have the capability, the replacement value. Via a variable table, you can control outputs on the modules with the "Isolate PQ" function, even in STOP mode.

During startup, the output modules remain disabled. Only when the cyclic scan begins are the output modules enabled. The signal states in a module's memory (not the process image!) are then applied to the outputs.

While the output modules are disabled, module memory can be set either with direct access (Transfer statements with address area PQ) or via the transfer of the process-image output table. If the CPU removes the Disable signal, the signal states in module memory are applied to the external outputs.

On a complete restart (OB 100), the process images and the module memory are cleared. If you want to scan inputs in OB 100, you must load the signal states from the module using direct access. You can then set the inputs (transfer them, for instance, with Load statements from address area PI to address area I), then work with the inputs. If you want certain outputs to be set on a transition from a complete restart to the cyclic program (prior to calling OB 1), you must use direct access to address the output modules. It is not enough to just set the outputs (in the process image), as the process-image output table is not transferred at the end of the complete restart routine.

On a warm restart, the "old" process-image input and process-image output tables, which were valid prior to power-down or STOP, are used in OB 101 and in the remainder of the cycle. At the end of that cycle, the process-image output table is transferred to module memory (but not yet switched through to the external outputs, since the output modules are still disabled). You now have the option of parameterizing the CPU to clear the process-image output table and the module memory at the end of the warm restart. Before switching to OB 1, the CPU revokes the Disable signal so that the signal states in the module memory are applied to the external outputs.

### 22.1.4 Start-Up Organization Blocks

On a *complete restart,* the CPU calls organization block OB 100 on a single-shot basis prior to processing the main program. If there is no OB 100, the CPU begins cyclic program scanning immediately.

On a *warm restart,* the CPU calls organization block OB 101 on a single-shot basis before processing the main program. If there is no OB 101, the CPU begins scanning at the point of interruption.

The start information in the temporary local data has the same format for both organization blocks; Table 22.1 shows the start information for OB 100. The reason for the restart is shown in the restart request.

- B#16#81 Manual complete restart (OB 100)
- B#16#82 Automatic complete restart (OB 100)
- B#16#83 Manual warm restart (OB 101)
- B#16#84 Automatic warm restart (OB 101)

**Table 22.1** Start Information for the Start-up OBs

| Byte | Name | Data Type | Description |
|---|---|---|---|
| 0 | OB100_EV_CLASS | BYTE | Event class |
| 1 | OB100_STRTUP | BYTE | Restart request |
| 2 | OB100_PRIORITY | BYTE | Priority class |
| 3 | OB100_OB_NUMBR | BYTE | OB number |
| 4 | OB100_RESERVED_1 | BYTE | Reserved |
| 5 | OB100_RESERVED_2 | BYTE | Reserved |
| 6..7 | OB100_STOP | WORD | Number of the stop event |
| 8..11 | OB100_STRT_INFO | DWORD | Additional information on the current restart |
| 12..19 | OB100_DATE_TIME | DT | Date and time event occurred |

The number of the stop event and the additional information define the restart more precisely (tells you, for example, whether a manual complete restart was initiated via the mode switch). With this information, you can develop an appropriate event-related start-up routine.

## 22.2 Power-Up

### 22.2.1 STOP Mode

The CPU goes to STOP in the following instances:

- When the CPU is switched on
- When the mode selector is set from RUN to STOP
- When an "unrecoverable" error occurs during program scanning
- When system function SFC 46 STP is executed
- When requested by a communication function (stop request from the programming device or via communication function blocks from another CPU)

The CPU enters the reason for the STOP in the diagnostic buffer. In this mode, you can also read the CPU information with a programming device in order to localize the problem.

In STOP mode, the user program is not scanned. The CPU retrieves the settings – either the values which you entered in the hardware configuration data when you parameterized the CPU or the defaults – and sets the modules to the specified initial state.

In STOP mode, the CPU can receive global data via GD communication and carry out passive unilateral communication functions. The real-time clock keeps running.

You can parameterize the CPU in STOP mode, for instance you can also set the MPI address, transfer or modify the user program, and execute a CPU memory reset.

### 22.2.2 Memory Reset

A memory reset sets the CPU to the "initial state". You can initiate a memory reset with the mode selector or with a programming device only in STOP mode.

The CPU erases the entire user program both in work memory and in RAM load memory. System memory (for instance bit memory, timers and counters) is also erased, regardless of retentivity settings.

The CPU sets the parameters for all modules, including its own, to their default values. The MPI parameters are an exception. They are not changed so that a CPU whose memory has been reset can still be addressed on the MPI bus. A memory reset also does not affect the diagnostic buffer, the real-time clock, or the run-time meters.

If a memory card with Flash EPROM is inserted, the CPU copies the user program from the memory card to work memory. The CPU also copies any configuration data it finds on the memory card.

### 22.2.3 Retentivity

A memory area is retentive when its contents are retained even when the mains power is switched off as well as on a transition from STOP to RUN following power-up. Retentive memory areas may be those for bit memory, timers, counters and, on the S7-300, data areas. The number of bytes in which areas can be made retentive depends on the CPU. You can specify the retentive areas via the "Retentivity" tab when you parameterize the CPU.

The settings for retentivity are in the system data blocks (SDBs) in load memory, that is, on the memory card. If the memory card is a RAM card, you must operate the programmable controller with a backup battery to save the retentivity settings permanently.

When you use a backup battery, the signal states of the memory bits, timers and counters specified as being retentive are retained. The user program and the user data remain unchanged. It makes no difference whether the memory card is a RAM or a Flash EPROM card.

If the memory card is a Flash EPROM card and there is no backup battery, an S7-300 and an S7-400 respond differently. In S7-300 controllers, the signal states of the retentive memory bits, timers and counters are retained, in S7-400 controllers they are not. The contents of retentive data blocks also remain unchanged in S7-300 controllers. The remaining data blocks in an S7-300 controller and all data blocks in an S7-400 controller are copied from the memory card to work memory, as are the code blocks. The only data blocks whose contents are retained are those on the memory card. Data blocks generated with system function SFC 22 CREAT_DB are not retentive. After startup, the data blocks have the contents that are on the memory card, that is, the contents with which they were programmed.

## 22.3 Types of Start-up

### 22.3.1 STARTUP Mode

The CPU executes a start-up in the following cases:

- When the mains power is switched on (via STOP)
- When the mode selector is set from STOP to RUN or RUN-P
- On a request from a communication function (request from a programming device or via communication function blocks from another CPU)

A *manual* startup is initiated via the keyswitch or a communication function, an *automatic* startup by switching on the mains power.

The start-up routine may be as long as required, and there is no time limit on its execution; the scan cycle monitor is not active. During the execution of the start-up routine, no interrupts will be serviced (exceptions being errors, which are handled as they would be in RUN mode). In the start-up routine, the CPU updates the timers, the run-time meters and the real-time clock. Figure 22.2 shows the activities carried out by the CPU during a startup.

A start-up routine can be aborted, for instance when the mode selector is actuated or when there is a power failure. The aborted start-up routine is then executed from the beginning when the power is switched on. If a complete restart is aborted, it must be executed again. If a warm restart is aborted, it is possible to execute either a complete restart or a warm restart.

### 22.3.2 Complete Restart

On a complete restart, the CPU sets both itself and the modules to the programmed initial state, erases the non-retentive data, calls OB 100, and then executes the main program in OB 1 from the beginning.

#### Complete manual restart

A complete manual restart is initiated in the following instances:

- Via the mode switch on the CPU on a transition from STOP to RUN or RUN-P; on the S7-400, the start-up switch is in the CRST position.
- Via a communication function from a PG or with an SFB from another CPU; the mode selector must be in the RUN or RUN-P position.

A complete manual restart can always be initiated unless the CPU requests a memory reset.

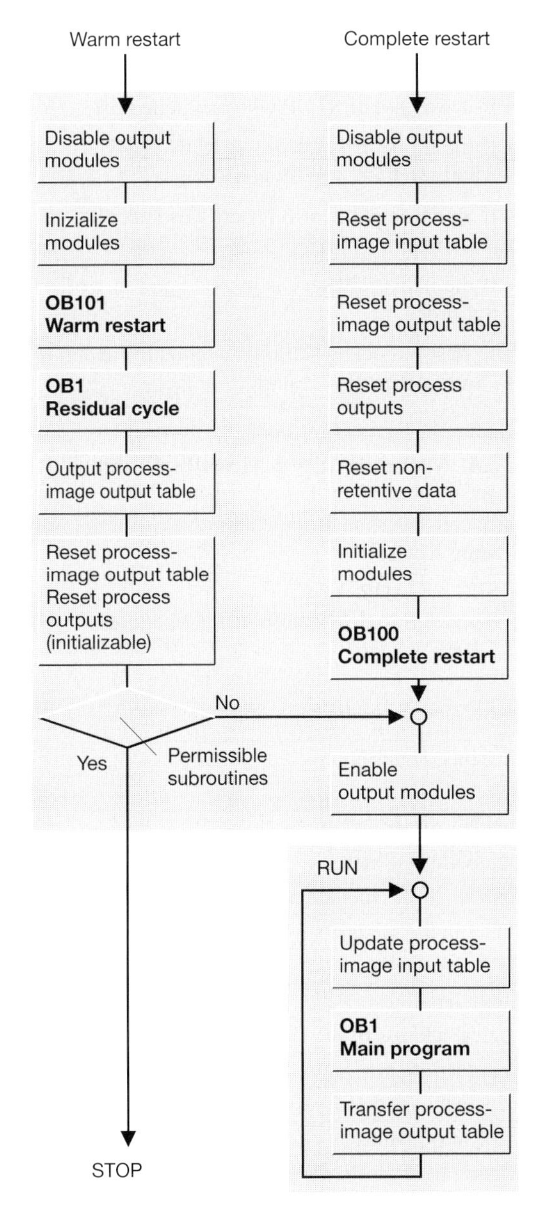


**Figure 22.2** CPU Activities During Start-up

### Complete automatic restart

An automatic complete restart is initiated by switching on the mains power. The restart is executed if

- the CPU was not at STOP when the power was switched off
- the mode switch is at RUN or RUN-P
- the CPU was interrupted by a power outage while executing a complete restart
- no automatic warm restart was specified (S7-400 only)

In all other cases, the CPU remains at STOP when the power is switched on. The startup switch (S7-400) has no effect in the case of an automatic complete restart.

When operated without a backup battery, the CPU executes an automatic non-retentive complete restart. The CPU starts the memory reset automatically, then copies the user program from the memory card to work memory. The memory card must be a Flash EPROM.

## 22.3.3 Warm Restart

A warm restart is possible only on an S7-400.

On a STOP or power outage, the CPU saves all interrupts as well as the internal CPU registers that are important to the processing of the user program. On a warm restart, it can therefore resume at the location in the program at which the interruption occurred. This may be the main program, or it may be an interrupt or error handling routine. All ("old") interrupts are saved and will be serviced.

The so-called "residual cycle", which extends from the point at which the CPU resumes the program following a warm restart to the end of the main program, counts as part of the restart. No (new) interrupts are serviced. The output modules are disabled, and are in their initial state.

A warm restart is permitted only when there have been no changes in the user program while the CPU was at STOP, such as modification of a block.

By parameterizing the CPU accordingly, you can specify how long the interruption may be for the CPU to still be able to execute a warm restart (from 100 milliseconds to 1 hour). If the interruption is longer, only a complete restart is allowed. The length of the interruption is the amount of time between exiting of the RUN mode (STOP or power-down) and reentry into the RUN mode (following execution of OB 101 and the residual cycle).

### Manual warm restart

A manual warm restart is initiated

- By moving the mode selector from STOP to RUN or RUN-P when the startup switch is at WRST
- Via a communication function from a PG or with an SFB from another CPU; the mode selector must be at RUN or RUN-P.

A manual warm restart is possible only when the warm restart disable was revoked in the "Startup" tab when the CPU was parameterized. The cause of the STOP must have been a manual activity, either via the mode selector or through a communication function; only then can a manual warm restart be executed while the CPU is at STOP.

### Automatic warm restart

An automatic warm restart is initiated by switching on the mains power. The CPU executes an automatic warm restart only in the following instances:

- If it was not at STOP when switched off
- If the mode selector was at RUN or RUN-P when the CPU was switched on
- If an automatic warm restart was specified per parameter
- If the backup battery is inserted and in working order

The position of the startup switch is irrelevant to an automatic warm restart.

## 22.4 Ascertaining a Module Address

You can ascertain module addresses with the following SFCs:

- SFC 5 GADR_LGC
  Ascertain logical dual-port RAM address

**Table 22.2** Parameters for the SFCs Used to Ascertain the Module Address

| SFC | Parameter | Declaration | Data Type | Contents, Description |
|---|---|---|---|---|
| 5 | SUBNETID | INPUT | BYTE | Area identifier |
| | RACK | INPUT | WORD | Number of the rack |
| | SLOT | INPUT | WORD | Number of the slot |
| | SUBSLOT | INPUT | BYTE | Number of the submodule |
| | SUBADDR | INPUT | WORD | Offset in the module's user data address area |
| | RET_VAL | OUTPUT | INT | Error information |
| | IOID | OUTPUT | BYTE | Area identifier |
| | LADDR | OUTPUT | WORD | Logical address of the channel |
| 50 | IOID | INPUT | BYTE | Area identifier |
| | LADDR | INPUT | WORD | A logical module address |
| | RET_VAL | OUTPUT | INT | Error information |
| | PEADDR | OUTPUT | ANY | WORD field for the PI addresses |
| | PECOUNT | OUTPUT | INT | Number of PI addresses returned |
| | PAADDR | OUTPUT | ANY | WORD field for the PQ addresses |
| | PACOUNT | OUTPUT | INT | Number of PQ addresses returned |
| 49 | IOID | INPUT | BYTE | Area identifier |
| | LADDR | INPUT | WORD | A logical module address |
| | RET_VAL | OUTPUT | INT | Error information |
| | AREA | OUTPUT | BYTE | Area identifier |
| | RACK | OUTPUT | WORD | Number of the rack |
| | SLOT | OUTPUT | WORD | Number of the slot |
| | SUBADDR | OUTPUT | WORD | Offset in the module's user data address area |

- SFC 50 RD_LGADR
  Ascertain all logical module addresses
- SFC 49 LGC_GADR
  Ascertain slot

Table 22.2 shows the parameters for these SFCs.

The SFCs have IOID and LADDR as common parameters for the logical address (= address in the I/O area). IOID is either B#16#54, which stands for the peripheral inputs (PIs) or B#16#55, which stands for the peripheral outputs (PQs). LADDR contains an I/O address in the PI or PQ area which corresponds to the specified channel. If the channel is 0, it is the module start address.

The hardware configuration data must specify an allocation between logical address (module start address) and slot address (location of the module in a rack or a station for distributed I/O) for the addresses ascertained with these SFCs.

- Ascertain the logical address of a channel
  System function SFC 5 GADR_LGC returns the logical address of a channel when you specify the slot address ("geographic" address). Enter the number of the subnet in the SUBNETID parameter if the module belongs to the distributed I/O or B#16#00 if the module is plugged into a rack. The RACK parameter specifies the number of the rack or, in the case of distributed I/O, the number of the station. If the module has no submodule slot, enter B#16#00 in the SUBSLOT parameter. SUBADDR contains the address offset in the module's user data (W#16#0000, for example, stands for the module start address).
- Ascertain all logical addresses for a module
  On the S7-400, you can assign addresses for a module's user data bytes which are not contiguous (under development). SFC 50 RD_LGADR returns all logical addresses for a module when you specify an arbitrary address from the user data area. Use the PEADDR and PAADDR parameters to define an area of WORD components (a word-based ANY pointer, for example *P#DBzDBXy..x* WORD *nnn*). SFC 50 then shows you the number of entries returned in these areas in the RECOUNT and PACOUNT parameters.
- Ascertain the slot address of a module
  SFC 49 LGC_GADR returns the slot address of a module when you specify an arbitrary logical module address. The value in the AREA parameter specifies the system in which the module is operated (Table 22.3).

## 22.5 Parameterizing Modules

The following system functions are available for parameterizing modules:

- SFC 54 RD_DPARM
  Read predefined parameters
- SFC 55 WR_PARM
  Write dynamic parameters
- SFC 56 WR_DPARM
  Write predefined parameters
- SFC 57 PARM_MOD
  Parameterize module
- SFC 58 WR_REC
  Write data record
- SFC 59 RD_REC
  Read data record

**Table 22.3** Description of SFC 49 LGC_GADR's Output Parameters

| AREA | System | Meaning of RACK, SLOT and SUBADDR |
|---|---|---|
| 0<br>1 | S7-400<br>S7-300 | RACK = Number of the rack<br>SLOT = Number of the slot<br>SUBADDR = Address offset to the start address |
| 2 | Distributed I/O | RACK, SLOT and SUBADDR irrelevant |
| 3<br>4<br>5<br>6 | S5 P area<br>S5 Q area<br>S5 IM3 area<br>S5 IM4 area | RACK = Number of the rack<br>SLOT = Slot number of the adapter casing<br>SUBADDR = Address in the S5 area |

**Table 22.4** Parameters for System Functions Used for Data Transfer

| Parameter for SFC | | | | | | Parameter Name | Declaration | Data Type | Contents, Description |
|---|---|---|---|---|---|---|---|---|---|
| – | 55 | 56 | 57 | 58 | 59 | REQ | INPUT | BOOL | "1" = Write request |
| 54 | 55 | 56 | 57 | 58 | 59 | IOID | INPUT | BYTE | B#16#54 = Peripheral inputs (PIs)<br>B#16#55 = Peripheral outputs (PQs) |
| 54 | 55 | 56 | 57 | 58 | 59 | LADDR | INPUT | WORD | Module start address |
| 54 | 55 | 56 | – | 58 | 59 | RECNUM | INPUT | BYTE | Data record number |
| – | 55 | – | – | 58 | – | RECORD | INPUT | ANY | Data record |
| 54 | 55 | 56 | 57 | 58 | 59 | RET_VAL | OUTPUT | INT | Error information |
| – | 55 | 56 | 57 | 58 | 59 | BUSY | OUTPUT | BOOL | Transfer still in progress if "1" |
| 54 | – | – | – | – | 59 | RECORD | OUTPUT | ANY | Data record |

Use system functions SFC 58 WR_REC and SFC 59 RD_REC to transfer arbitrary data records from a module with the appropriate capability. The parameters for these and a number of other system functions are listed in Table 22.4.

Table 22.5 shows the available data records which can be transferred with the aforementioned system functions. Depending on the data record, the information transferred may comprise up to 240 bytes.

### General remarks on parameterizing modules

Some S7 modules can be parameterized, that is to say, values may be set on the module which deviate from the default. To specify parameters, open the module in the hardware configuration and fill in the tabs in the dialog box. When you transfer the *System Data* object in the *Blocks* container to the PLC, you are also transferring the module parameters.

The CPU transfers the module parameters to the module automatically in the following cases

- On start-up (complete restart or warm restart)
- When a module has been plugged into a configured slot (S7-400)
- Following the "return" of a rack or a distributed I/O station.

The module parameters are subdivided into static parameters and dynamic parameters. You can set both parameter types off-line in the hardware configuration. You can also modify the dynamic parameters at runtime using SFC calls. In the start-up routine, the parameters set on the modules using SFCs are overwritten by the parameters set (and stored on the CPU) via the hardware configuration.

The parameters for the signal modules are in two data records: the static parameters in data record 0 and the dynamic parameters in data record 1 in the system data blocks (SDBs). You can transfer both data records to the module with SFC 57 PARM_MOD, data record 0 or 1 with SFC 56 WR_DPARM, and only data record 1 with SFC 55 WR_PARM. The data records must be in the SDBs.

After parameterization of an S7-400 module, the specified values do not go into force until bit 2 ("Operating mode") in byte 2 of diagnostic data record 0 has assumed the value "RUN" (can be read with SFC 59 RD_REC).

As far as addressing for data transfer is concerned, use the *lowest* module start address (LADDR parameter) together with the identifier indicating whether you have defined this address as input or output (IOID parameter). If you as-

**Table 22.5** Available Data Records

| Record No. | Contents for Read | Contents for Write |
|---|---|---|
| 0 | Diagnostic data | Parameters |
| 1 | Diagnostic data | Parameters |
| 2 to 127 | User data | User data |
| 128 to 255 | Diagnostic data | Parameters |

signed the same start address to both the input and output area, use the identifier for input. Use the I/O identifier regardless of whether you want to execute a Read or a Write operation.

Use the RECORD parameter with the data type ANY to define an area of BYTE components. This may be a variable of type ARRAY, STRUCT or UDT, or an ANY pointer of type BYTE (for example P#DB*z*DBX*y.x* BYTE *nnn*). If you use a variable, it must be a "complete" variable; individual array or structure components are not permissible.

### Reading predefined parameters

System function SFC 54_RD_PARAM transfers the data record with the number specified at the RECNUM parameter from the relevant system data block SDB to the destination area specified at the RECORD parameter.

You can now change this data record, for example, and write it to the module with SFC 58 WR_REC.

### Writing dynamic parameters

System function SFC 55 WR_PARM transfers the data record addressed by RECORD to the module specified by the IOID and LADDR parameters. Specify the number of the data record in the RECNUM parameter. The prerequisite is that the data record be in the correct SDB system data block, and that the data record contains only dynamic parameters.

When the job is initiated, the SFC reads the entire data record; the transfer may be distributed over several program scan cycles. The BUSY parameter is "1" during the transfer.

### Writing predefined parameters

System function SFC 56 WR_DPARM transfers the data record with the number specified in the RECNUM parameter from the relevant SDB system data block to the module identified by the IOID and LADDR parameters.

The transfer may be distributed over several program scan cycles; the BUSY parameter is "1" during the transfer.

### Parameterizing a module

System function SFC 57 PARM_MOD transfers all the data records programmed when the module was parameterized via the hardware configuration.

The transfer may be distributed over several program scan cycles; the BUSY parameter is "1" during the transfer.

### Writing a data record

SFC 58 WR_REC transfers the data record addressed by the RECORD parameter and the number RECNUM to the module defined by the IOID and LADDR parameters. A "1" in the REQ parameter starts the transfer. When the job is initiated, the SFC reads the complete data record.

The transfer may be distributed over several program cycles; the BUSY parameter is "1" during the transfer.

### Reading a data record

When the REQ parameter is "1", SFC 59 RD_REC reads the data record addressed by the RECNUM parameter from the module and places it in destination area RECORD. The destination area must be longer than or at least as long as the data record. If the transfer is completed without error, the RET_VAL parameter contains the number of bytes transferred.

The transfer may be distributed over several program scan cycles; the BUSY parameter is "1" during the transfer.

S7-300s delivered prior to February 1997: the SFC reads as much data from the specified data record as the destination area can accommodate. The size of the destination area may not exceed that of the data record.

# 23 Error Handling

The CPU reports errors or faults detected by the modules or by the CPU itself in different ways:

- Errors in arithmetic operations (overflow, invalid REAL number) by setting status bits (status bit OV, for example, for a numerical overflow)
- Errors detected while processing STL functions (synchronous errors) by calling organization blocks OB 121 and OB 122
- Errors in the programmable controller which do not relate to program scanning (asynchronous errors) by calling organization blocks OB 80 to OB 87.

The CPU signals the occurrence of an error or fault, and in some cases the cause, by setting error LEDs on the front panel. In the case of unrecoverable errors (such as invalid OP code), the CPU goes directly to STOP.

With the CPU in STOP mode, you can use a programming device and the CPU information functions to read out the contents of the block stack (B stack), the interrupt stack (I stack) and the local data stack (L stack) and then draw conclusions as to the cause of error.

The system diagnostics can detect errors/faults on the modules, and enters these errors in a diagnostic buffer. Information on CPU mode transitions (such as the reasons for a STOP) are also placed in the diagnostic buffer. The contents of this buffer are retained on STOP, on a memory reset, and on power failure, and can be read out following power recovery and execution of a start-up routine using a programming device.

## 23.1 Synchronous Errors

The CPU's operating system generates a synchronous error when an error occurs in immediate conjunction with program scanning. In the case of a programming error, OB 121 is called, while OB 122 is called in the event of an access

**Table 23.1** Start Information for the Synchronous Error OBs

| Variable Name | Data Type | Description, Contents |
|---|---|---|
| OB12x_EV_CLASS | BYTE | B#16#25 = Call programming error OB 121<br>B#16#29 = Call access error OB 122 |
| OB12x_SW_FLT | BYTE | Error code (see text) |
| OB12x_PRIORITY | BYTE | Priority class in which the error occurred |
| OB12x_OB_NUMBR | BYTE | OB number (B#16#79 or. B#16#80) |
| OB12x_BLK_TYPE | BYTE | Type of block interrupted (S7-400 only)<br>OB: B#16#88, FB: B#16#8E, FC: B#16#8C |
| OB121_RESERVED_1<br>OB122_MEM_AREA | BYTE | Addition to error code (see text) |
| OB121_FLT_REG<br>OB122_MEM_ADDR | WORD | OB 121: Error source (see text)<br>OB 122: Address at which the error occurred |
| OB12x_BLK_NUM | WORD | Number of the block in which the error occurred (S7-400 only) |
| OB12x_PRG_ADDR | WORD | Error address in the block that caused the error (S7-400 only) |
| OB12x_DATE_TIME | DT | Time at which programming error was detected |

error. If a synchronous error OB has not been programmed, the CPU goes to STOP. Figure 23.1 shows the start information for both synchronous error organization blocks.

A synchronous error OB has the same priority as the block in which the error occurred. It is for this reason that it is possible to access the registers of the interrupted block in the synchronous error OB, and it is also for this reason that the program in the synchronous error OB (under certain circumstances with modified content) can return the registers to the interrupted block. Note that when a synchronous error OB is called, its 20 bytes of start information are also pushed onto the L stack for the priority class that caused the error, as are the other temporary local data for the synchronous error OB and for all blocks called in this OB.

In the case of a CPU 416, another synchronous error OB can be called in an error OB. The block nesting depth for a synchronous error OB is 3 for S7-400 CPUs and 4 for S7-300 CPUs.

You can disable and enable a synchronous error OB call with system functions SFC 36 MSK_FLT and SFC 37 DMSK_FLT.

### 23.1.1 Programming Errors

If the operating system detects a programming error at runtime, it calls OB 121.

The error code in variable OB121_SW_FLT is described in detail in Section 23.2.1, "Error Filters". Bits 4 to 7 of variables OB121_RESERVED_1 and OB122_MEM_AREA (in OB 122) contain the access mode in the event of a read or write error.

1 Bit access
2 Byte access
3 Word access
4 Doubleword access

and bits 0 to 3 contain the address area

0 I/O area PI or PQ
1 Process-image input area I
2 Process-image output area Q
3 Bit memory area M
4 Global data block DB
5 Instance data block DI
6 Temporary local data L
7 Temporary local data for the preceding block V

Example: Error code B#16#24 means that a synchronous error occurred while accessing a data byte located in a data block opened via the DB register.

The variable OB121_FLT_REG contains the error source as per the error code.

- The errored address (in the event of a read/write error)
- The errored area (in the event of a range error)
- The errored number of the block or the timer/counter

### 23.1.2 Access Errors

The operating system calls OB 122 if a runtime access error is detected.

The error code in variable OB122_SW_FLT is described in detail in Section 23.2.1, "Error Filters". The contents of variable OB122_MEM_AREA correspond to the contents of variable OB121_RESERVED_1 (see above under "Programming Errors").

## 23.2 Synchronous Error Handling

The following system functions are provided for handling synchronous errors:

- SFC 36 MSK_FLT
  Mask synchronous errors (disable OB call)
- SFC 37 DMSK_FLT
  Unmask synchronous error (reenable OB call)
- SFC 38 READ_ERR
  Read error register

The operating system enters the synchronous error in the diagnostic buffer without regard to the use of system functions SFC 37 to SFC 39. The parameters for these system functions are listed in Table 23.2.

### 23.2.1 Error Filters

The error filters are used to control the system functions for synchronous error handling. In the programming error filter, one bit stands for each programming error detected; in the access error filter, one bit stands for each access error de-

**Table 23.2** SFC Parameters for Synchronous Error Handling

| SFC | Parameter Name | Declaration | Data Type | Contents, Description |
|---|---|---|---|---|
| 36 | PRGFLT_SET_MASK | INPUT | DWORD | New (additional) programming error filter |
| | ACCFLT_SET_MASK | INPUT | DWORD | New (additional) access error filter |
| | RET_VAL | OUTPUT | INT | W#16#0001 = The new filter overlaps the existing filter |
| | PRGFLT_MASKED | OUTPUT | DWORD | Complete programming error filter |
| | ACCFLT_MASKED | OUTPUT | DWORD | Complete access error filter |
| 37 | PRGFLT_RESET_MASK | INPUT | DWORD | Programming error filter to be reset |
| | ACCFLT_RESET_MASK | INPUT | DWORD | Access error filter to be reset |
| | RET_VAL | OUTPUT | INT | W#16#0001 = The new filter contains bits that are not set (in the current filter) |
| | PRGFLT_MASKED | OUTPUT | DWORD | Remaining programming error filter |
| | ACCFLT_MASKED | OUTPUT | DWORD | Remaining access error filter |
| 38 | PRGFLT_QUERY | INPUT | DWORD | Programming error filter to be queried |
| | ACCFLT_QUERY | INPUT | DWORD | Access error filter to be queried |
| | RET_VAL | OUTPUT | INT | W#16#0001 = The query filter contains bits that are not set (in the current filter) |
| | PRGFLT_CLR | OUTPUT | DWORD | Programming error filter with error messages |
| | ACCFLT_CLR | OUTPUT | DWORD | Access error filter with error messages |

tected. When you define an error filter, you set the bit that stands for the synchronous error you want to mask, unmask or query. The error filters returned by the system functions show a “1” for synchronous errors that are still masked or which have occurred.

The programming error filter is shown in Table 23.3, the Error Code column shows the contents of variable OB121_SW_FLT in the start information for OB 121.

The access error filter is shown in Table 23.4; the Error Code column shows the contents of variable OB122_SW_FLT in the start information for OB 122.

The S7-400 CPUs distinguish between two types of access error: access to a non-existent module and invalid access attempt to an existing module. If a module fails during operation, a time-out is signaled approximately 150 µs after an access attempt. At the same time, that module is marked “non-existent” and an I/O access error is reported on every subsequent attempt to access the module. The CPU also reports an I/O access error when an attempt is made to access a non-existent module, regardless of whether the attempt was direct (via the I/O area) or indirect (via the process image).

The error filter bits not listed in the tables are not relevant to the handling of synchronous errors.

### 23.2.2 Masking Synchronous Errors

System function SFC 36 MSK_FLT disables synchronous error OB calls via the error filters. A “1” in the error filters indicates the synchronous errors for which the OBs are not to be called (the synchronous errors are “masked”). The masking of synchronous errors in the error filters is in addition to the masking stored in the operating system’s memory. SFC 36 returns a function value indicating whether a (stored) masking already exists on at least one bit for the masking specified at the input parameters (W#16#0001).

SFC 36 returns a “1” in the output parameters for all currently masked errors.

If a masked synchronous error event occurs, the respective OB is not called and the error is entered in the error register. The Disable applies

**Table 23.3** Programming Error Filter

| Bit | Error Code | Contents |
|---|---|---|
| 1 | B#16#21 | BCD conversion error (pseudo-tetrad detected during conversion) |
| 2 | B#16#22 | Area length error on read (address not within area limits) |
| 3 | B#16#23 | Area length error on write (address not within area limits) |
| 4 | B#16#24 | Area length error on read (wrong area in area pointer) |
| 5 | B#16#25 | Area length error on write (wrong area in area pointer) |
| 6 | B#16#26 | Invalid timer number |
| 7 | B#16#27 | Invalid counter number |
| 8 | B#16#28 | Address error on read (bit address 0 in conjunction with byte, word or doubleword access and indirect addressing) |
| 9 | B#16#29 | Address area on write (bit address 0 in conjunction with byte, word or doubleword access and indirect addressing) |
| 16 | B#16#30 | Write error, global data block (write-protected block) |
| 17 | B#16#31 | Write error, instance data block (write-protected block) |
| 18 | B#16#32 | Invalid number of a global data block (DB register) |
| 19 | B#16#33 | Invalid number of an instance data block (DI register) |
| 20 | B#16#34 | Invalid number of a function (FC) |
| 21 | B#16#35 | Invalid number of a function block (FB) |
| 26 | B#16#3A | Called data block (DB) does not exist |
| 28 | B#16#3C | Called function (FC) does not exist |
| 30 | B#16#3E | Called function block (FB) does not exist |

to the current priority class (priority level). For example, if you were to disable a synchronous error OB call in the main program, the synchronous error OB would still be called if the error were to occur in an interrupt service routine.

### 23.2.3 Unmasking Synchronous Errors

System function SFC 37 DMSK_FLT enables the synchronous error OB calls via the error filters. You enter a "1" in the filters to indicate the synchronous errors for which the OBs are once again to be called (the synchronous errors are

**Table 23.4** Access Error Filter

| Bit | Error Code | Contents |
|---|---|---|
| 3 | B#16#42 | I/O access error on read<br>S7-300: Module does not exist or does not acknowledge<br>S7-400: An existing module does not acknowledge after first access operation (time-out) |
| 4 | B#16#43 | I/O access error on write<br>S7-300: Module does not exist or does not acknowledge<br>S7-400: An existing module does not acknowledge after first access operation (time-out) |
| 5 | B#16#44 | S7-400 only:<br>I/O access error on attempt to write to non-existent module or on repeated access to modules which do not acknowledge |
| 6 | B#16#45 | S7-400 only:<br>I/O access error on attempt to write to non-existent module or on repeated access to modules which do not acknowledge |

"unmasked"). The entries corresponding to the specified bits are deleted in the error register. SFC 37 returns W#16#0001 as function value if no (stored) masking already exists on at least one bit for the unmasking specified at the input parameters.

SFC 37 returns a "1" in the output parameters for all currently masked errors.

If an unmasked synchronous error occurs, the respective OB is called and the event entered in the error register. The Enable applies to the current priority class (priority level).

### 23.2.4 Reading the Error Register

System function SFC 38 READ_ERR reads the error register. You must enter a "1" in the error filters to indicate the synchronous errors whose entries you want to read. SFC 38 returns W#16#0001 as function value when the selection specified in the input parameters included at least one bit for which no (stored) masking exists.

SFC 38 returns a "1" in the output parameters for the selected errors when these errors occurred, and deletes these errors in the error register when they are queried. The synchronous errors that are reported are those in the current priority class (priority level).

### 23.2.5 Entering a Substitute Value

SFC 44 REPL_VAL allows you to enter a substitute value in accumulator 1 from within a synchronous error OB. Use SFC 44 when you can no longer read any values from a module (for instance when a module is defective). When you program SFC 44, OB 122 ("access error") is called every time an attempt is made to access the module in question. When you call SFC 44, you can load a substitute value into the accumulator; the program scan is then resumed with the substitute value. Table 23.5 lists the parameters for SFC 44.

You may call SFC 44 in only one synchronous error OB (OB 121 or OB 122).

## 23.3 Asynchronous Errors

Asynchronous errors are errors which can occur independently of the program scan. When an asynchronous error occurs, the operating system calls one of the organization blocks listed below:

OB 80 Time error
OB 81 Power supply fault
OB 82 Diagnostic interrupt
OB 83 Insert/remove interrupt
OB 84 CPU hardware fault
OB 85 Program sequence error
OB 86 Rack failure
OB 87 Communication error

The OB 82 call (diagnostic interrupt) is described in detail in section 23.4, "System Diagnostics".

The asynchronous error OB call can be disabled and enabled with system functions SFC 39 DIS_IRT and SFC 40 EN_IRT, and delayed and enabled with system functions SFC 41 DIS_AIRT and SFC 42 EN_AIRT.

#### Time error

The operating system calls organization block OB 80 when one of the following errors occurs:

- Cycle monitoring time exceeded
- OB request error (the requested OB is still executing or an OB was requested too frequently within a given priority class)
- Time-of-day interrupt error (TOD interrupt time past because clock was set forward or after transition to RUN)

If no OB 80 is available and a timing error occurs, the CPU goes to STOP. The CPU also goes to STOP if the OB is called a second time in the same program scan cycle.

**Table 23.5** Parameters for SFC 44 REPL_VAL

| SFC | Parameter Name | Declaration | Data Type | Contents, Description |
|---|---|---|---|---|
| 44 | VAL | INPUT | DWORD | Substitute value |
| | RET_VAL | OUTPUT | INT | Error information |

### Power supply fault

The operating system calls organization block OB 81 if one of the following errors occurs:

- At least one backup battery in the central controller or in an expansion unit is empty
- No battery voltage in the central controller or in an expansion unit
- 24 V supply failed in central controller or in an expansion unit

OB 81 is called for incoming and outgoing events. If there is no OB 81, the CPU continues functioning when a power supply error occurs.

### Insert/remove interrupt

The operating system monitors the module configuration once per second. An entry is made in the diagnostic buffer and in the system status list each time a module is inserted or removed in RUN, STOP or STARTUP mode. In addition, the operating system calls organization block OB 83 if the CPU is in RUN mode. If there is no OB 83, the CPU goes to STOP on an insert/remove module interrupt.

As much as a second can pass before the insert/remove module interrupt is generated. As a result, it is possible that an access error or an error relating to the updating of the process image could be reported in the interim between removal of a module and generation of the interrupt.

If a suitable module is inserted into a configured slot, the CPU automatically parameterizes that module, using data records already stored on that CPU. Only then is OB 83 called.

### CPU hardware fault

The operating system calls organization block OB 84 when an interface error (MPI network, PROFIBUS-DP) occurs or disappears. If there is no OB 84, the CPU goes to STOP on a CPU hardware fault.

### Program sequence error

The operating system calls organization block OB 85 when one of the following errors occurs:

- Start request for an organization block which has not been loaded
- Error occurred while the operating system was accessing a block (for instance no instance data block when a system function block (SFB) was called)
- I/O access error while executing a full process image update

If there is no OB 85, the CPU goes to STOP on a program execution error.

### Rack failure

The operating system calls organization block OB 86 if it detects the failure of a rack (power failure, line break, defective IM), a subnet, or a distributed I/O station. OB 86 is called for both incoming and leaving errors.

In multiprocessor mode, OB 86 is called in all CPUs if a rack fails.

If there is no OB 86, the CPU goes to STOP if a rack failure occurs.

### Communication error

The operating system calls organization block OB 87 when a communication error occurs. Some examples of communication errors are

- Invalid frame identification or frame length detected during global data communication
- Sending of diagnostic entries not possible
- Clock synchronization error
- GD status cannot be entered in a data block

If there is no OB 87, the CPU goes to STOP when a communication error occurs.

## 23.4 System Diagnostics

### 23.4.1 Diagnostic Events and Diagnostic Buffer

System diagnostics is the detection, evaluation and reporting of errors occurring in programmable controllers. Examples are errors in the user program, module failures or wirebreaks on signaling modules. These *diagnostic events* may be:

- Diagnostic interrupts from modules with this capability
- System errors and CPU mode transitions
- User messages via system functions.

Modules with diagnostic capabilities distinguish between programmable and non-programmable diagnostic events. Programmable diagnostic events are reported only when you have set the parameters necessary to enable diagnostics. Non-programmable diagnostic events are always reported, regardless of whether or not diagnostics have been enabled. In the event of a reportable diagnostic event,

- The fault LED on the CPU goes on
- The diagnostic event is passed on to the CPU's operating system
- A diagnostic interrupt is generated if you have set the parameters enabling such interrupts (by default, diagnostic interrupts are disabled).

All diagnostic events reported to the CPU operating system are entered in a *diagnostic buffer* in the order in which they occurred, and with date and time stamp. The diagnostic buffer is a battery-backed memory area on the CPU which retains its contents even in the event of a memory reset. The diagnostic buffer is a ring buffer whose size depends on the CPU. When the diagnostic buffer is full, the oldest entry is overwritten by the newest.

You can read out the diagnostic buffer with a programming device at any time. In the CPU's *System Diagnostics* parameter block you can specify whether you want expanded diagnostic entries (all OB calls). You may also specify whether the last diagnostic entry made before the CPU goes to STOP should be sent to a specific node on the MPI bus.

### 23.4.2 Writing User Entries in the Diagnostic Buffer

System function SFC 52 WR_USMSG writes an entry in the diagnostic buffer which may be sent to all nodes on the MPI bus. Table 23.6 lists the parameters for SFC 52.

The entry in the diagnostic buffer corresponds in format to that of a system event, for instance the start information for an organization block. Within the permissible boundaries, you may choose your own event ID (EVENTN parameter) and additional information (INFO1 and INFO2 parameters).

The event ID is identical to the first two bytes of the buffer entry (Figure 23.1). Permissible for a user entry are the event classes 8 (diagnostic entries for signal modules), 9 (standard user events), A and B (arbitrary user events).

Additional information (INFO1) corresponds to bytes 7 and 8 of the buffer entry (one word) and additional information 2 (INFO2) to bytes 9 to 12 (one doubleword). The contents of both variables may be of the user's own choice.

Set SEND to "1" to send the diagnostic entry to the relevant node. Even if sending is not possible (because no node is logged in or because the Send buffer is full, for example), the entry is still made in the diagnostic buffer (when bit 9 of the event ID is set).

### 23.4.3 Evaluating Diagnostic Interrupts

When a diagnostic interrupt is incoming or outgoing, the operating system interrupts scanning of the user program and calls organization block OB 82. If OB 82 has not been programmed, the CPU goes to STOP on a diagnostic interrupt. You can disable or enable OB 82 with system function SFC 39 DIS_IRT or SFC 40 EN_IRT, and delay or enable it with system function SFC 41 DIS_AIRT or SFC 42 EN_AIRT.

In the first byte of the start information, B#16#39 stands for an incoming diagnostic in-

**Table 23.6** Parameters for SFC 52 WR_USMSG

| SFC | Parameter | Declaration | Data Type | Contents, Description |
|---|---|---|---|---|
| 52 | SEND | INPUT | BOOL | If "1": Send is enabling |
| | EVENTN | INPUT | WORD | Event ID |
| | INFO1 | INPUT | ANY | Additional information 1 (one word) |
| | INFO2 | INPUT | ANY | Additional information 2 (one doubleword) |
| | RET_VAL | OUTPUT | INT | Error information |

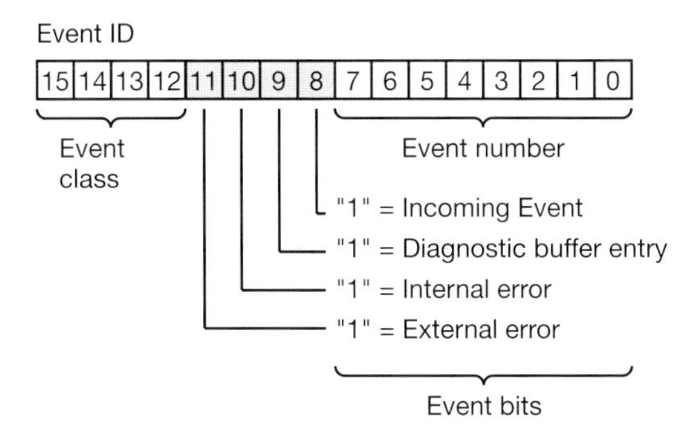


**Figure 23.1** Event ID for Diagnostic Buffer Entries

terrupt and B#16#38 for a leaving diagnostic interrupt. The sixth byte gives the address identifier (B#16#54 stands for an input, B#16#55 for an output); the subsequent INT variable contains the address of the module that generated the diagnostic interrupt. The next four bytes contain the diagnostic information provided by that module.

You can use system function SFC 59 RD_REC (read data record) in OB 82 to obtain detailed error information. The diagnostic information are consistent until OB 82 is exited, that is, they remain "frozen". Exiting of OB 82 acknowledges the diagnostic interrupt on the module.

A module's diagnostic data are in data records DS 0 and DS 1. Data record DS 0 contains four bytes of diagnostic data describing the current status of the module. The contents of these four bytes are identical to the contents of bytes 8 to 11 of the OB 82 start information. Data record DS 1 contains the four bytes from data record DS 0 and, in addition, the module-specific diagnostic data.

### 23.4.4 Reading the System Status List

The system status list (SZL) describes the current status of the programmable controller. Using information functions, the list can be read but not modified. you can read part of the list (that is, a sublist) with system function SFC 51 RDSYSST. Sublists are virtual lists, which means that they are made available by the CPU operating system only on request. The parameters for SFC 51 are listed in Table 23.7.

REQ = "1" initiates the read operation, and BUSY = "0" tells you when it has been completed. The operating system can execute several asynchronous read operations quasi simultaneously; how many depends on the CPU being used. If SFC 51 reports a lack of resources via the function value (W#16#8085), you must resubmit your read request.

The contents of parameters SZL_ID and INDEX are CPU-dependent. The SZL_HEADER parameter is of data type STRUCT, with variables LENGTHDR (data type WORD) and N_DR (WORD) as components. LENGTHDR contains the length of a data record, N_DR the number of data records read.

Use the DR parameter to specify the variable or data area in which SFC 51 is to enter the data records. For example, P#DB200.DBX0.0 WORD 256 would provide an area of 256 data words in data block DB 200, beginning with DBB 0. If the area provided is of insufficient capacity, as many data records as possible will be entered. Only complete data records are transferred. The specified area must be able to accommodate at least one data record.

**Table 23.7** Parameters for SFC 51 RDSYSST

| SFC | Parameter Name | Declaration | Data Type | Contents, Description |
|---|---|---|---|---|
| 51 | REQ | INPUT | BOOL | If "1": Submit request |
| | SZL_ID | INPUT | WORD | Sublist ID |
| | INDEX | INPUT | WORD | Type or number of the sublist object |
| | RET_VAL | OUTPUT | INT | Error information |
| | BUSY | OUTPUT | BOOL | If "1": Read not yet completed |
| | SZL_HEADER | OUTPUT | STRUCT | Length and number of data records read |
| | DR | OUTPUT | ANY | Field for data records read |

# Variable Handling

This section provides information on handling complex variables. Knowledge of the structure of data types, mastery of indirect addressing and the ability to determine the addresses of the variables at runtime are all requirements here.

- Variables if elementary **data types** can be accessed direct with STL statements, whether you are dealing with binary logic operations, memory functions or load and transfer operations. With complex data types and user-defined data types, only the individual *components* can currently be accessed direct. If you still want to access variables of these data types, you must know the inner structure of the variables.
- **Indirect addressing** allows you to access operands whose addresses are not known until runtime. You can choose between memory-indirect and register-indirect addressing. You can even wait until runtime to use the operand area. Indirect addressing allows you to access variables of complex and user-defined data types using absolute addressing.
- **Direct variable access** loads the current address of a local variable. When you have determined the address, you can process local variables (and so also block parameters) of any data types. The two preceding chapters contain the information required for this purpose.

Several extensive examples – collected in Chapter 26.4 "Brief Description of the Message Frame Example" – explain the handling of complex variables. The examples "Message Frame Data", "Preparing a Message Frame" and "Clock Check" deal with handling user-defined data types and the use of variables of complex data types in conjunction with system functions and standard functions. The examples "Checksum" and "Data Item Conversion" describe how to access parameters of complex data types with the help of indirect addressing. The example "Save Message Frame" shows how to use the system function SFC 20 BLKMOV to transfer data areas whose addresses are not known until runtime.

**24 Data Types**
Elementary, complex and user-defined data types; declaration and structure of the data types

**25 Indirect Addressing**
Area pointers, DB pointers and ANY pointers; memory-indirect and register-indirect addressing, area-internal and area-crossing; working with address registers

**26 Direct Variable Access**
Addresses of local variables; data storage of variables; data storage with parameter transfer; 'Variable' ANY pointer; Message Frame example

# 24 Data Types

Data types determine the properties and characteristics of data, essentially the representation of the content of one or more related operands and the permissible areas. STEP 7 provides pre-defined data types that you can compile in addition to user-defined data types. The data types are globally available; they can be used in any block.

This chapter gives detailed information on elementary data types and complex data types and shows the structure of the relevant variables. You will learn how user-defined data types are created and used.

You can find examples of the data types on the diskette accompanying the book under the "Variable Handling" program in function blocks FB 101, FB 102 and FB 103 or source file Chap_24.

## 24.1 Elementary Data Types

Variables of elementary data types have a maximum length of one doubleword; they can therefore be processed with load and transfer functions or with binary logic operations.

### 24.1.1 Declaration of Elementary Data Types

Elementary data types can occupy one bit, one byte, one word or one doubleword.

*Declaration*

| *varname* : *datatype* := *pre-assignment*; |
|---|

*varname* is the name of the variable
*datatype* is an elementary data type
*pre-assignment* is a fixed value

The identifiers of the data types (for example, BOOL, REAL) are keywords; they can also be written in lower case. A variable of elementary data type can be declared globally in the symbol table or locally in the declaration section.

*Pre-assignment*
The variable can be pre-assigned when it is declared (not as a block parameter in a function or as a temporary variable). The pre-assignment must be of the same data type as the variable.

*Application*
You can apply variables of elementary data type at the correspondingly declared block parameter (of the same data type POINTER or ANY) or you can access them with "normal" STL statements (for example, binary checks, load functions).

*Storing the variables*
A variable of elementary data type is stored in the same way as the relevant operand. All operand areas including block parameters are permissible.

### 24.1.2 BOOL, BYTE, WORD, DWORD, CHAR

A variable of data type BOOL represents a bit value (for example, input I 1.0). Variables of data types BYTE, WORD and DWORD are bit strings of 8, 16 or 32 bits. The individual bits are not evaluated. Chapter 3 "STL Programming Language" shows possible representations as constants.

Special forms of these data types include the BCD numbers and the counter value as used in conjunction with counter functions, as well as the data type CHAR that represents a character in ASCII code (Figure 24.1).

#### BCD numbers

BCD numbers have no special identifier in STL. You enter a BCD number with data type 16# (hexadecimal) and use only digits 0 to 9.

BCD numbers occur in coded loading of timer and counter values and in conjunction with con-

**Data type CHAR**

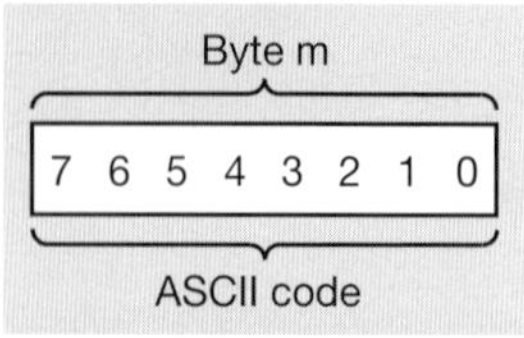


**BCD number, 3 Decades**

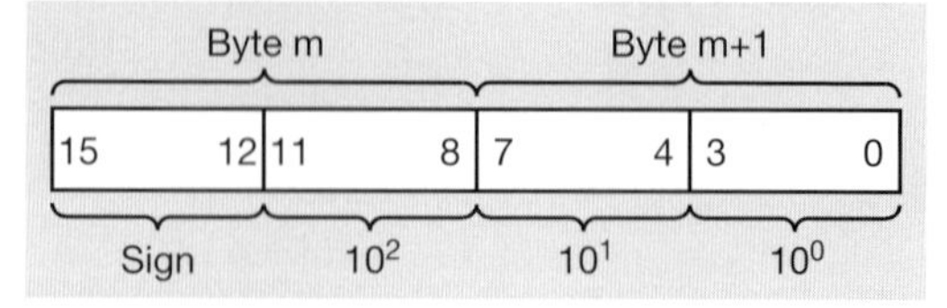


**BCD number, 7 Decades**

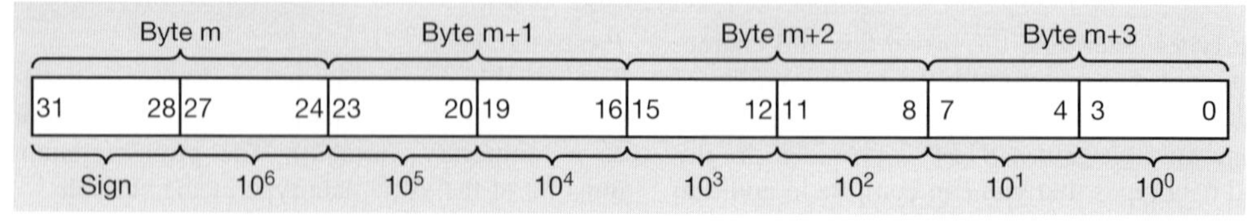


**Figure 24.1** Representation of BCD Numbers and CHAR

version functions. Data type S5TIME# is available for specifying a timer value when starting a timer function (see below), and for specifying a counter value there is data type 16# or C#. A counter value C# is a BCD number between 000 and 999, where the sign is always 0.

In general, BCD numbers are unsigned numbers. In conjunction with conversion functions, the sign of a BCD number is accommodated in the extreme-left (highest) decade. This results in the loss of one decade in the number range.

In the case of a BCD number stored in a 16-bit word, the sign is in the upper decade with only bit position 15 being relevant. Signal state "0" signifies that the number is positive and signal state "1" represents a negative number. The sign does not affect the assignment of the individual decades. An equivalent assignment applies for a 32-bit word.

The number range available is 0 to ± 999 for 16-bit BCD numbers and 0 to ± 9 999 999 for 32-bit BCD numbers.

**Table 24.1** Special Characters for CHAR

| CHAR | Hex | Meaning |
|---|---|---|
| $$ | 24hex | Dollar sign |
| $' | 27hex | Single inverted comma |
| $L or $l | 0Ahex | Line feed (LF) |
| $P or $p | 0Chex | Form feed (FF) |
| $R or $r | 0Dhex | Carriage return (CR) |
| $T or $t | 09hex | Tabulator |

### CHAR

A variable of data type CHAR (character) occupies one byte. The data type CHAR represents a single character stored in ASCII format. Example: "A".

You can use every printable character in single inverted commas. Some special characters take the notation shown in Table 24.1. Example: L "$$" loads a dollar sign in ASCII code.

You can use special forms of the data type CHAR when loading ASCII-coded characters into the accumulator. L "a" loads one character (in this case, an a) right-justified into the accumulator, L "aa" loads two characters and L "aaaa" loads 4 characters.

## 24.1.3 Number Representations

The data types INT, DINT and REAL are summarized in this section. Figure 24.2 shows the bit assignments of these data types.

### INT

A variable of data type INT represents an integer (whole number) that is stored as a 16-bit fixed-point number. The data type INT has no special identifier.

A variable of data type INT occupies one word. The signal states of bits 0 to 14 represent the positional weight of the number; the signal state of bit 15 represents the sign (SI). Signal state "0" means that the number is positive. Signal state

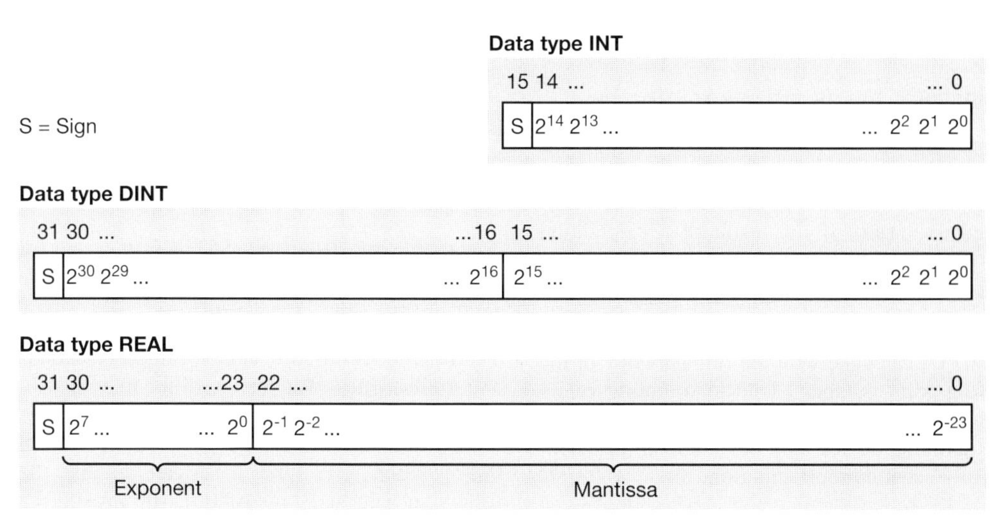


**Figure 24.2** Bit Assignments of the Data Types INT, DINT and REAL

"1" represents a negative number. Negative numbers are represented in two's complement. The number range is from +32,767 ($7FFF_{hex}$) to –32,768 ($8000_{hex}$).

## DINT

A variable of data type DINT represents an integer that is stored as a 32-bit fixed-point number. An integer is stored as a DINT variable if it is greater than 32,767 or less than –32,768 or if an L# precedes the number as the type identifier. Examples:

```
L#150  DINT variable (32-bit fixed-
       point number)
150    INT variable (16-bit fixed-
       point number)
```

A variable of data type DINT occupies a doubleword. The signal states of bits 0 to 30 represent the positional weights of the number; the sign is stored in bit 31. This bit contains "0" for a positive number and "1" for a negative number. Negative numbers are stored in two's complement. The number range is
from +2,147,483,647 ($7FFF\ FFFF_{hex}$)
to –2,147,483,648 ($8000\ 0000_{hex}$).

## REAL

A variable of data type REAL represents a fraction that is stored as a 32-bit floating-point number. An integer is stored as a REAL variable if the decimal point is followed by a zero. Examples:

```
1.0e+02  REAL variable in exponent
         representation
100.0    REAL variable (32-bit float-
         ing-point number)
L#100    DINT variable (32-bit fixed-
         point number)
100      INT variable (16-bit fixed-
         point number)
```

In exponent representation, you can specify an integer or a fraction with 7 significant digits with sign before the "e" or "E". The specification following the "e" or "E" is the exponent to base 10. Conversion of the REAL variable into the internal representation of a floating-point number is handled by STEP 7. With REAL numbers, a distinction is made between numbers that can be represented with total accuracy ("normalized" floating-point numbers) and numbers with restricted accuracy ("denormalized" floating-point numbers). The value range of a normalized floating-point number lies between the limits:

$-3.402\,823 \times 10^{+38}$ to $-1.175\,494 \times 10^{-38}$
$\pm 0$
$+1.175\,494 \times 10^{-38}$ to $+3.402\,823 \times 10^{+38}$

A denormalized floating-point number can lie within the following limits:

$-1.175\,494 \times 10^{-38}$ to $-1.401\,298 \times 10^{-45}$
and
$+1.401\,298 \times 10^{-45}$ to $+1.175\,494 \times 10^{-38}$

The S7-300 CPUs cannot perform calculations with denormalized floating-point numbers. The bit pattern of a denormalized number is interpreted as a zero. If the result of a calculation falls within this range, it is represented as a zero, with the status bits OV and OS being set (number range violation).

The CPUs calculate with the full accuracy of the floating-point numbers. Due to rounding errors in the conversion, the results displayed on the programming device may deviate from the theoretically accurate representation.

A variable of data type REAL consists internally of three components: the sign, the 8-bit exponent to base 2 and the 32-bit mantissa. The sign can assume the values "0" (positive) or "1" (negative). The exponent is stored incremented by one constant (bias, +127), so that it has a value range of 0 to 255. The mantissa represents the fraction component. The integer component of the mantissa is not stored since it is either always 1 (in the case of normalized floating-point numbers) or always 0 (in the case of denormalized floating-point numbers). Table 24.2 shows the internal range limits of a floating-point number.

### 24.1.4 Time Representations

The data types S5TIME, DATE, TIME and TIME_OF_DAY are summarized in this section. Figure 24.3 shows the bit assignments of these data types. A data type that fits into this category (DATE_AND_TIME) belongs to the complex data types since it occupies 8 bytes.

#### S5TIME

A variable of data type S5TIME is used for initializing the timer functions. It occupies one 16-bit word with 1 + 3 decades.

The time is specified in hours, minutes, seconds and milliseconds. Conversion to the internal representation is handled by STEP 7. The number is represented internally as a BCD number from 000 to 999. The time base can assume the following values: 10 ms (0000), 100 ms (0001), 1 s (0010) and 10 s (0011). The time is the product of the time base and the time value. Examples:

```
S5TIME#500ms        (= W#16#0050)
S5T#2h46m30s        (= W#16#3999)
```

#### DATE

A variable of data type DATE is stored in a word as a un-signed fixed-point number. The contents of the variable correspond to the number of days since 01.01.1990. The representation contains the year, the month and the day, each separated by a hyphen. Examples:

DATE#1990-01-01 (= W#16#0000)
D#2168-12-31 (= W#16#FF62)

**Table 24.2** Range Limits of a Floating-Point Number

| Sign | Exponent | Mantissa | Meaning |
|---|---|---|---|
| 0 | 255 | not equal to 0 | Not a valid floating-point number (not a number) |
| 0 | 255 | 0 | + infinite |
| 0 | 1 ... 254 | any | Positive normalized floating-point number |
| 0 | 0 | not equal to 0 | Positive denormalized floating-point number |
| 0 | 0 | 0 | + zero |
| 1 | 0 | 0 | – zero |
| 1 | 0 | not equal to 0 | Negative denormalized floating-point number |
| 1 | 1 ... 254 | any | Negative normalized floating-point number |
| 1 | 255 | 0 | – infinite |
| 1 | 255 | not equal to 0 | Not a valid floating-point number (not a number) |

### TIME

A variable of data type TIME occupies one doubleword. The representation contains the specifications for days (d), hours (h), minutes (m), seconds (s) and milliseconds (ms); individual specifications can be omitted. The contents of the variable are interpreted as milliseconds (ms) and stored as a 32-bit fixed-point number with sign. Examples:

```
TIME#24d20h31m23s647ms
(= DW#16#7FFF_FFFF)
```

```
TIME#0ms
(= DW#16#0000_0000)
```

```
T#-24d20h31m23s648ms
(= DW#16#8000_0000)
```

### TIME_OF_DAY

A variable of data type TIME_OF_DAY occupies one doubleword. It contains the number of milliseconds since the start of the day (0:00 hours) as an unsigned fixed-point number. The representation contains the specifications for hours, minutes and seconds, each separated by a colon. Specification of the milliseconds, following the seconds and separated by a dot, can be omitted. Examples:

```
TIME_OF_DAY#00:00:00
(= DW#16#0000_0000)
```

```
TOD#23:59:59.999
(= DW#16#0526_5BFF)
```

## 24.2 Complex Data Types

Complex data types are data types which (in their totality) cannot be processed direct by STL statements (Table 24.3). The data types are predefined but the length of the data type STRING (character string) and the composition and size of data types ARRAY (field) and STRUCT (structure) can be defined by the user.

You can declare variables of complex data types only in global data blocks, in instance data blocks, as temporary local data or as block parameters.

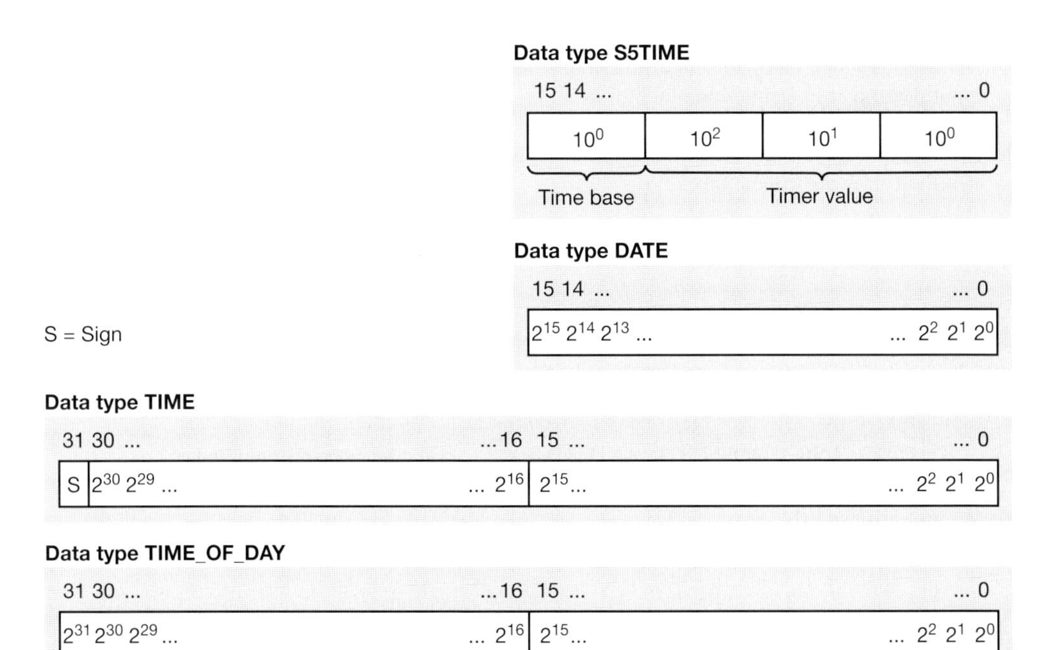


**Figure 24.3** Bit Assignments of the Data Types S5TIME, DATE, TIME and TIME_OF_DAY

**Table 24.3** Complex Data Types

| Data Type | Length | Description |
|---|---|---|
| DATE_AND_TIME | 8 bytes | Date and time |
| STRING | (n+2) bytes | Character string with n characters |
| ARRAY | n bytes | Field (combination of variables of the same type) |
| STRUCT | n bytes | Structure (combination of variables of different types) |

### 24.2.1 DATE_AND_TIME

The data type DATE_AND_TIME represents a time consisting of the date and the time of day. You can also use the abbreviation DT in place of DATE_AND_TIME.

*Declaration*

*varname* : DATE_AND_TIME := *Pre-assignment*;

*varname* : DT := *Pre-assignment*;

DATE_AND_TIME or DT are keywords; they can also be written in lower case.

*Pre-assignment*
At the declaration stage, the variable can be pre-assigned (not as a block parameter in a function, as an in/out parameter in a function block or as a temporary variable). The pre-assignment must be of the type DATE_AND_TIME or DT and must have the following appearance:

Keyword#Year-Month-Day-Hours:Minutes:Seconds.Milliseconds

Specification of the milliseconds can be omitted (Table 24.4).

*Application*
Variables of data type DT can be applied at block parameters of data type DT or ANY; for example, they can be copied with the system function SFC 20 BLKMOV. There are standard function blocks available for processing these variables ("IEC functions").

*Structure of the variables*
A variable of data type DATE_AND_TIME occupies 8 bytes (Figure 24.4). The variable begins at a word boundary (at a byte with an even address). All specifications are available in BCD format.

### 24.2.2 STRING

The data type STRING represents a character string consisting of up to 254 characters.

*Declaration*

*varname* : STRING[*maxNumber*] := *Pre-assignment*;

**Table 24.4** Examples of the Declaration of DT Variables and STRING Variables

| Name | Type | Initial Value | Comment |
|---|---|---|---|
| Date1 | DT | DT#1990-01-01-00:00:00 | DT variable minimum value |
| Date2 | DATE_AND_TIME | DATE_AND_TIME#<br>2089-12-31-23:59:59.999 | DT variable maximum value |
| FirstName | STRING[10] | 'Jack' | STRING variable, 4 characters out of 10 occupied |
| Surname | STRING[7] | 'Daniels' | STRING variable, all 7 characters occupied |
| NewLine | STRING[2] | '$R$L' | STRING variable, occupied by special characters |
| EmptyString | STRING[16] | '' | STRING variable without entry |

**Data format DT**

| | | |
|---|---|---|
| Byte n | Year | 0 to 99 |
| Byte n+1 | Month | 1 to 12 |
| Byte n+2 | Day | 1 to 31 |
| Byte n+3 | Hour | 0 to 23 |
| Byte n+4 | Minute | 0 to 59 |
| Byte n+5 | Second | 0 to 59 |
| Byte n+6 | ms | 0 to 999 |
| Byte n+7 | | Week-day |

Week-day
from 1 = Sunday
to 7 = Saturday

**Data format STRING**

| | | |
|---|---|---|
| Byte n | Maximum length | (k) |
| Byte n+1 | Current length | (m) |
| Byte n+2 | 1st character | current length |
| Byte n+3 | 2nd character | |
| Byte ... | ... | |
| Byte n+m+1 | mth character | |
| Byte ... | ... | |
| Byte n+k+1 | ... | |

(Bytes n+2 to n+k+1: maximum length)

**Figure 24.4** Structure of a DT and a STRING Variable

STRING is a keyword and can also be written in lower case.

*maxNumber* specifies the number of characters that a variable declared in this way can have (from 0 to 254). This specification can also be omitted; the Editor then uses a length of 254 bytes. With functions FCs, the Editor does not permit length specifications or it demands the standard length of 254.

*Pre-assignment*
At the declaration stage, the variable can be pre-assigned at the declaration stage (not as a block parameter in a function, as an in/out parameter in a function block or as a temporary variable). The pre-assignment is made with ASCII-coded characters enclosed in single inverted commas or with a preceding dollar sign in the case of certain characters (see data type CHAR).

If the pre-assignment value is shorter than the declared maximum length, the remaining character positions are not occupied. When further processing a variable of data type STRING, only the currently occupied character positions are taken into account. Pre-assignment as "EmptyString" is possible.

*Application*
Variables of data type STRING can be applied at block parameters of data type STRING or ANY; for example, they can be copied with the system function SFC 20 BLKMOV. There are standard function blocks available for processing these variables ('IEC functions').

*Structure of the variables*
A variable of data type STRING (character string) has a maximum length of 256 characters with 254 bytes of net data. It starts at a word boundary (at a byte with an even address).

When the variables are applied, their maximum length is defined. The current length (the actual used length of the character string = Number of valid characters) is entered when pre-assigning or when processing a character string. The first byte of the character string contains the maximum length and the second byte contains the current length; these are followed by the characters in ASCII format (Figure 24.4).

### 24.2.3 ARRAY

The data type ARRAY represents a field consisting of a fixed number of components of the same data type.

*Declaration*

| | |
|---|---|
| *fieldname* : | ARRAY [*minIndex..maxIndex*] OF *datatype* := *pre-assignment*; |
| *fieldname* : | ARRAY [$minIndex_1..maxIndex_1$, .., $minIndex_6..maxIndex_6$] OF *datatype* := *pre-assignment*; |

ARRAY and OF are keywords and can also be written in lower case.

*fieldname* is the name of the field

*minIndex* is the lower limit of the field and *maxIndex* is the upper limit. Both limits are INT numbers in the range –32768 to +32767; *maxIndex* must be greater than or equal to *minIndex*. A field can have up to 6 dimensions whose limits can be specified separated by a comma.

*datatype* can be any data type, even user-defined data types, except ARRAY itself

*Pre-assignment*
At the declaration stage, you can pre-assign values to individual field components (not as a block parameter in a function, as an in/out parameter in a function block or as a temporary variable). The data type of the pre-assignment value must match the data type of the field. You do not require to pre-assign all field components; if the number of pre-assignment values is less than the number of field components, only the first components are pre-assigned. The number of pre-assignment values must not be greater than the number of field components. The pre-assignment values are each separated by a comma. Multiple pre-assignment with the same values is specified within round brackets with a preceding repetition factor.

*Application*
You can apply a field as a complete variable at block parameters of data type ARRAY with the same structure or at a block parameter of data type ANY. For example, you can copy the contents of a field variable using the system function SFC 20 BLKMOV. You can also specify individual field components at a block parameter if the block parameter is of the same data type as the components.

If the individual field components are of elementary data types, you can process them with "normal" STL statements. A field component is accessed with the field name and an index in square brackets. The index is a fixed value and cannot be modified at runtime (no variable indexing possible).

*Multi-dimensional fields*
Fields can have up to 6 dimensions. Multi-dimensional fields are analogous to one-dimensional fields. At the declaration stage, the ranges of the dimensions are written in square brackets, each separated by a comma. When accessing the field components of multi-dimensional fields, you must always specify the indices of all dimensions.

*Structure of the variables*
An ARRAY variable always begins at a word boundary, that is, at a byte with an even address. ARRAY variables occupy the memory up to the next word boundary.

Components of data type BOOL begin in the least significant bit; components of data type BYTE and CHAR begin in the right-hand byte (Figure 24.5 left). The individual components are listed in order.

In multi-dimensional fields, the components are stored line-wise (dimension-wise) starting with the first dimension (Figure 24.5 right). With bit and byte components, a new dimension always

**Table 24.5** Examples of a Field Declaration

| Name | Type | Initial Value | Comment |
|---|---|---|---|
| MeasVal | ARRAY[1..24] | 0.4, 1.5, 11 (2.6, 3.0) | Field variable with 24 REAL components |
| | REAL | | |
| Time_array | ARRAY[-10..10] | 21 (TOD#08:30:00) | Time of day field with 21 components |
| | TIME_OF_DAY | | |
| Result | ARRAY[1..24,1..4] | 96 (L#0) | Two-dimensional field with 96 components |
| | DINT | | |
| Character | ARRAY[1..2,3..4] | 2 ('a'), 2 ('b') | Two-dimensional field with 4 components |
| | CHAR | | |

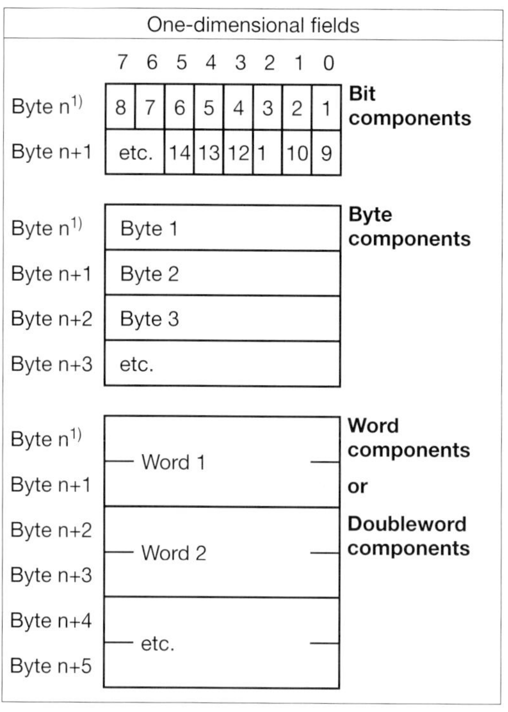


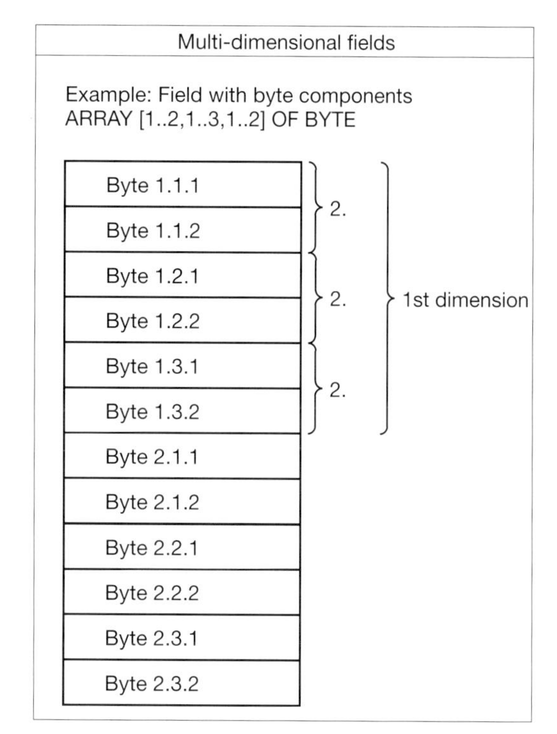


[1] n = even

**Figure 24.5** Structure of an ARRAY Variable

starts in the next byte, and with components of other data types a new dimension always starts in the next word (in the next even byte).

### 24.2.4 STRUCT

The data type STRUCT represents a data structure consisting of a fixed number of components that can each be of a different data type.

*Declaration*

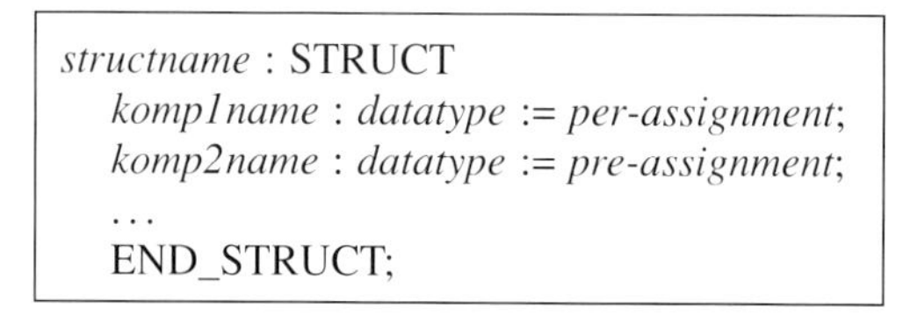
*structname* : STRUCT
*komp1name* : *datatype* := *per-assignment*;
*komp2name* : *datatype* := *pre-assignment*;
...
END_STRUCT;

STRUCT and END_STRUCT are keywords that can also be written in lower case.

*structname* is the name of the structure.

*komp1name, komp2name* etc. are the names of the individual structure components.

*datatype* is the data type of the individual components. All data types can be used, including further structures.

*Pre-assignment*
At the declaration stage, you can pre-assign values to the individual structure components (not as a block parameter in a function, as an in/out parameter in a function block or as a temporary variable). The data types of the pre-assignment values must match the data types of the components.

*Application*
You can apply a complete variable at block parameters of data type STRUCT with the same structure or at a block parameter of data type ANY. For example, you can copy the contents of a STRUCT variable with the system function SFC 20 BLKMOV. You can also specify an individual structure component at a block parameter if the block parameter is of the same data type as the component.

If the individual structure components are of elementary data types, you can process them

**Table 24.6** Example of Declaring a Structure

| Name | | Type | Initial Value | Comment |
|---|---|---|---|---|
| MotCont | | STRUCT | | Simple structure variable with 4 components |
| | On | BOOL | FALSE | Variable MotCont.On of type BOOL |
| | Off | BOOL | TRUE | Variable MotCont.Off of type BOOL |
| | Delay | S5TIME | S5TIME#5s | Variable MotCont.Delay of type S5TIME |
| | maxSpeed | INT | 5000 | Variable MotCont.maxSpeed of type INT |
| | | END_STRUCT | | |

with 'normal' STL statements. A structure component is accessed with the structure name and the component name separated by a dot.

*Structure of the variables*

A STRUCT variable always begins at a word boundary, that is, at a byte with an even address; following this, the individual components are located in the memory in the order of their declaration. STRUCT variables occupy the memory up to the next word boundary.

Components of data type BOOL begin in the least significant bit of the next byte; components of data type BYTE and CHAR begin in the next byte (Figure 24.6). Components of other data types begin at a word boundary.

A nested structure is a structure as a component of another structure. A nesting depth of up to 6 structures is possible. All components can be accessed individually with STL statements provided they are of elementary data type. The individual names are each separated by a dot.

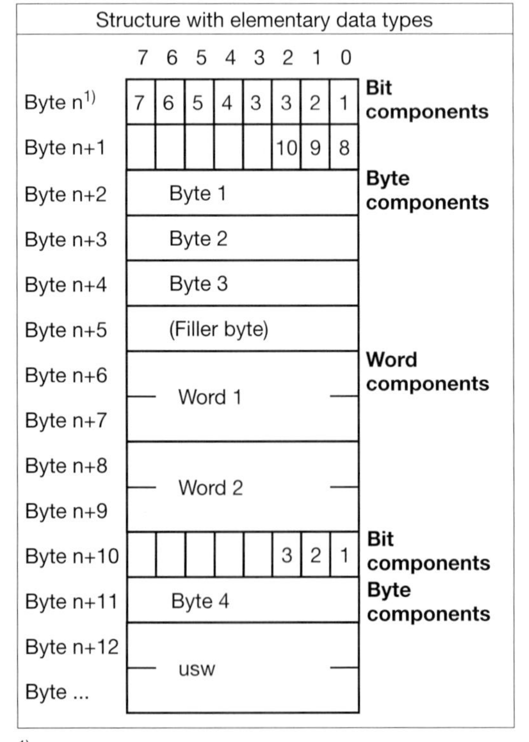


1) n = even

**Figure 24.6** Structure of a STRUCT Variable

## 24.3 User-Defined Data Types

If a data structure occurs frequently in your program, or if you want to give a name to a data structure, you can specify a user-defined data type (UDT). You create UDTs with the incremental Editor or with the Text Editor as source file. You assign a name in the symbol table. UDTs have global validity; that is, once declared, they can be used in all blocks.

*Declaration*

```
TYPE udtname
   STRUCT
   komp1name : datatype := pre-assignment;
   komp2name : datatype := pre-assignment;
   ...
   END_STRUCT
END_TYPE
```

TYPE, END_TYPE, STRUCT and END_STRUCT are keywords that can also be written in lower case.

*udtname* is the name of the user-defined data type. In place of *udtname,* you can also use the absolute address UDT*n*; it is located in the range from UDT 0 to UDT 65535.

*komp1name, komp2name* etc. are the names of individual structure components.

*datatype* is the data type of the individual components. All data types can be used.

User-defined data types are pre-assigned and used like structures; the structure is the same as for structures. You assign the name of an absolute address in the symbol table. The data type of a UDT (in the symbol table) is identical with the absolute address.

The example "Message Frame Data" in Section 26.4 shows you how to work with user-defined data types.

**Table 24.7** Example of a User-Defined Data Type UDT

| Name | | Type | Initial Value | Comment |
|---|---|---|---|---|
| | | STRUCT | | |
| | Identifier | WORD | W#16#F200 | UDT component Identifier of type WORD |
| | Number | INT | 0 | UDT component Number of type INT |
| | Time1 | TIME_OF_DAY | TOD#0:0:0.0 | UDT component Time1 of type TOD |
| | | END_STRUCT | | |

# 25 Indirect Addressing

Indirect addressing gives you the ability to address operands whose addresses are not known until runtime. With indirect addressing, you can also effect multiple processing of program sections, for example, in a loop, and with each pass, you can assign a different address to the operands used. This chapter shows how the STL programming language supports you here.

Since the addresses are not calculated until runtime in the case of indirect addressing, there is a danger that memory areas could be inadvertently overwritten. *The programmable controller might then respond unpredictably! Please exercise the utmost caution when using indirect addressing!*

The examples in this chapter are also presented on the diskette accompanying the book under the "Variable Handling" program in function block FB 125 or source file Chap_25.

## 25.1 Pointers

The address for indirect addressing must be structured in such a way that it contains the bit address, the byte address and, if applicable, also the operand area. It therefore has a special format, called *Pointer*. A pointer is used for "pointing" to an operand.

STL recognizes three types of pointers:

- Area pointers; these are 32 bits long and contain a specific operand or its address
- DB pointers; these are 48 bits long and also contain the number of the data block in addition to the area pointer
- ANY pointer; these are 80 bits long and contain further specifications such as the data type of the operand in addition to the DB pointer

Only the area pointer is significant for indirect addressing, the DB pointer and the ANY pointer are used when transferring block parameters. Since these pointer types contain the area pointer, this chapter also describes the structure of the DB pointer and the ANY pointer.

**Table 25.1** Area Coding in the Area Pointer

| Area | | Coding | | |
|---|---|---|---|---|
| P | Peripheral I/O | 0 | 0 | 0 |
| I | Inputs | 0 | 0 | 1 |
| Q | Outputs | 0 | 1 | 0 |
| M | Memory bits | 0 | 1 | 1 |
| DBX | Global data | 1 | 0 | 0 |
| DIX | Instance data | 1 | 0 | 1 |
| L[1)] | Temporary local data | 1 | 1 | 0 |
| V[2)] | Temporary local data of the preceding block | 1 | 1 | 1 |

[1)] Not with area-crossing addressing

[2)] Used at block parameter transfer

### 25.1.1 Area Pointer

The area pointer contains the operand address and possibly also the operand area. Without the operand area, it is an *area-internal* pointer; if the pointer also contains the operand area, it is referred to as an *area-crossing* pointer.

You can address an area pointer direct and load it into the accumulator or into the address register, since it is 32 bits long. The notation for constant representation is as follows:

P#y.x for an area-internal pointer (for example P2.0) and

P#Zy.x for an area-crossing pointer (for example P#M22.0)

where x = bit address, y = byte address and Z = area. As the area, you specify the operand identifier (Table 25.1). The assignment of bit 31 differentiates between the two pointer types (Figure 25.1).

**Area-internal pointer**

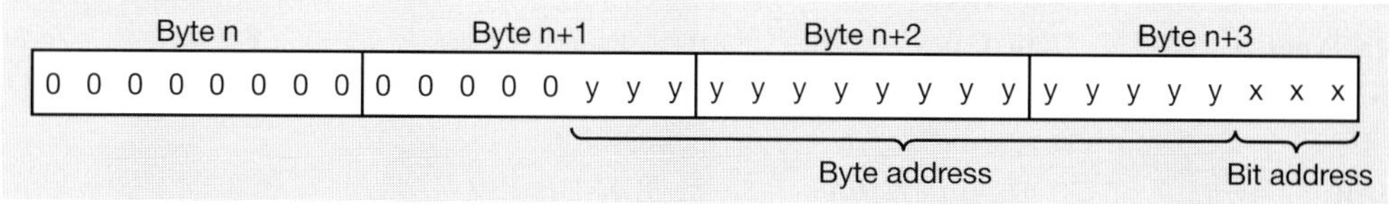


**Area-crossing pointer**

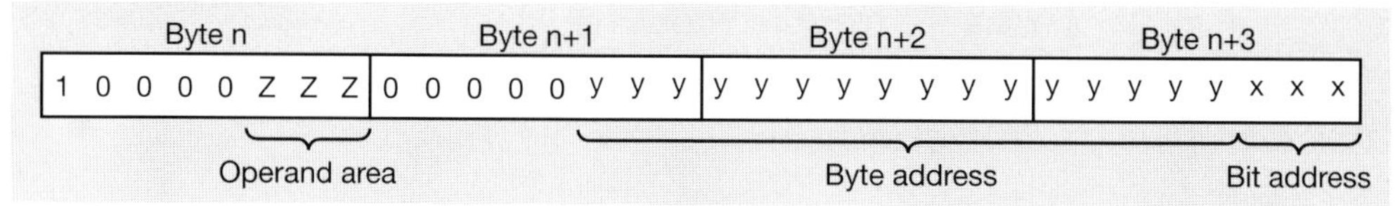


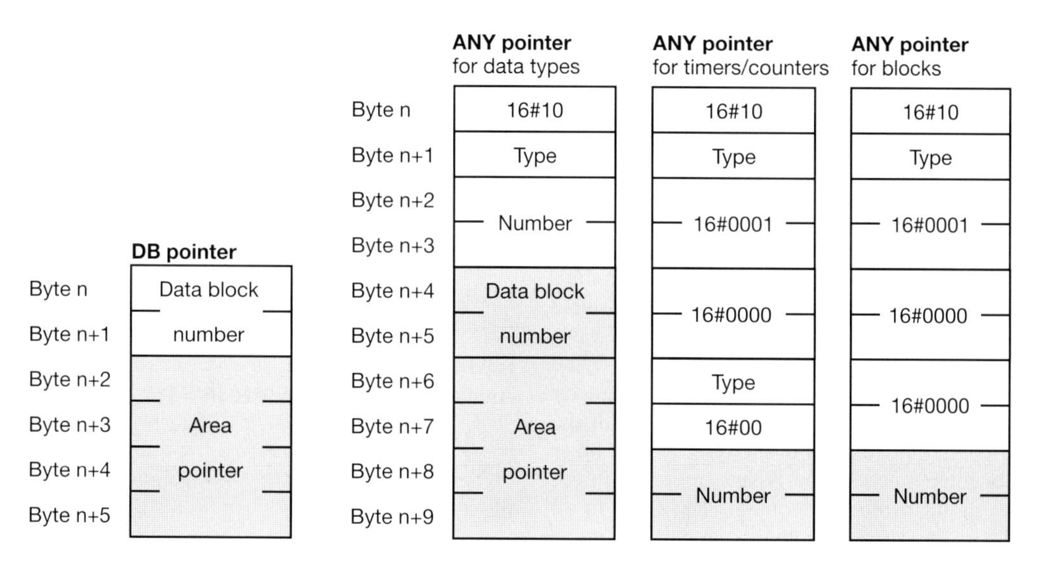


**Figure 25.1** Structure of the Pointer for Indirect Addressing

The area pointer has a bit address that must always be specified even for digital operands; with digital operands, you specify 0 as the bit address. With the area pointer P#M22.0, you can address memory bit M 22.0 but also memory byte MB 22, memory word MW 22 or memory doubleword MD 22.

## 25.1.2 DB Pointer

A DB pointer also contains a data block number as a positive INT number in addition to the area pointer. It specifies the data block if the area pointer contains the operand areas global data or instance data. In all other cases, the first two bytes contain zero.

You are familiar with the notation of the pointer from the full addressing of data operands. Here too, the data block and the data operand are specified separated by a dot:

P#DataBlock.DataOperand

Example: P#DB 10.DBX 20.5

You cannot load this pointer; however, you can apply it at a block parameter of parameter type POINTER in order to point to a data operand. STL uses this pointer type internally in order to transfer actual parameters.

**Table 25.2** Data Type Coding in ANY Pointers

| Data Types | Type | Data Types | Type | Data Types | Type |
|---|---|---|---|---|---|
| BOOL | 01 | REAL | 08 | BLOCK_FB | 17 |
| BYTE | 02 | DATE | 09 | BLOCK_FC | 18 |
| CHAR | 03 | TOD | 0A | BLOCK_DB | 19 |
| WORD | 04 | TIME | 0B | BLOCK_SDB | 1A |
| INT | 05 | S5TIME | 0C | COUNTER | 1C |
| DWORD | 06 | DT | 0E | TIMER | 1D |
| DINT | 07 | STRING | 13 | | |

### 25.1.3 ANY Pointer

The ANY pointer also contains the data type and a repetition factor in addition to the DB pointer. This makes it possible to also point to a data area.

The ANY pointer is available in two versions: for variables with data types and for variables with parameter types. If you point to a variable with a data type, the ANY pointer contains a DB pointer, the type and a repetition factor. If the ANY pointer points to a variable with parameter type, it contains only the number instead of the DB pointer, in addition to the type. With a timer or counter function, the type is repeated in the byte (n+6); byte (n+7) contains B#16#00. In all other cases, these two bytes contain the value W#16#0000.

The first byte of the ANY pointer contains the syntax ID; in STL it is always $10_{hex}$. The type specifies the data type of the variables for which the ANY pointer applies. Variables of elementary data types, DT and STRING receive the type shown in Table 25.2 and the quantity 1.

If you apply a variable of data type ARRAY or STRUCT (also UDT) to an ANY parameter, the Editor generates an ANY pointer to the field or the structure. This ANY pointer contains the identifier for BYTE ($02_{hex}$) as the type and the number of bytes making up the length of the variable as the quantity. The data type of the individual field or structure components are insignificant here. An ANY pointer thus points to a WORD field with double the quantity of bytes. Exception: A pointer to a field consisting of components of data type CHAR is also applied with CHAR type ($03_{hex}$).

You can apply an ANY pointer at a block parameter of parameter type ANY if you want to point to a variable or an operand area. The constant representation for data types is as follows:

[DataBlock.]Operand Type Quantity

Examples:

P#DB 11.DBX 30.0 INT 12
Area with 12 words in DB 11 from DBB 30

P#M 16.0 BYTE 8
Area with 8 bytes from MB 16

P#I 18.0 WORD 1
Input word IW 18

P#I 1.0 BOOL 1
Input I 1.0

With parameter types, you write the pointer as follows:

L# Number Type Quantity

Examples:

L# 10 TIMER 1
Timer function T10

L# 2 COUNTER 1
Counter function C2

The Editor then applies an ANY pointer that agrees in type and in quantity with the specifications in the constant representation. Please note that the operand address in the ANY pointer for data types must always be a bit address.

Specification of a constant ANY pointer is meaningful if you want to access a data area for which you have not declared a variable. In principle, you can also apply variables or operands at an ANY parameter. For example, the representation "P#I 1.0 BOOL 1" is identical to "I 1.0" or the corresponding symbol address.

With parameter type ANY, you can also declare variables in the temporary local data. You use these variables to create an ANY pointer that can be modified at runtime (see Section 26.3.3 "Variable ANY Pointer").

If you do not specify any pre-assignment when declaring an ANY parameter in a function block, the Editor assigns $10_{hex}$ to the syntax ID and $00_{hex}$ to the remaining bytes. It then represents these (empty) ANY pointers (in the data view) thus: P#P0.0 VOID 0.

## 25.2 Types of Indirect Addressing

### 25.2.1 General

Indirect addressing is only possible with absolute addressing. You cannot indirectly address variables with symbolic addresses (you must also access the components of a field individually and directly). If you want to access a variable indirectly, you must know the absolute address of the variable. STL supports you here with direct variable access (see next chapter). Absolute addressing recognizes the following

- immediate addressing,
- direct addressing and
- indirect addressing.

Addressing via block parameters is a special form of indirect addressing: By specifying the actual parameter at the block parameter, you determine the operand to be processed at runtime.

We refer to *immediate addressing* when the number value is specified together with the operation. Examples of immediate addressing include loading a constant value into the accumulator, shifting by a fixed value and also setting and resetting the result of the logic operation with SET or CLR.

With *direct addressing*, you access the operand direct, for example A I 1.2 or L MW 122. The value you want to combine or load into the accumulator is located in an operand, that is, in a memory cell. You address this memory cell by specifying the address direct in the STL statement.

With *indirect addressing*, the STL statement indicates where the address is to be found instead of containing the address itself. We distinguish between two types of indirect addressing depending on the type of the indicator:

- *Memory-indirect addressing* uses an operand from the system memory to accommodate the address. Example: In the statement T QW [MD 220], the address of the output word to which the transfer is to be made, is located in the memory doubleword MD 220.
- *Register-indirect addressing* uses an address register to determine the address of the operand. Example: With the statement T QW [AR1,P.0] a transfer is made to the output word whose address is 2 (bytes) higher than the address located in the address register AR1.

You can use register-indirect addressing in two variants: With *area-internal* register-indirect addressing, you program in the statement the operand area for which the address in the address register is to apply. The address in the address register therefore moves within an operand area (example: L MW[AR1,P#0.0], load the memory word whose address is located in AR1). With *area-crossing* register-indirect addressing, you specify only the operand width (bit, byte, word or doubleword) in the statement; the operand area is located in the address register and can be modified dynamically (example: L W[AR1,P#0.0], load the word whose operand area and address are located in the AR1).

### 25.2.2 Indirect Addressable Operands

The indirect addressable operands can be divided into two categories:

- Operands that can be assigned an elementary data type, and
- Operands that can be assigned a parameter type.

You can use memory-indirect and register-indirect addressing with the former, but only memory-indirect addressing with the latter (Table 25.3). The operands that cannot have a bit address also require no bit address in the pointer, so that a 16-bit wide number is sufficient as the address (unsigned INT number).

The areas of the pointers have a theoretical size of 0 to 65535 (byte address or number). In prac-

**Table 25.3** Indirect Addressable Operands

| Indirect Addressable Operands | Addressing | Pointer |
|---|---|---|
| Peripheral I/O, inputs, outputs, memory bits, global data, instance data, temporary local data | Memory-indirect and register-indirect | Area pointers, either area-internal or area crossing |
| Timers, counters, functions, function blocks, data blocks | Memory-indirect | 16-bit number |

tice, the addresses are restricted by the operand volume of the CPU in each case. The bit address lies in the range from 0 to 7.

### 25.2.3 Memory-Indirect Addressing

With memory-indirect addressing, the address is located in an operand. This (address) operand has doubleword width if an area pointer is required, or word width, if indirect addressing via a number is used.

The address operand can be within one of the following operand areas:

- Bit memory
  as absolute addressed operand or symbolically addressed variable
- L stack (temporary local data)
  as absolute addressed operand or symbolically addressed variable
- Global data block
  as absolute addressed operand
  When using global data operands, please ensure that the 'correct' data block is opened via the DB register. If, for example, you address a global data operand indirect via a global data doubleword, both operations must be located in the same data block.
- Instance data block
  as absolute addressed operand or as symbolically addressed variable
  There are restrictions to the use of instance data as address operands; see below.

If you use instance data as address operands in functions, treat them in exactly the same way as global data operands; you use only the DI register in place of the DB register. Symbolic addressing is not permissible in this case. You can use instance data as address operands in function blocks only if you compile the blocks as CODE_VERSION1 block (no multi-instance capability).

#### Indirect addressing with an area pointer

The area pointer required for memory-indirect addressing is always an area-internal pointer; that is, it consists of byte and bit address. If you want to address digital operands, you must always specify 0 as the bit address.

Example: Memory doubleword MD 10 contains the pointer P#30.0. The statement A M [MD 10] accesses the memory bit whose address is located in memory doubleword MD 10; memory bit M 30.0 is therefore checked (Figure 25.2). With the statement L MW [MD 10], you load memory word MW 30 into the accumulator.

You can use memory-indirect addressing for all binary operands in conjunction with the binary logic operations and the memory functions and for all digital operands in conjunction with the load and transfer functions.

#### Indirect addressing with a number

The number for indirect addressing of timers, counters and blocks is 16 bits wide and can be accommodated in one word.

Example: Memory word MW 20 contains the number 133. The statement OPN DB [MW 20] opens the global data block whose number is located in memory word MW 20. With the statement SP T [MW 20] you start timer T 133 as a pulse.

You can use all timer and counter operations together with indirect addressing. You can open a data block either via the DB register (OPN DB [..]) or via the DI register (OPN DI [..]). If the address word contains zero, the CPU executes a NOP operation.

You can address the call of code blocks indirectly with UC FC [..] and CC FC [..] or UC FB [..] and CC FB [..]. Please note that the blocks

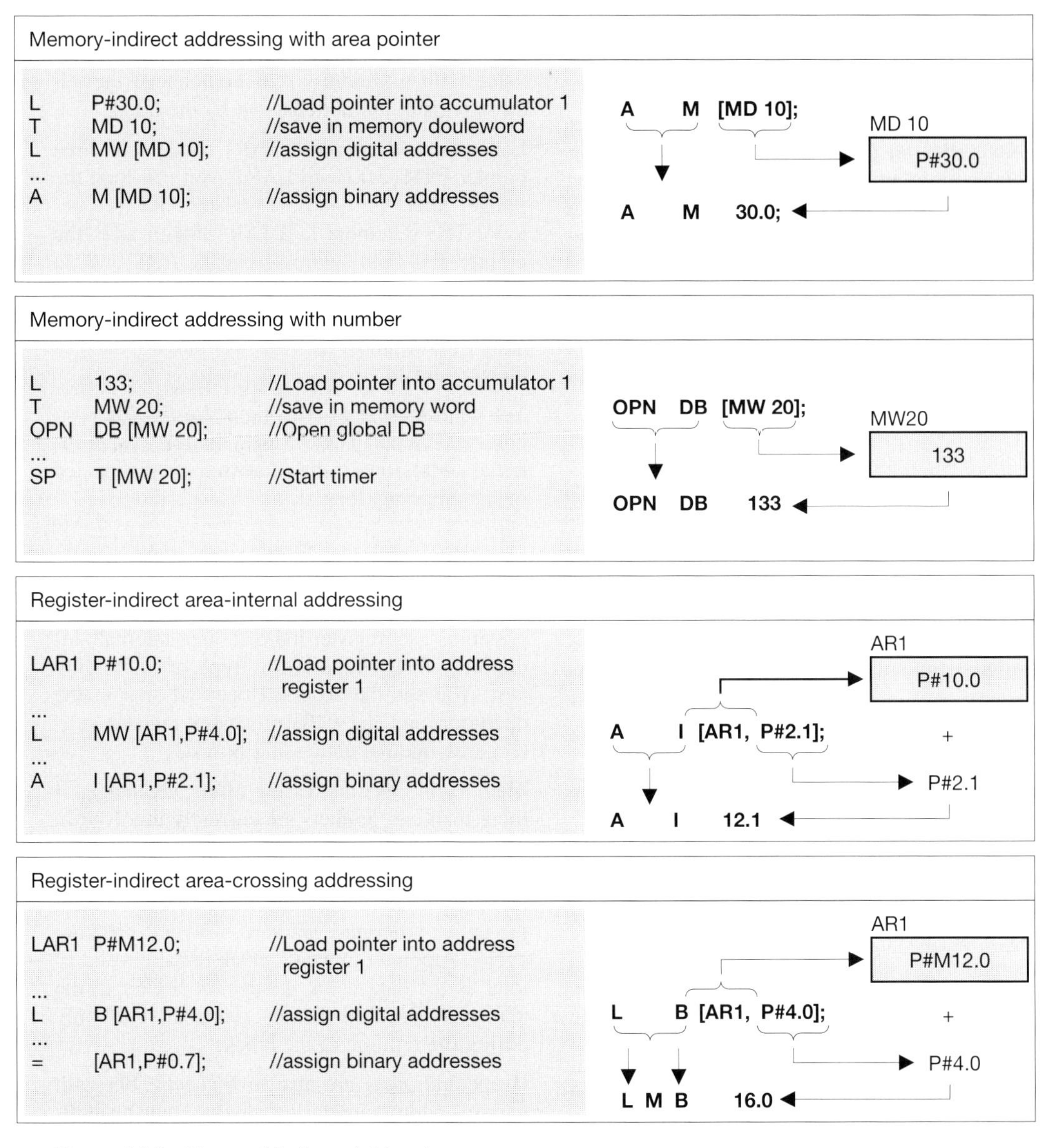


**Figure 25.2** Types of Indirect Addressing

must have no block parameters; the function block called via UC or CC also has no instance data block.

### 25.2.4 Register-Indirect Area-Internal Addressing

With register-indirect area-internal addressing, the address is located in one of the two address register. The contents of the address register is an area-internal pointer.

With register-indirect addressing, an offset is specified in addition to the address register. This offset is added to the contents of the address register when the operation is executed (without changing the contents of the address register). This offset has the format of an area-internal pointer. You must always specify it and you can only specify it as a constant. With indirect addressed digital operands, this offset must have the bit address 0. The maximum value is P#4095.7.

Example: Address register AR1 contains the area pointer P#10.0 (with LAR1, you can load the pointer direct into address register AR1, see below). The statement A I [AR1,P#2.1] adds the pointer P.1 to address register AR1 and so forms the address of the input to be checked. With the statement L MW [AR1,P#4.0], you load memory word MW 14 into the accumulator.

### Area-internal addressing with area-crossing pointers

If the address register contains an area-crossing pointer and if you use this address register in conjunction with area-internal operations, the operand area in the address register is ignored.

Example: The following statements load an area-crossing pointer to the global data bit DBX 20.0 into address register AR1 and then execute area-internal addressing via AR1 on a memory doubleword. When the load statement is executed, memory doubleword MD 20 is loaded.

```
LAR1    P#DBX20.0;
L       MD[AR1,P#0.0];
```

## 25.2.5 Register-Indirect Area-Crossing Addressing

With register-indirect area-crossing addressing, the address is located in one of the two address registers. The contents of the address register is an area-crossing pointer.

With area-crossing addressing, you write the operand area in conjunction with the area pointer into the address register. If you use indirect addressing you only specify an ID for the operand width as the operand: no specification for a bit, "B" for a byte, "W" for a word and "D" for a doubleword.

As with area-internal addressing, you work here with an offset that you specify with as a fixed value with bit address. The contents of the address register are not changed by the offset.

Example: Address register AR1 contains the area pointer P#M12.0 (with LAR1, you can load the pointer direct into address register AR1, see below). The statement L B [AR1,P#4.0] adds the pointer P#4.0 to address register AR1 and so forms the address of the memory byte to be loaded. With the statement = [AR1,P#0.7], you assign the result of the logic operation (RLO) to memory bit M 12.7.

You cannot use area-crossing addressing on temporary local data (CPU Stop). Switch to area-internal addressing, if the addressed area is located in the temporary local data.

## 25.2.6 Summary

When do you use which type of addressing? If possible, use register-indirect area-internal addressing. STL supports this type of addressing best. You see the accessed operand area in the operation and the CPU processes register-indirect area-internal addressing fastest.

Memory-indirect addressing offers advantages if more than two pointers are currently involved in program execution. However, please note the "validity period" of a pointer: A pointer in the bit memory area is available unrestrictedly during the entire program even across several program cycles. A pointer in a data block remains valid as long as the data block is open. A pointer in the temporary local data area remains valid only during the runtime of the block.

If operand areas are also to be accessible with variable addressing during runtime, register-indirect area-crossing addressing is the right choice.

Table 25.4 shows a comparison of indirect addressing types. All statement sequences shown lead to the same result, the setting of output Q 4.7.

**Table 25.4** Comparison of Indirect Addressing Types

| Memory-Indirect | | Register-Indirect Area-Internal | | Register-Indirect Area-Crossing | |
|---|---|---|---|---|---|
| L | P#4.7 | LAR1 | P#4.7 | LAR1 | P#Q4.7 |
| T | MD 24 | | | | |
| S | Q [MD 24] | S | Q [AR1,P#0.0] | S | [AR1,P#0.0] |

## 25.3 Working the Address Registers

### 25.3.1 Overview

The listing below shows the statements that are possible in conjunction with the address registers. Figure 25.3 gives a graphical representation of the overview. All statements are executed unconditionally and do not affect the status bits.

```
LAR1   –        Load address register AR1
LAR2   –        Load address register AR2
       P#Zy.x           with an area-crossing pointer
       P#y.x            with an area-internal pointer

LAR1   –        Load address register AR1 with the content of
LAR2   –        Load address register AR2 with the contents of
       MD y             a memory doubleword
       LD y             a local data doubleword
       DBD y            a global data doubleword
       DID y            an instance data doubleword 1)

LAR1            Load address register AR1 with the contents of accumulator 1
LAR2            Load address register AR2 with the contents of accumulator 1
LAR1   AR2      Load address register AR1 with the contents of address register AR 2

TAR1   –        Transfer the contents of address register AR1 to
TAR2   –        Transfer the contents of address register AR2 to
       MD y             a memory doubleword
       LD y             a local data doubleword
       DBD y            a global data doubleword
       DID y            an instance data doubleword1)

TAR1            Transfer the contents of address register AR1 to accumulator 1
TAR2            Transfer the contents of address register AR2 to accumulator 1
TAR1   AR2      Transfer the contents of address register AR1 to accumulator AR 2

CAR             Swap the contents of the address registers

+AR1            Add the contents of accumulator 1 to address register AR 1
+AR2            Add the contents of accumulator 1 to address register AR 2
+AR1   P#y.x    Add a pointer to the contents of address register AR1
+AR2   P#y.x    Add a pointer to the contents of address register AR2
```

[1)] There are restrictions to the use of these operands (see "Special Features of Indirect Addressing" below).

### 25.3.2 Loading into an Address Register

The statement LAR*n* loads an area pointer into address register AR*n*. As the source, you can select an area-internal or area-crossing pointer or a doubleword from the operand areas bit memory, temporary local data, global data and instance data. The contents of the doubleword must correspond to the format of an area pointer.

If you do not specify an operand, LAR*n* loads the contents of accumulator 1 into address register AR*n*.

With the statement LAR1 AR2, you copy the contents of address register AR2 into address register AR1.

Examples:

```
LAR2 P#20.0;  //Load AR2 with P0.0
L    P#24.0;
LAR1 ;        //Load AR1 with <Accum 1>
LAR1 MD 120;  //Load AR1 with <MD 120>
LAR1 AR2;     //Load AR1 with <AR2>
```

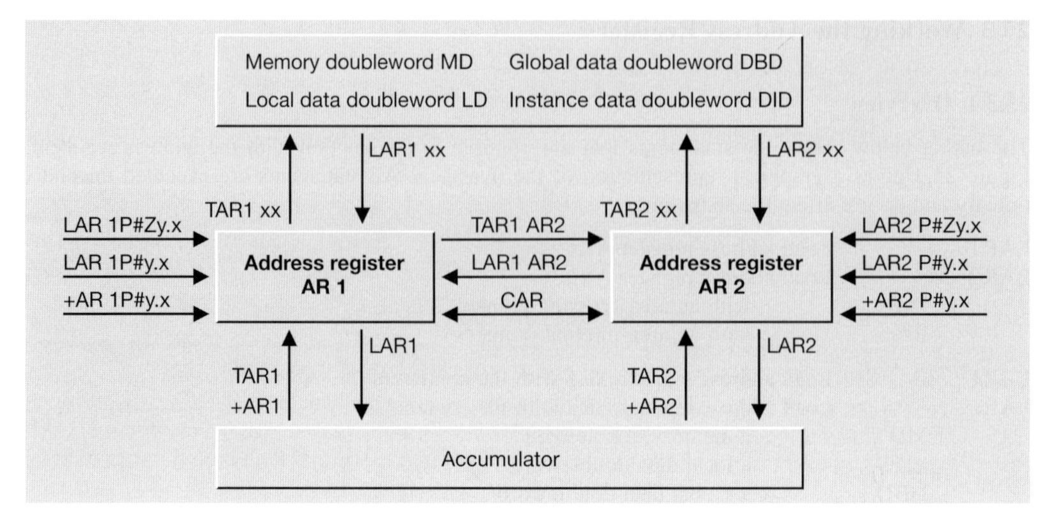


**Figure 25.3** Operations in Conjunction with Address Registers

### 25.3.3 Transferring from an Address Register

The statement TAR*n* transfers the complete area pointer from address register AR*n*. As the destination, you can specify a doubleword from the operand areas bit memory, temporary local data, global data and instance data.

If you do not specify an operand, TAR*n* transfers the contents of address register AR*n* into accumulator 1. In so doing, the previous contents of accumulator 1 are shifted into accumulator 2; the previous contents of accumulator 2 are lost. Accumulators 3 and 4 remain unaffected.

With the statement TAR1 AR2, you copy the contents of address register AR1 into address register AR2.

Examples:

```
TAR2    MD 140; //Transfer  to MD 140
TAR1    ;       //Transfer  to Accum 1
TAR1    AR2;    //Transfer  to AR2
```

### 25.3.4 Swap Address Registers

The statement CAR swaps the contents of address registers AR1 and AR2.

Example: 8 Bytes of data are transferred between the bit memory area from MB 100 and the data area from DB20.DBB 200. The direction of transfer is determined by memory bit M 126.6. If M 126.6 has signal state "0", the contents of the address registers are swapped. If you want to transfer data between two data blocks in this way, load the two data block registers (with OPN DB and OPN DI) together with the address registers and swap with the statement TDB.

```
      LAR1 P#M100.0;
      LAR2 P#DBX200.0;
      OPN  DB 20;
      A    M 126.6;
      JC   OV;
      CAR  ;
OV:   L    D[AR1,P#0.0];
      T    D[AR2,P#0.0];
      L    D[AR1,P#4.0];
      T    D[AR2,P#4.0];
```

Note: System function SFC 20 BLKMOV is available for transferring larger data areas.

### 25.3.5 Adding to the Address Register

You can add a value to the address registers in order, for example, to increment the address of an operand at each loop pass in program loops. You either specify the value as a constant (as an area-internal pointer) in the statement, or the value is located in the right-hand word of accumulator 1. The type of pointer in the address register (area-internal or area-crossing) and the operand area are retained.

## Adding with pointers

The statements +AR1 P#y.x and +AR2 P#y.x add a pointer to the address register specified. Please note, that with these statements, the area pointer has a maximum size of P#4095.7. If the accumulator contains a value greater than P#4095.7, the number is interpreted as a fixed-point number in two's complement and subtracted (see below).

Example: A data area is to be compared word-wise with a value. If the comparison value is greater than the value in the data field, a memory bit is to be set to "1", otherwise it is to be set to "0".

```
      OPN  DB 14;
      LAR1 P#DBX20.0;
      LAR2 P#M10.0;
      L    Quantity_Data;
Loop: T    LoopCounter;
      L    ComparisonValue;
      L    W[AR1,P#0.0];
      >I   ;
      =    [AR2,P#0.0];
      +AR1 P#2.0;
      +AR2 P#0.1;
      L    LoopCounter;
      LOOP Loop;
```

## Adding with value in the accumulator

The statements +AR1 and +AR2 interpret the value in accumulator 1 as a number in INT format, extend it with the correct sign to 24 bits and add it to the contents of the address register. In this way, a pointer can also be reduced. Violation of the maximum range of the byte address (0 to 65535) has no further effects: The uppermost bits are "cut" (Figure 25.4).

Please note, that the bit address is located in bits 0 to 2. If you want to increment the byte address already in accumulator 1, you must add from bit 3 (shift the value by 3 to the left).

Example: In data block DB 14, the 16 bytes whose addresses are calculated from the pointer in memory doubleword MD 220 and a (byte) offset in memory byte MB 18 are to be deleted. Before adding to AR1, the contents of MB 18 must be adjusted (SLW 3).

```
OPN   DB 14;
LAR1  MD 220;
L     MB 18;
SLW   3;
+AR1  ;
L     0;
T     DBD[AR1,P#0.0];
T     DBD[AR1,P#4.0];
T     DBD[AR1,P#8.0];
T     DBD[AR1,P#12.0];
```

Note: System function SFC 21 FILL is available for filling larger data areas with bit patterns.

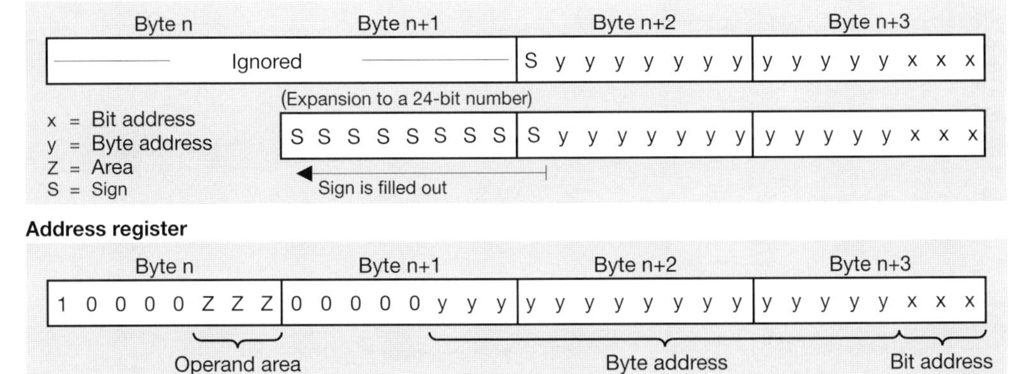


**Figure 25.4** Adding to the Address Register

## 25.4 Special Features of Indirect Addressing

### 25.4.1 Using Address Register AR1

STL uses address register AR1 to access block parameters that are transferred as DB pointers. In the case of functions, these include all block parameters of complex data type and in the case of function blocks, it means in/out parameters of complex data type.

When you access a block of this kind, in order, for example, to check a bit component of a structure or to write an INT value to a field component, the contents of address register AR1 are changed and so, incidentally, are the contents of the DB register. This also applies when you 'pass on' block parameters of this data type to called blocks.

If you use address register AR1, there must be no block parameter access as described above between loading the address register and indirect addressing. Otherwise, you must save the contents of AR1 before the access, and load them again following the access.

Example: You load a pointer into AR1 and use this address register for indirect addressing. In the meantime, you want to load the value of the structure component *Motor.Act*. Before loading *Motor.Act*, you save the contents of the DB register and address register AR1; after loading, you restore the contents of the registers (Figure 25.5).

### 25.4.2 Using Address Register AR2

With "multi-instance-capable" function blocks (block version 2), STEP 7 uses address register AR2 as the "Base address register" for instance data. When an instance is called, AR2 contains P#DBX0.0 and all accesses to block parameters or static local data in the FB use register-indirect area-internal addressing with the operand area DI via this register. A call of a local instance increments the "base address" with +AR2 P#y.x, so that access can be made relative to this address within the called function block that uses the instance data block of the calling function block. In this way, function blocks can be called both as autonomous instances and as local instances (and here at any point in a function block, even several times).

If you program a function block with block version 1 (no "multi-instance capability"), STEP 7 does not use address register AR2.

So if you want to use address register AR2 in a function block with multi-instance capability, you must first save the contents and then restore them after use.You must not program any block

```
VAR_TEMP
  AR1_Memory    : DWORD;
  DB_Memory     : WORD;
END_VAR
...
//Indirect addressing with AR1 and DB register
LAR1    P#y.x;
OPN     DB z;
...
//Save the register contents
L       DBNO;
T       DB_Memory;
TAR1    AR1_Memory;
//Access block parameters of complex data types
L       Motor.Act;
//Restore the register contents
OPN     DB [DBMemory];
LAR1    AR1_Memory;
T       DBW[AR1,P#0.0];                // store loaded value
```

**Figure 25.5** Example: Saving Address Register AR1

```
VAR_TEMP
  AR2_Memory    : DWORD;
  DI_Memory     : WORD;
END_VAR
...

//Save the register contents
L    DINO;
T    DI_Memory;
TAR2 AR2_Memory;
//Indirect addressing with AR2 and the DI register
LAR2 P#y.x;
OPN  DI z;
...
L    DIW[AR2, P#0.0];
...
//Restore the register contents
OPN  DI [DI_Memory];
LAR2 AR2_Memory;
```

**Figure 25.6** Example: Saving Address Register AR2

parameter or static local data accesses in the area in which you work with address register AR2.

Within functions, there are no restrictions on working with address register AR2.

Example: You want to perform indirect addressing in a function block with AR2 and the DI register. First, you save their contents. You must not access block parameters or static local data again until you have restored the contents of AR2 and the DI register (Figure 25.6).

**Table 25.5** Different Programming in the Case of Static Local Data

| In the FB with CODE_VERSION1 (no "multi-instance capability") | In function block with "multi-instance capability" |
|---|---|
| VAR<br>PTR : DWORD;<br>END_VAR | VAR<br>PTR : DWORD;<br>END_VAR<br>VAR_TEMP<br>tPTR : DWORD;<br>END_VAR |
| L MW[PTR]; | L PTR;<br>T tPTR;<br>L MW[tPTR]; |
| LAR1 PTR; | L PTR;<br>LAR1; |
| TAR1 PTR; | TAR1;<br>T PTR; |

### 25.4.3 Restrictions with Static Local Data

With function blocks compiled with CODE_VERSION1 (no "multi-instance capability"), you can use all statements described in this chapter without restriction.

In the case of function blocks with "multi-instance capability", the Editor accesses instance data via address register AR2 (see above); that is, all accesses are indirect. This also applies in conjunction with indirect addressing or when handling address registers. If you use absolute addressing for the instance data in which you store area pointers, the Editor adopts the absolute address. However, as soon as you use symbolic addressing, the Editor rejects this programming as 'double indirect addressing'.

Table 25.5 gives two examples of this: If you are using memory-indirect addressing in the case of function blocks with "multi-instance capability", you cannot use direct a pointer that you want to store in the static local data. You copy the pointer into a temporary local data item and then you can work with it. You cannot load the pointer in the static local data direct into an address register and you cannot transfer the contents of an address register direct to the pointer (second example).

# 26 Direct Variable Access

This chapter shows you how to access the absolute addresses of local variables direct. The "normal" STL statements are available for local variables of elementary data types. Local variables of complex data types or block parameters of type POINTER or ANY cannot be handled "as a whole". To process these variables, you first calculate the starting address at which the variable is stored and you then process parts of the variable with indirect addressing. In this way, you can also process block parameters of complex data types.

The examples in this chapter are also given on the diskette accompanying the book under the "Variable Handling" program in function block FB 126 or source file Chap_26.

## 26.1 Loading the Variable Address

The following statements give the starting address of a local variable

```
L     P#name;
LAR1  P#name;
LAR2  P#name;
```

with *name* as the name of the local variable. These statements load an area-crossing pointer into accumulator 1 or into address register AR1 or AR2. The area pointer contains the address of the first byte of the variable. If *name* cannot be identified uniquely as the local variable, insert a "#" before the name so that the statement becomes, for example: L P##*name*. Depending on the block, the variable areas listed in Table 26.1 are permissible for *name*.

With functions, the address of a block parameter cannot be loaded direct into an address register. You can take the route via accumulator 1 here (for example: L P#name; LAR1;).

In function blocks compiled with CODE_VERSION1 (no "multi-instance capability"), the absolute address of the instance variable is loaded.

In "multi-instance capable" functions blocks, the absolute address *relative to address register AR2* is loaded in the case of the static local data and the block parameters. If you want to calculate the absolute address of the variable in the instance data block, you must add the *area-internal pointer* (address only) of AR2 to the loaded variable address

Example 1:
Load variable address into address register AR1

```
TAR2   ;
UD     DW#16#00FF_FFFF;
LAR1   P#name;
+AR1   ;
```

**Table 26.1** Loading Permissible Operands for Variable Addresses

| Operation | *name* is s | OB | FC | FB V1 | FB V2 |
|---|---|---|---|---|---|
| L P#*name* | temporary local datum | x | x | x | x |
| | static local datum | – | – | x | x[1] |
| | block parameter | – | x | x | x[1] |
| LAR*n* P#*name* | temporary local datum | x | x | x | x |
| | static local datum | – | – | x | x[1] |
| | block parameter | – | – | x | x[1] |

[1] Variable address relative to the address register AR2

With the first two statements, the address in AR2 is loaded into the accumulator and added to the contents of AR1 with +AR1. As a result, AR1 contains the address of the variable #name.

Example 2:
Load variable address into accumulator 1

```
TAR2 ;
UD   DW#16#00FF_FFFF;
L    P#name;
+D   ;
```

Similarly to example 1, the result of this is that accumulator 1 then contains the address of the variable #name.

The addition of the area-internal pointer can be omitted if it has the value P#0.0. This is the case if you do not use the function block as a local instance.

Please note that "LAR2 P#name" overwrites address register AR2 that is used in the case of 'multi-instance capable' function blocks as the "base address register" for addressing the instance data!

You can only access one overall variable with these load statements and not individual components of fields, structures or local instances. You cannot reach variables in global data blocks or in the operand areas inputs, outputs, peripheral I/O and bit memory with these load statements.

Table 26.2 shows you how to calculate the address of an INT and a STRING variables in the static local data and how to work with this address. If you use the example program in a function block that you call as a local instance, you must add the base address to the variable address as shown above.

## 26.2 Data Storage of Variables

### 26.2.1 Storage in Global Data Blocks

The Editor stores the individual variables in the data block in the order of their declaration. Essentially, the following rules apply here:

- The first bit variable of an uninterrupted declaration sequence is located in bit 0 of the next byte followed by the next bit variables.
- Byte variables are stored in the next byte.
- Word and doubleword variables always start at a word boundary, that is, at a byte with an even address.

**Table 26.2** Load Variable Address (Examples)

```
//Variable declaration (function block is not local instance!)
//Variable assignment begins at address P#0.0
VAR
  Field     : ARRAY [1..22] OF BYTE;      //ARRAY variable, occupies 22 bytes
  Number    : INT := 123;                 //INT variable, occupies 2 bytes
  FirstName : STRING[12] := 'Elisabeth'; //STRING variable, occupies 14 bytes
END_VAR
```

| Statement | Description |
|---|---|
| LAR1 P#Number; | Loads the starting address of Number into AR1<br>AR1 now contains P#DIX22.0 |
| L W[AR1,P#0.0]; | Corresponds to the statement L DIW 22 or L Number |
| LAR1 P#FirstName; | Loads the starting address of FirstName into AR1<br>AR1 now contains P#DIX24.0 |
| L B[AR1,P#0.0]; | Loads the first byte (maximum length of the character string) into accumulator 1 |
| L B[AR1,P.0]; | Loads the third byte (first relevant byte) into accumulator 1 |
| L 'John';<br>T D[AR1,P.0]; | Writes "John" into the character string |
| L 4;<br>T B[AR1,P#1.0]; | Corrects the current length of the character string to 4<br>The variable FirstName now contains 'John' |

```
DATA_BLOCK StorageNonOptimized
STRUCT
  Bit1           : BOOL;
  Bit2           : BOOL;
  Bit3           : BOOL;
  Real1          : REAL;
  Byte1          : BYTE;
  Bit_field      : ARRAY [1..3] OF BOOL;
  Structure      : STRUCT
    S_Bit1       : BOOL;
    S_Bit2       : BOOL;
    S_Bit3       : BOOL;
    S_Int1       : INT;
    S_Byte       : BYTE;
    END_STRUCT;
  Character      : STRING[3];
  Date           : DATE;
  Byte2          : BYTE;
END_STRUCT
BEGIN
END_DATA_BLOCK
```

```
DATA_BLOCK StorageOptimized
STRUCT
  Bit1           : BOOL;
  Bit2           : BOOL;
  Bit3           : BOOL;
  Byte1          : BYTE;
  Real1          : REAL;
  Bit_field      : ARRAY [1..3] OF BOOL;
  Structure      : STRUCT
    S_Bit1       : BOOL;
    S_Bit2       : BOOL;
    S_Bit3       : BOOL;
    S_Byte       : BYTE;
    S_Int1       : INT;
   END_STRUCT;
  Character      : STRING[3];
  Byte2          : BYTE;
  Date           : DATE;
END_STRUCT
BEGIN
END_DATA_BLOCK
```

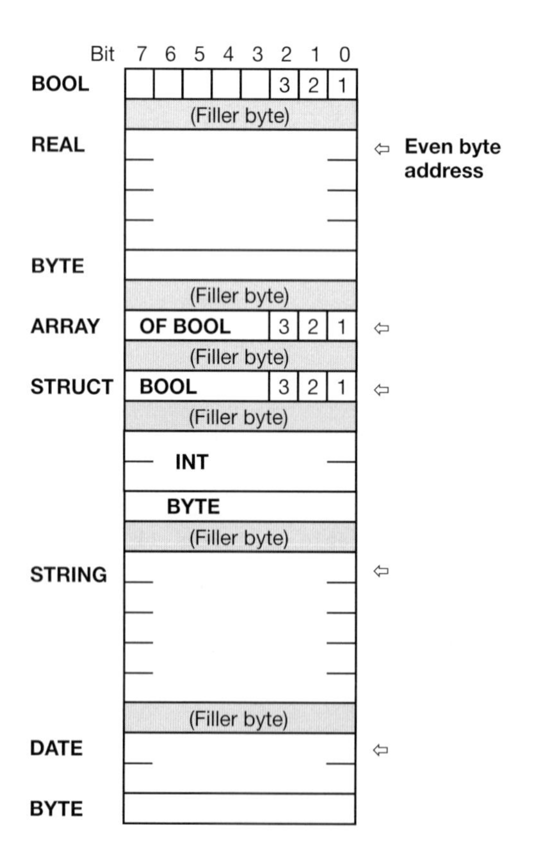


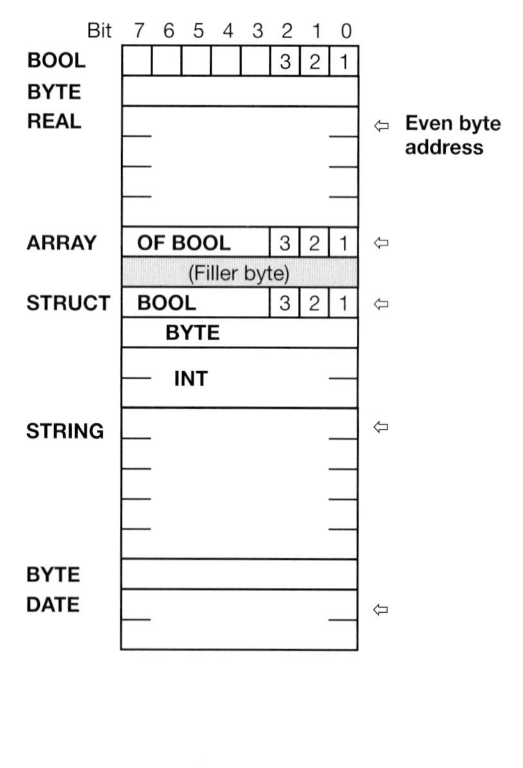


**Figure 26.1** Example of Data Block Assignment

- DT and STRING variables start at a word boundary.
- ARRAY variables start at a word boundary and are "filled" up to the next word boundary. This applies also for bit and byte fields. Field components of elementary data types are stored as described above. Field components of higher data types start at word boundaries. Each dimension of a field is aligned like an autonomous field.
- STRUCT variables begin at a word boundary and are 'filled' up to the next word boundary. This applies also for purely bit and byte structures. Structure components of elementary data types are stored as described above. Structure components of higher data types begin at word boundaries.

By combining bit variables and arranging byte variables in pairs, you can optimize data storage in a data block.

In Figure 26.1, you can see one example each of non-optimized and optimized data storage. Please note that the Editor always 'fills' ARRAY and STRUCT variables up to the next word, that is, no bit or byte variables can be stored in byte gaps. However, you can have optimized arrangement of the variables within the structure.

### 26.2.2 Storage in Instance Data Blocks

The Editor stores the variables in an instance data block in the following order:

- Input parameters
- Output parameters
- In/out parameters
- Local variables (including local instances)

Each variable is stored in the order of its declaration. The declaration areas each begin at a word boundary, that is, at a byte with an even address. Within the declaration areas, the individual variables are arranged as described in the previous chapter (as in a global data block).

Figure 26.2 shows an example for the assignment of an instance data block.

```
FUNCTION_BLOCK StorageExample
VAR_INPUT
  E_Bit1     : BOOL;
  E_Bit2     : BOOL;
  E_Bit3     : BOOL;
  E_Real1    : REAL;
END_VAR
VAR_OUTPUT
  A_Byte1    : BYTE;
  A_Byte2    : BYTE;
  A_Byte3    : BYTE;
END_VAR
VAR_IN_OUT
  D_Byte1    : BYTE;
  D_Bit1     : BOOL;
  D_Bit2     : BOOL;
  D_Bit3     : BOOL;
END_VAR
VAR
  Date       : DATE;
  Character  : STRING [3];
  Bit_feld   : ARRAY [1..3] OF BYTE;
END_VAR
BEGIN
//...
END_FUNCTION_BLOCK
```

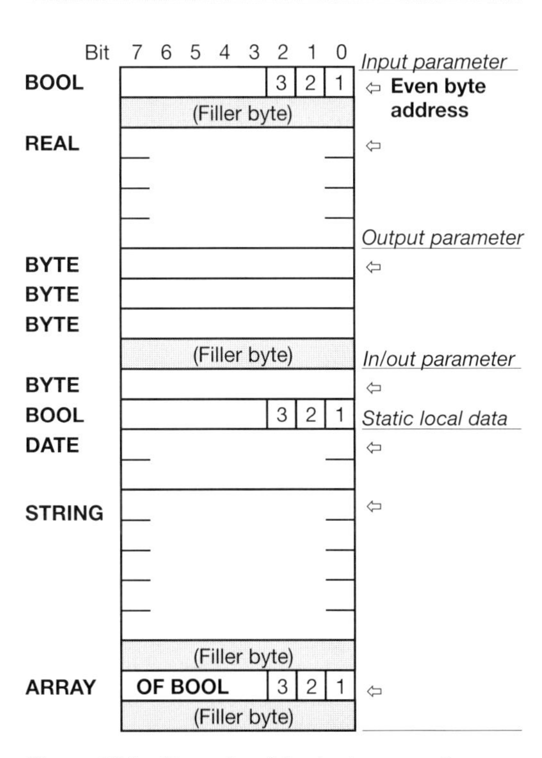


**Figure 26.2** Example of the Assignment of an Instance Data Block

### 26.2.3 Storage in the Temporary Local Data

Storage of the variables in the temporary local data (L stack) corresponds to storage in a global data block. The assignment always begins at (relative) byte 0. Please note, that in organization blocks, the first 20 bytes are occupied by the start information. Even if you do not use the start information, the first 20 bytes must be declared (even if only with a field of 20 bytes).

The Editor itself also uses local data, for example when transferring parameters in a block call. The Editor applies the symbolically declared temporary local data and the temporary local data it uses itself in the order of their declaration or their use. The absolutely addressed temporary local data are not taken into account here, so that overlaps can occur if you do not know which local data the Editor is applying. If you want or have to access local data with absolute addressing, you can, for example, declare at the first location of the temporary local data declaration a field that keeps free the required number of bytes (words, doublewords). You can then make absolute accesses in this field area. With organization blocks, you define the field after the 20 bytes for the start information.

The example in Figure 26.3 shows the assignment of the temporary local data of an organization block. The field "Ldata" starts immediately following the start information, at byte LB 20 and stretches in this example to byte LB 35. The Editor does not occupy this area with its own temporary data so you can use this area for absolute addressing.

The start information is omitted in functions and function blocks. If you require the temporary local data for absolute addressing, apply the field as the first variable in these blocks; it then begins at byte LB 0.

## 26.3 Data Storage when Transferring Parameters

The block parameters are stored differently in functions and function blocks. You as the user need not be concerned with this; you program the block parameters for both block types in the same way. However, this difference is extremely important for direct block parameter access.

```
Organization_BLOCK Cycle
VAR_TEMP
  SInfo      : ARRAY [1..20] OF BYTE;
  LData      : ARRAY [1..16] OF BYTE;
  Temp1      : STRING [36];
  Temp2      : BOOL;
  Temp3      : BOOL;
  Temp4      : BOOL;
  Temp5      : BOOL;
  Temp6      : BOOL;
END_VAR
BEGIN
//Access via absolute addresses
  ...
  T    LW 20;
  ...
  =    L  22.2;
//Access symbolic
  T    Temp6;
  =    Temp3;
  T    LDaten[16];
//Load variable address
  L    P#Temp1;
  LAR1 P#Temp2;
//...
END_Organization_BLOCK
```

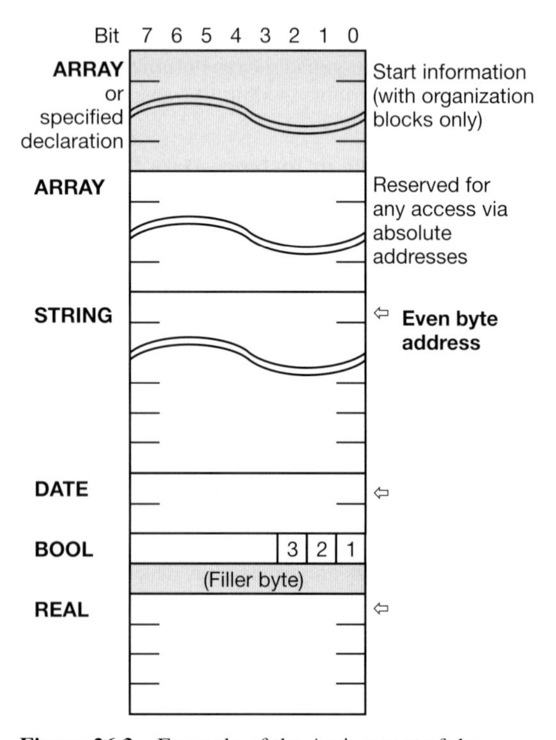


**Figure 26.3** Example of the Assignment of the L Stack in Organization Blocks

**Table 26.3** Parameter Storage in Functions

| Data Type | INPUT | IN_OUT | OUTPUT |
|---|---|---|---|
| | The parameter is an area pointer to | | |
| Elementary | a value | a value | a value |
| Complex | a DB pointer | a DB pointer | a DB pointer |
| TIMER, COUNTER, BLOCK | a number | not possible | not possible |
| POINTER | a DB pointer | a DB pointer | a DB pointer |
| ANY | an ANY pointer | an ANY pointer | an ANY pointer |

### 26.3.1 Parameter Storage in Functions

The Editor stores a block parameter of a function as area-crossing area pointer in block code in accordance with its own call statement so that each block parameter requires one doubleword of memory. Depending on data type and declaration type, the pointer points to the actual parameter itself, to a copy of the actual parameter in the temporary local data of the calling block (set up by the Editor), or to a pointer in the temporary local data of the calling block that in turn points to the actual parameter (Table 26.3). Exception: With the parameter types TIMER, COUNTER and BLOCK_xx the pointer is a 16-bit number located in the left-hand word of the block parameter.

With elementary data types, the block parameter points direct to the actual operand (Figure 26.4). However, with the area-pointer as block parameter, you cannot reach any constants or operands located in data blocks. For this reason, at the compiling stage, the Editor copies a constant or a (fully-addressed) actual operand located in a data block into the temporary local data of the calling block and directs the area pointer to this. This operand area is called V (temporary local data of the preceding block, V area).

Copying to the V area takes place before the actual FC call with input parameters and with in/out parameters and after the call with in/out and output parameters. For this reason, the rule that you can only check input parameters and write to output parameters also applies. If, for example, you transfer a value to an input parameter with a fully-addressed data operand, the value will be stored in the temporary local data of the preceding block and it will be forgotten because no more copying take place to the "actual" variable in the data block. It is a similar story for loading a corresponding output parameter: Since no copying takes place from the "actual" variable from the data block to the V area, you load an (indeterminate) value from the V area in this case.

The copy process makes it *essential* to assign an output parameter with an elementary data type in the block with a value, if a completely addressed data operand is or could be provided as current parameter. If you do not assign a value to the output parameter (e.g. because you exit the block beforehand or jump over the program point), the local date will not be provided. It then remains at the "random" value that it had before the block call. The output parameter is then assigned with this "undefined" value.

With complex data types (DT, STRING, ARRAY, STRUCT as well as UDT), the actual parameters are located either in a data block or in the V area. Since an area pointer cannot reach an actual operand in a data block, the Editor creates a DB pointer in the V area at the compiling stage. Since pointer then points to the actual operand in the data block (DB No. <> 0) or to the V area (DB No. = 0). The DB pointers for all declaration types in the V area are created before the 'actual' FC call.

With the parameter types TIMER, COUNTER and BLOCK_xx the block parameter contains a number (16 bits left-justified in the 32-bit parameter) instead of the area pointer. The parameter type POINTER is handled in exactly the same way as a complex data type.

With the parameter type ANY, the Editor creates an ANY pointer with a length of 10 bytes in the V area that can then point to any variable. The

**Pointer to the actual operand or its value**

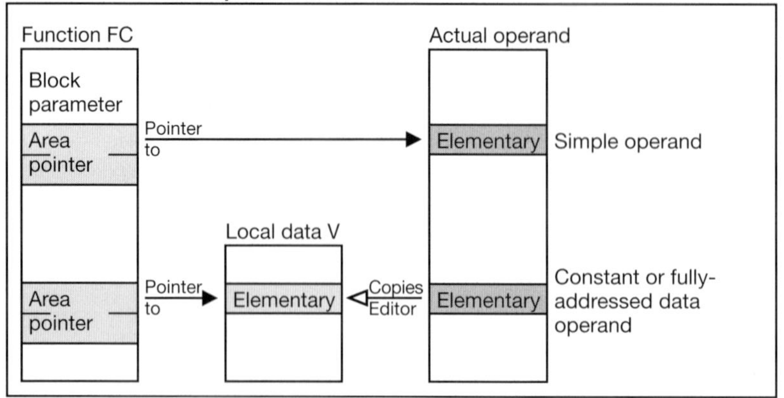


**Pointer to another pointer**

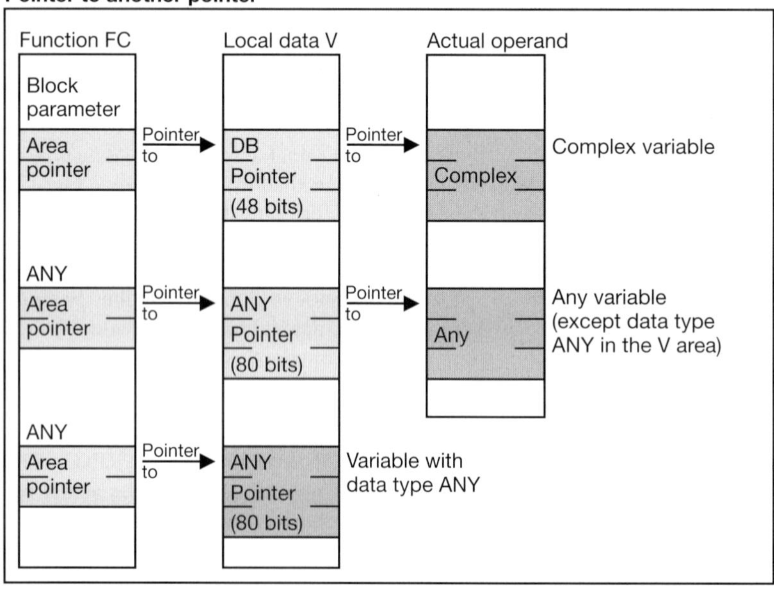


**Figure 26.4** Transferring Parameters in Functions

principle is the same as with complex data types. The Editor makes an exception if you apply at a block parameter of type ANY an actual parameter that is located in the temporary local data and is of data type ANY. The Editor then does not create any more ANY pointers, but instead directs the area pointer (the block parameter) straight to the actual parameter (in this case, the ANY pointer can be modified at runtime, see Section 26.3.3 'Variable ANY Pointer').

### 26.3.2 Storing Parameters in Function Blocks

The Editor stores the block parameters of a function block in the function block's instance data block. At the function block call, the Editor generates statement sequences that copy the values of the actual parameters into the instance data block before the actual call and then copy them back from the instance data block to the actual

**Table 26.4** Storing Parameters in the Case of Function Blocks

| Data Type | INPUT | IN_OUT | OUTPUT |
|---|---|---|---|
| Elementary | Value | Value | Value |
| Complex | Value | DB pointer | Value |
| TIMER, COUNTER, BLOCK | Number | not possible | not possible |
| POINTER | DB pointer | DB pointer | not possible |
| ANY | ANY pointer | ANY pointer | not possible |

parameters following the call. You do not see these statement sequences when you look at the compiled block. You only notice it indirectly in the occupied memory space.

In the instance data block, the block parameters are stored either as a value, as a 16-bit number or as a pointer to the actual parameter (Table 26.4). When stored as a value, the memory required depends on the data type of the block parameter; the number occupies 2 bytes, the pointers occupy 6 bytes (DB pointers) or 10 bytes (ANY pointers).

The relationships between block parameters, instance data assignments and actual parameters are shown in Figure 26.5. When copying actual parameters of complex data types into the instance data block (input parameters) or back to the actual parameter (output parameter), the Editor uses the system function SFC 20 BLKMOV, whose parameters it builds up in the temporary local data area of the calling block.

Copying block parameters that are stored as a value in the instance data block is carried out using statement sequences before the "actual" FB call in the case of input parameters and in/out parameters, and after the call in the case of in/out and output parameters. For this reason, the rule that you can only check input parameters and write to output parameters also applies. If, for example, you transfer a (new) value to an input parameter, the current value of the actual pa-

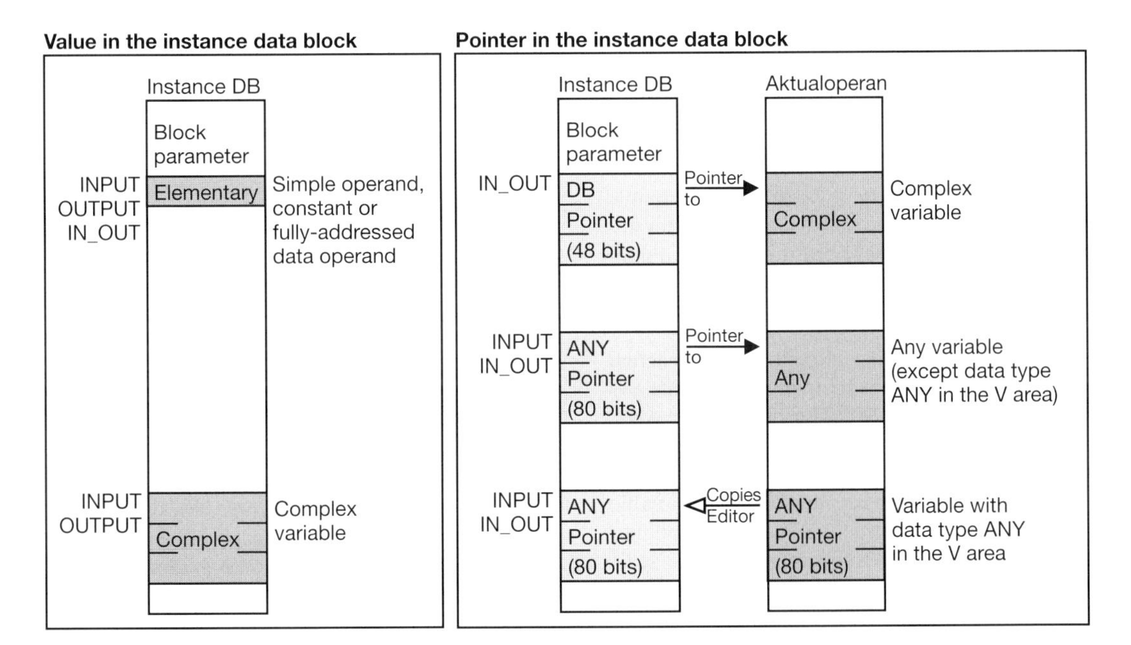


**Figure 26.5** Transferring Parameters in the Case of Function Blocks

rameter is lost. If you load an output parameter, you load the (old) value in the instance data block and not that of the actual parameter.

Because the block parameters are stored in the instance data block, they need not be initialized every time the function block is called. When no initialization is made, the program works with the "old" value of the input or in/out parameter or you fetch the value of the output parameter from another subsequent position in the program. Outside the function block, you can access variables in the instance data block in the same way as you do variables in a global data block (with the symbolic name of the data block and the name of the block parameter). The same applies also for the static local data.

If you apply a temporary local variable of data type ANY at an ANY parameter, the Editor copies the contents of this variable into the ANY pointer (into the block parameter) in the instance data block.

### 26.3.3 "Variable" ANY Pointer

ANY parameters can only be parameterized with data areas or variables that must be defined already at the compiling stage. Example: Copying a variable into a data area with SFC 20 BLKMOV

```
CALL SFC 20 (
  SRCBLK  := "Rec_mailb".Data,
  DSTBLK  := P#DB63.DBW0.0 BYTE 8,
  RET_VAL := SFC20Error);
```

It is possible to modify or re-define the variable or the data area at runtime because the Editor applies a fixed ANY pointer to the actual parameter in the temporary local data (see below in this chapter).

The Editor makes an exception to this if the actual parameter itself is in the temporary local data and the is of data type ANY. Then no further ANY pointer is set up, instead the Editor interprets these ANY variables as ANY pointers to the actual parameter. This means that the ANY variable must have the same structure as an ANY pointer (see Section 25.1.3 "ANY Pointers").

You can now modify these ANY variables in the temporary local data at runtime and so specify another actual parameter for an ANY parameter.

To apply this "variable" ANY pointer, proceed as follows:

- Applying a temporary local variable of data type ANY:

```
VAR_TEMP
  ANY_POINTER : ANY;
END_VAR
```

The name of the ANY variable can be selected freely within the permissible framework for block-local variables.

- Initialize the ANY variable with values:

a) The ANY variable is located at a fixed known address, for example, from address LB 0

```
L    W#16#1002;
T    LW 0;
L    16;
T    LW 2;
L    63;
T    LW4;
L    P#DBB0.0;
T    LD 6;
```

b) The address of the ANY variable is not fixed

```
LAR1 P#ANY_POINTER;
L    W#16#1002;
T    W[AR1,P#0.0];
L    16;
T    W[AR1,P#2.0];
L    63;
T    W[AR1,P#4.0];
L    P#DBB0.0;
T    D[AR1,P#6.0];
```

- Initialize the ANY parameter, for example, at an SFC 20

```
CALL SFC 20 (
  SRCBLK  := "Rec_mailb".Data,
  DSTBLK  := ANY_POINTER,
  RET_VAL := SFC20Error);
```

This procedure is not restricted to SFC 20 BLKMOV; you can use it on all ANY parameters of any blocks.

Example: We want to write a copy block that is to copy data areas between data blocks. The source and destination area is to be parameterizable. We use SFC 20 BLKMOV for copying.

The block – a function FC – has the following parameters:

```
VAR_INPUT
  QDB  : INT; //Source data block
  SSTA : INT; //Source starting
                address
  NUMB : INT; //Number of bytes
  DDB  : INT; //Destination data
                block
  DSTA : INT; //Destination starting
                address
END_VAR
```

The function value is to contain the error message of SFC 20 and can then also be evaluated as if we were using SFC 20 direct. In addition, the status bit BR is set to "0" in the event of an error.

Two ANY variables, one as a pointer for the source area and one as a pointer for the destination area, are sufficient for the block-local data:

```
VAR_TEMP
  SANY : ANY; //ANY pointer source
  DANY : ANY; //ANY pointer destina-
                tion
END_VAR
```

Since we know the addresses of the ANY pointers in the temporary local data, we can program them with absolute addresses, for example, the preparation of a source pointer:

```
L    W#16#1002; //Type BYTE
T    LW 0;
L    NUMB;      //Number of bytes
T    LW 2;
L    QDB;       //Source DB
T    LW 4;
L    SSTA;      //Start of the source
SLD  3;
OD   DW#16#8400_0000;
T    LD 6;
```

The destination pointer starting at address LB 10 is prepared in the same way.

It only remains to initialize SFC 20:

```
CALL SFC 20 (
  SRCBLK  := SANY,
  DSTBLK  := DANY,
  RET_VAL := RET_VAL);
```

The function value RET_VAL of SFC 20 is initialized with the function value RET_VAL of our function FC.

You can find this little example in full on the diskette accompanying the book (function FC 47 in the program "General Examples").

In this way, an ANY pointer can be assigned any value, for example, you can vary the type in word 2 or the area pointer so that in principle, you can address any variables and data areas, for example also the bit memory area.

## 26.4 Brief Description of the Message Frame Example

The following examples will deepen your understanding of how to handle complex variables. the program of the different blocks each emphasizes a specific aspect of this topic. The declared technological function of the examples such as "Generate_Frame" and "Checksum", are intended only to make things clearer and, where necessary, are dealt with only briefly.

At this point, the examples are described with text and figures. You can find the program on the diskette accompanying the book under the program "Message Frame Example". This example consists of the following sections:

- Message frame data (UDT 51, UDT 52, DB 61, DB 62, DB 63)
  shows how to handle self-defined data structures
- Clock_check (FC 61)
  shows how to handle system blocks and standard blocks
- Checksum (FC 62)
  shows how to use direct variable access
- Generate_Frame (FB 51)
  shows how to use SFC 20 BLKMOV with fixed addresses
- Store_Frame (FB 52)
  Shows how to use the 'variable' ANY pointer
- DT_Conv (date conversion) (FC 63)
  shows the processing of variables of complex data types

### Message frame data example

The example shows how you can define frequently occurring data structures as your own data type and how to use this data type in variable declaration and parameter declaration.

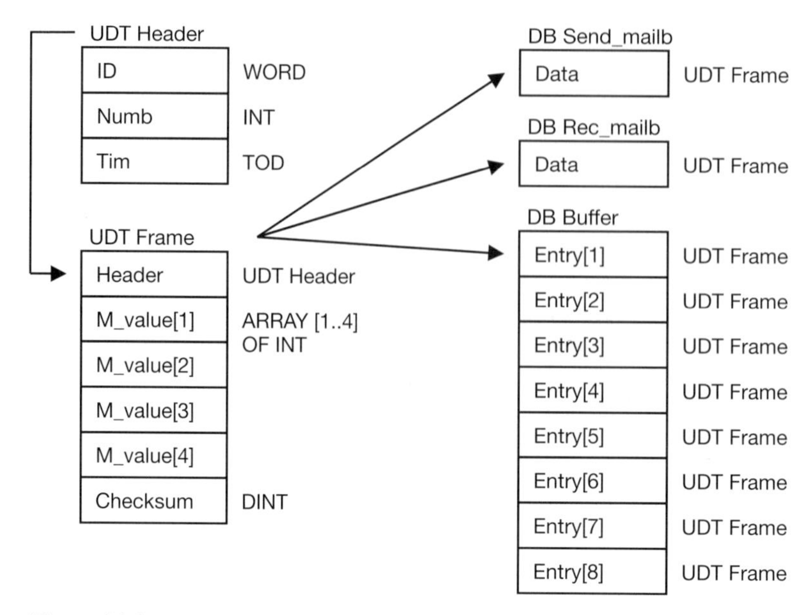


**Figure 26.6** Data Structure for the Message Frame Data Example

We establish a database for incoming and outgoing frames: A send mailbox with the structure of a message frame, a receive mailbox with the same structure and a (receive) ring buffer for intermediate storage of the incoming message frames (Figure 26.6). Since the data structure of the message frame occurs frequently, we want to make it a user-defined data type (UDT) *Frame*. The frame contains a frame header; we also want to give a name to the structure of the frame header. The send mailbox and the receive mailbox are to be data blocks each containing a variable with the structure of *Frame*. Finally, there is the ring buffer, a data block with a field of eight components that also have the data structure of *Frame*.

First, we define the UDT *Header*, then the UDT *Frame*. *Frame* consists of a structure *Header*, a field *M_value* with 4 components and a variable *Checksum*. All components are initialized with zero. In the data blocks *Send_mailb* and *Rec_mailb*, a variable *Data* with the structure *Frame* is defined in each case. The variables can now be individually initialized in the initialization section of the data block. In the example, the component *ID* receives a value in each case that deviates from the initialization in the UDT. The data block *Buffer* contains the variable *Entry* as a field with 8 components of the structure *Frame*. Here too, the individual components can be initialized with different values in the initialization section (for example, Entry[1].Header.Numb := 1).

| | |
|---|---|
| UDT 51 | User-defined data type Header |
| UDT 52 | User-defined data type Frame |
| DB 61 | Send mailbox (Send_mailb) |
| DB 62 | Receive mailbox (Rec_mailb) |
| DB 63 | Buffer |

## Clock check example

The example shows how to handle system blocks and standard blocks (error evaluation, copying from the library, renaming).

The function *Clock_check* is to output the time-of-day in the CPU-integrated real-time clock as a function value. For this purpose, we require the system function SFC 1 READ_CLK that reads the date and time-of-day of the real-time clock in data format DATE_AND_TIME or DT. Since we only want to read the time-of-day, we also require the IEC function FC 8 DT_TOD. This function fetches the time-of-day in format TIME_OF_DAY or TOD from the data format DT (Figure 26.7).

The time specification of the real-time clock is stored in data block *Data66* since we still require

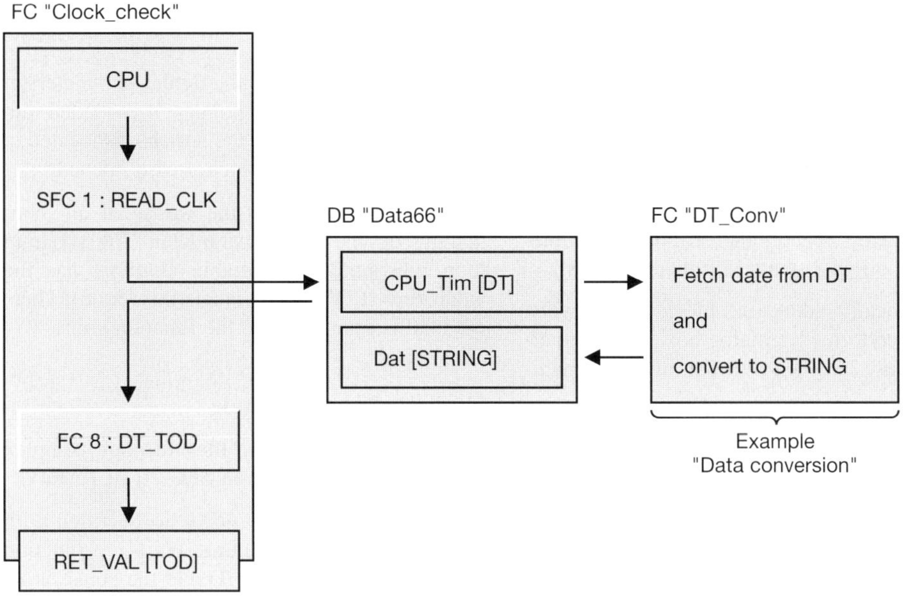

FC "Clock_check"

CPU

SFC 1 : READ_CLK

DB "Data66"

CPU_Tim [DT]

Dat [STRING]

FC "DT_Conv"

Fetch date from DT

and

convert to STRING

Example
"Data conversion"

FC 8 : DT_TOD

RET_VAL [TOD]

**Figure 26.7** Clock Check Example

this information for the "Date conversion" example. Without this additional use, we could have also declared a temporary local variable instead of the variable *CPU_Tim*.

*Error evaluation*
The system functions signal an error via the binary result BR and the via the function value RET_VAL. An error exists if the binary result BR = "0"; the function value is then also negative (bit 15 is set). The IEC standard functions signal an error only via the binary result. Both types of error evaluation are shown in the example. In the *Clock_check* function, the binary result is first set to "1"; if an error exists, the binary result is set to "0" by the relevant block. Then an invalid value is output for the time-of-day. After the *Clock_check* function has been called, you can also check for an error via the binary result.

*Off-line programming of system functions*
Before compiling the example program or before calls in incremental program input, the off-line user program must contain the system function SFC 1 and the standard function FC 8. Both functions are included in the scope of supply of STEP 7. You can find these functions in the block libraries supplied. (For the system functions integrated in the CPU, the library contains an interface description instead of the program of the system functions. The function can be called off-line via this interface description; the interface description is not transferred to the CPU. The loadable functions such as the IEC functions are stored in the library as executable programs.)

With FILE → OPEN in the SIMATIC Manager, you select the library "Standard Library V3.x" and open the library "System Function Blocks". Under *Blocks* here you will find all interface descriptions for the system functions. If you have not opened the project window of your project, you can arrange both windows next to each other with WINDOW → ARRANGE → VERTICALLY and drag the selected system functions into your program with the mouse (mark the SFC with the mouse, "hold" it, "drag" to *Blocks* or to its open window and "drop"). You can copy the standard function FC 8 in the same way. You will find it in the library 'IEC Converting Blocks'. FC 8 is a loadable function; it therefore occupies user memory, in contrast to SFC 1.

If a standard block is called from the Editor's Program Element Catalog under "Libraries" during incremental programming, it is automatically copied to *Blocks* and entered in the symbol table.

*Renaming standard functions*
You can rename a loadable standard function. You mark the standard function (for example, FC 8) in the project window and click (again) on the identifier. A frame appears around the name and you can specify a new address (for example, FC 98). If you now press F1 while the standard function (renamed to FC 98) is marked, you will still nevertheless receive the on-line Help function for the original standard function (FC 8).

If an identically addressed block exists when copying is performed, a dialog box appears to allow to choose between overwriting and renaming.

*Symbol address*
In the symbol table, you can assign names to the system functions and the standard functions, so that you can also access these functions symbolically. You have a free choice of names within the framework of the permissible definitions for block names. In the example, a symbolic name has been selected for each block name (for improved identification).

## Checksum example

This example shows direct access to a block parameter of type ANY with calculation of the variable address and use of indirect addressing.

A checksum is to be generated from a data structure by simply adding all bytes with no account being taken of any carry (overflow, number range violation for DINT).

All data structures (STRUCT and UDT) are treated by the Editor like a field with byte components if they are applied to a block parameter of parameter type ANY. With this program, therefore, you can generate the checksum not only from a field with byte components (ARRAY OF BYTE), but also from structure variables. If you also want to use the program on variables of other data types, you must modify the relevant check (type ID in the ANY pointer).

The checksum function uses direct variable access to get the absolute address of the block parameter (more precisely: the address at which the Editor has stored the ANY pointer). First, a check is made to ensure that the type ID "Byte" and a repetition factor >1 has been entered. In the event of an error, the binary result is set to "0" and the function is exited with a function value equal to zero.

The starting address of the actual parameter (at runtime) is in the ANY pointer. It is loaded into address register AR1. If the variable is located in a data block, this data block is also opened.

The next network adds the values of all bytes making up the actual parameter. The program loop runs until the variable *Quantity* has the value zero (LOOP decrements this value). Then, the total is transferred to the function value.

## Generate frame example

The example shows you how to copy complex variables with the function SFC 20 BLKMOV.

The data block *Send_mailb* is to be filled with the data of a message frame. We use a function block that has stored the ID and the consecutive number in its instance data block. The net data are located in a global data block; they are copied to the send mailbox with the system function BLKMOV.

We get the time-of-day from the real-time clock in the CPU with the help of the function *Clock_check* (see previous example) and we generate the checksum by simply adding all bytes in the message frame header and the data (see "Checksum" examples). Figure 26.8 shows the program and the data structures.

The first network in the function block FB *Generate_Frame* transfers the ID stored in the instance data block to the frame header. The consecutive number is incremented by +1 and is also entered in the frame header.

The second network contains the call of the function *Clock_check* that fetches the time-of-day from the real-time clock and enters it in the format TIME_OF_DAY in the frame header.

In the third network, you can see a method of using system function SFC 20 BLKMOV to copy variables selected at runtime without using indirect addressing. It is therefore also not necessary to know the absolute address and the structure of the variable. The principle is extremely simple: The desired copy function is selected with the jump distributor. The numbers 1 to 4 are permissible as selection criteria. The example "Buffer entry" shows the same functionality this time

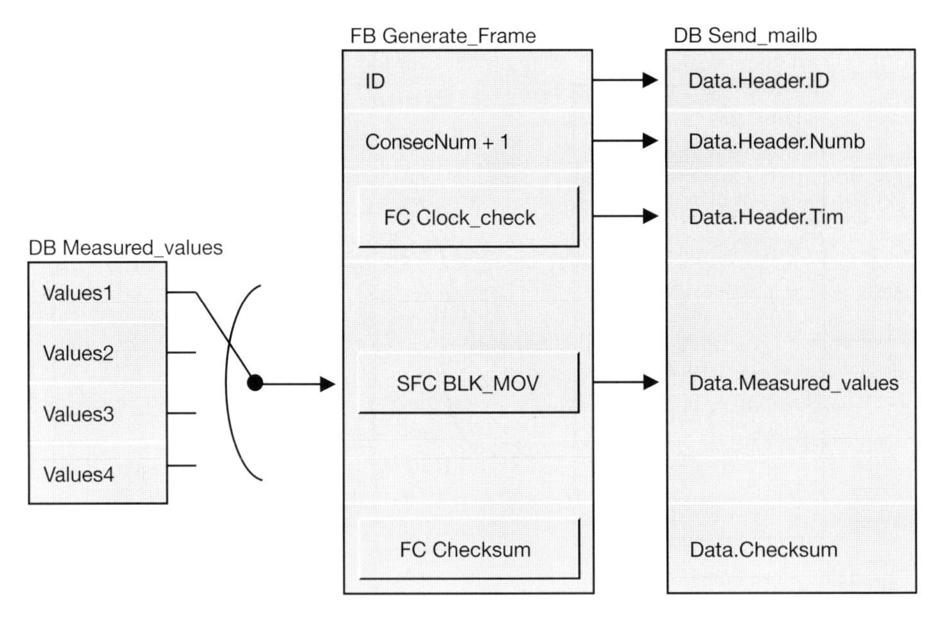


**Figure 26.8** Example: Generate Frame

with a variable destination range using a pointer calculated at runtime.

The next network generates the checksum via the frame header and the frame data. Since the function *Checksum* generates the checksum over a single data area, the frame header and the data are first combined in the temporary variable *Block*. The contents of *Block* are then added bytewise and stored in the checksum in the send frame.

The FB *Generate_Frame* is programmed in such a way that it can be called via a signal edge for the purpose of generating the frame.

### Store frame example

This example concentrates on showing you how to use an "variable" ANY pointer.

A frame in the data block *Rec_mailb* is to be entered at the next location in the data block *Buffer*. The block-local variable Entry determines the location in the ring buffer; the address of the ring buffer is calculated from the value in this location (Figure 26.9).

If the number of the frame in the receive mailbox has changed, the frame is to be written to a buffer at the next location. The buffer is to be a data block that can accommodate 8 frames. After the eighth frame has been entered, the next frame is to b entered at the first location again.

The FB *Store_Frame* compares the entered frame number with the stored number in the data block *Rec_mailb*. If the frame numbers are different, the stored number is corrected and the frame in the receive mailbox is copied to the data block *Buffer* in the next entry. The system function SFC 20 BLKMOV handles the copying. Since the destination can be different depending on the value of *Entry*, we calculate the absolute address of the destination area, generate an ANY pointer from this in the variable *ANY pointer* and transfer it to the SFC at the parameter DST BLK. Please note that you only use area-internal addressing for indirect addressing of a temporary local variable.

The data structure *Frame* has a length of 20 bytes (header: 8 bytes, Measured_values: 8 bytes, Checksum: 4 bytes). The variable *Data* in the data block *Rec_mailb* is therefore 20 bytes long, just as every component of the field *Entry* in the data block *Buffer* is also 20 bytes long. Consequently, the individual components *Entry[n]* begin at byte address *n* * 20, where *n* corresponds to the variable *Entry*.

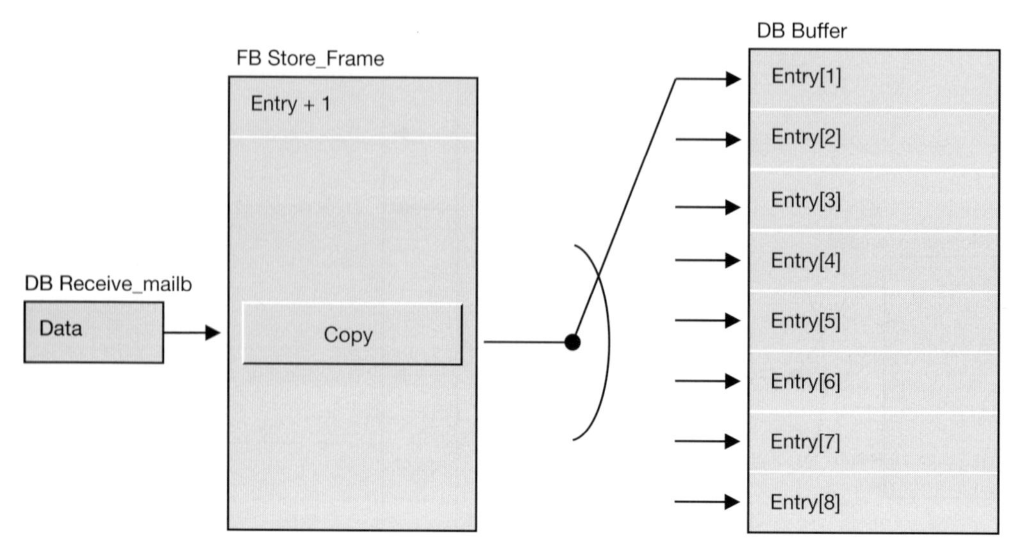


**Figure 26.9** Example: Store Frame

## Date conversion example

The example concentrates on showing the processing of variables of complex data types using direct variable access and indirect addressing with both address registers.

A global data block *Data66* contains the variables *CPU_Tim* (data type DATE_AND_ TIME) and *Dat* (data type STRING). The date is to be fetched from the variable *CPU_Tim* and stored as a character string with the format "JJMMTT" in the variable *Dat*.

The subsequent program in the function "DT_Conv" uses address register AR1 and the DB register for the pointer to the input parameter *Tim* and address register AR2 and the DI register for the pointer to the function value (corresponding to the STRING variable *Dat* in the data block "Data66"). The program is located in a function so that both data block registers and both address registers are available without restriction.

The program in the first network calculates the address for the actual parameter at the block parameter *Tim,* valid at runtime, and stores the address in the DB register in AR1. An actual parameter of complex data type can only be located in a data block (global or instance data) or in the temporary local data of the calling block (in the V area). If the actual parameter is in a data block, the data block number would be loaded into the DB register and the area pointer in AR1 would contain the operand area DB. If the actual parameter is in the V area, zero would be loaded into the DB register and the are pointer in AR1 would contain the operand area V.

The second network contains the equivalent program for the function value whose address is then located in address register AR2 and in the DI register. In order to be able to address indirect via the DI register as well, the operand area DI must be entered in AR2. However, depending on the memory occupied by the actual operand, either DB for data block or V for V area would be found here. By setting bit 24 in AR2 to "1", we change the operand area from DB to DI but we do not change any operand area V.

Prepared in this way, the maximum length fixed for the actual parameter at the function value can be checked in the next network. The length must be at least 6 characters. If it is less than 6, "0" in is entered in the binary result BR (otherwise "1") and block processing is terminated. In this way, you can check for processing errors via the binary result after calling the function DT_Conv.

The next network fetches the year and the month from *Tim* (in BCD form), converts the values into ASCII characters (precedes them with a 3) and writes them back to the function value. The same happens with the days.

The program is ended with the correction of the current length in the function value.

# Appendix

This section of the book contains instructions for converting a STEP 5 program into a STEP 7 program, an overview of the contents of the STEP 7 block libraries and an overview of all STL statements.

- With the optional package **S5/S7-Converter,** you can convert an existing STEP 5 program into a (STEP 7) STL program as a text file.
- The scope of supply of STEP 7 includes **Block Libraries** with loadable functions and function blocks and with block headers and interface descriptions of system blocks (SFCs and SFBs). The Editor needs interface description if you call system blocks off-line. You can also find templates for the start information of the organization blocks.
- The book ends with an **Operations Overview** of all STL operations.

The **diskette** accompanying the book contains the file STL_BOOK.ARJ that in turn contains an archived library. You can retrieve (de-archive) the library under the SIMATIC Manager with FILE → RETRIEVE. Select the archive (the diskette) from the dialog field shown. In the next dialog field, you determine the destination directory. Libraries are generally stored under ...\STEP7\S7LIBS; however, you can select any other directory, for example ...\STEP7\S7PROJ, that normally contains the projects. In the final dialog field "Retrieve – Options", you deactivate the option "Restore full path".

The library STL_Book contains eight programs that are essentially demonstration examples of STL representation. Two more comprehensive examples show the programming of functions, function blocks and local instances (Conveyor Example) and the handling of data (Message Frame example). The memory requirement for the library STL_Book is approximately 2.12 Mbytes.

To try out an example, set up a project corresponding to your own hardware configuration and copy the program including the symbol table from the library to the project. Now you can test the program on-line.

**27 S5/S7 Converter**
Conversion of STEP 5 programs to STEP 7 programs

**28 Block Libraries**
Organization Blocks, System Function Blocks, IEC Converting Blocks, S5-S7 Converting Blocks, TI-S7 Converting Blocks, PID Control Blocks and Communication Blocks

**29 Operation Overview**
All STL operations

# 27 S5/S7 Converter

With the S5/S7 Converter, you can convert a STEP 5 program into a STEP 7 STL source file. The Converter turns all directly convertible statements into the corresponding STEP 7 statements. STEP 5 statements that cannot be converted to STEP 7 statements, are commented out. The Converter takes over all comments. As an option, the assignment list can also be converted to an importable symbol table.

To convert a sequential control with GRAPH 5 to a STEP 7 program, you must create the program again with S7-GRAPH.

The S5/S7 Converter is included in the scope of supply of the STEP 7 Standard Package. You do not require authorization to use the Converter.

## 27.1 General

To convert a STEP 5 program, you require the program file *name*ST.S5D and the cross-reference list *name*XR.INI as well as the assignment list *name*Z0.SEQ if available. In addition, you can create a macro file. It contains statement sequences that the Converter can use in place of certain STEP 5 statements. From these files, the Converter generates a STEP 7 source file and, if required, a symbol table. All generated files are stored in the same directory as the STEP 5 files.

The Converter transfers organization blocks with user program into the corresponding STEP 7 organization blocks and all other code blocks into functions FCs. The block numbers of the FC blocks start at zero and are numbered consecutively; you can change the suggested block numbers in a dialog window.

The Converter detects standard blocks from the following Siemens block packages

- Floating-point arithmetic
- Signal functions
- Basic functions with analog functions
- Math functions

The library 'S5–S7 Converting Blocks', included in the scope of supply of STEP 7, contains replacement blocks for the standard blocks from these packages. You can also find standard blocks ("integral functions") in this library, that replace some of the function blocks integrated into the S5-115U-CPUs.

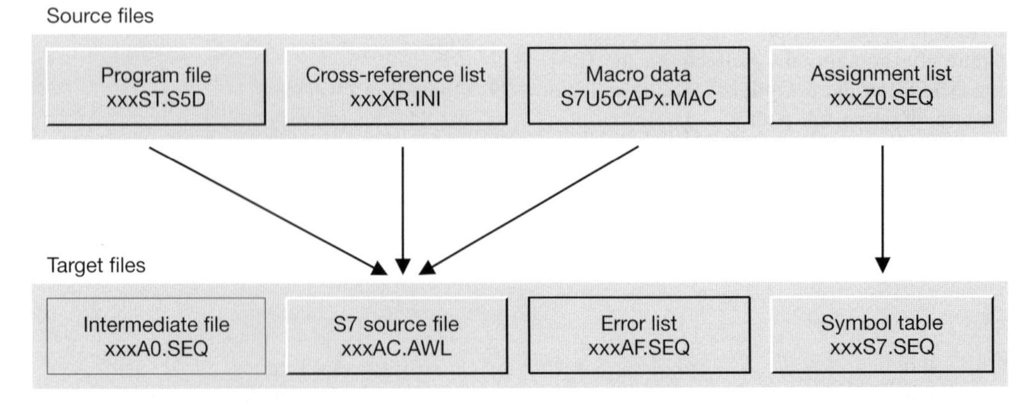


**Figure 27.1** Files for the Converter

If the STEP 5 program contains blocks from these packages, the Converter converts the call and signals which blocks occur in the program. You must then copy the relevant blocks from the library to your user program before compiling the converted program.

You can follow the procedure below when converting a STEP 5 program:

- Program executability check in the destination environment
- If necessary, preparation of the STEP 5 program (removal of non-convertible sections that are replaced, for example, by CPU parameterization)
- If necessary, creation of macros (replacement of STEP 5 statements with self-selected STEP 7 statement sequences when converting)
- Conversion (generation of a STEP 7 source program)
- Setting up of a STEP 7 project with importing of the source program and the symbol table into the STEP 7 project, if necessary, copying of the standard function blocks used
- If post editing is required, correct or supplement the STEP 7 source program
- Compilation

The conversion sequence is not fixed. You can, for example, convert a STEP 5 program without preparation and then make all the corrections in the STEP 7 source program.

## 27.2 Preparation

### 27.2.1 Checking Executability on the Destination System

If you want to use an existing STEP 5 program in a SIMATIC S7, you must first check that the program can execute on the destination system. For example:

- Does the destination CPU have the required properties? Do the required program execution characteristics exist?
- Which modules has the STEP 5 program worked with? Which modules are accessed in the STEP 7 program?
- Does the destination CPU have the required number of operands (for example, inputs, outputs, blocks)?

You can operate an S5 expansion unit via the IM 463-2 interface module or you can operate certain S5 modules in an adapter casing in an S7-400. It is also possible to connect SIMATIC S5 modules to SIMATIC S7 as distributed I/O via PROFIBUS-DP.

### 27.2.2 Checking Program Execution Characteristics

The program execution levels familiar to you from SIMATIC S5 generally correspond to the program execution levels in SIMATIC S7, now called priority classes. You replace the settings you have made in data block DB 1 or DX 0 or perhaps in the system data with the parameterization of the S7-CPU (for example, restart characteristics, watchdog interrupt handling).

The integral organization blocks and the integral function blocks in S5 correspond to the S7 system blocks. If you have used integral functions in S5, you must imitate this functionality in S7 with system blocks or with CPU parameterization.

#### Data block DB 1

On the S5-115U, the program execution characteristics are set in data block DB 1 or in the system data RS. The top of Table 27.1 shows how these characteristics can be implemented with SIMATIC S7.

#### System utilities

The CPUs of the S5-115U provide system utilities that you can use with organization block OB 250 (CPU 945) or via system datum RS 125 (CPU 941 to CPU 944). The middle section of Table 27.1 contains suggestions for converting these system utilities to SIMATIC S7.

#### Data block DX 0

With the CPUs of the upper performance range, the entries in data block DX 0 determine the program execution characteristics. The bottom section of Table 27.1 shows the conversion to SIMATIC S7.

The system data areas RI, RJ, RS and RT are omitted without replacement in S7. Any information you have buffered in these areas is stored in S7 in global data blocks or in memory bits. You now get system information from the RS area via system functions; you implement functions initiated via this area via system functions or CPU parameterization.

### 27.2.5 Preparing the STEP 5 Program

Before conversion, you can prepare your STEP 5 program for its future use as a STEP 7 program (but you do not have to do this; you can also carry out all corrections after conversion to the STEP 7 source file). With this adaptation, you can reduce the number of error messages and warnings. For example, you can make the following adaptations before conversion:

- Deletion of the data blocks with program characteristics DB 1 or DX 0
- Removal of all calls of integral blocks or accesses to the system data area RS whose functionality can be reached via the parameterization of the S7-CPU
- Adaptation of the operand areas inputs, outputs, peripheral I/O to the (new) module addresses (you should ensure here that the STEP 5 address range is not exceeded, otherwise an error will be signaled already in the first conversion run; these statements are then not converted)
- With unconvertible program sections that occur repeatedly, you can delete the sections down to one "unique" STEP 5 statement per program section. You assign a macro (a STEP 7 statement sequence) to this "unique" statement that is to replace the program section.
- If your program contains many (long) data blocks that have no data structure (those used, for example, as a data buffer), you can significantly reduce the number of statements to be compiled and therefore the source code if you delete all but one of the data words in this data block. After conversion (and before compiling) program the contents of these data blocks in the source file with a field declaration, for example `Buffer : ARRAY [1..256] OF WORD`.

You can use the Converter to convert not only whole programs but also individual blocks.

## 27.3 Converting

### 27.3.1 Creating Macros

You can create macros before conversion for replacing unconvertible STEP 5 statements or for making a change different to the standard conversion. You create conversion macros with the Converter. If a macro is defined twice, the first definition is used. Macros with the SIMATIC instruction set (German) are stored in the file S7U5CAPA.MAC; macros with the international instruction set (English) are stored in the file S7U5CAPB.MAC. The Converter distinguishes between instruction macros and OB macros. You can create 256 instruction macros and 256 OB macros.

*Instruction macros* replace a STEP 5 statement with a sequence of specified STEP 7 statements. General structure of an instruction macro:

```
$MACRO: <Step 5 statement>
<Step 7 statement sequence>
$ENDMACRO
```

The STEP 5 statement must be specified in full (with complete operands). The Converter then inserts the specified STEP 7 statement sequence in their place.

Example: You use a delay interrupt (organization block OB 6) in the STEP 5 program for the CPU 945. You have started this interrupt by calling special function OB 250:

```
L  KF +200
L  KB      1
JU OB   250
```

The first load statement contains the number of milliseconds by which the call of OB 6 is to be delayed. This statement can remain and you replace the remaining two statements with a STEP 5 statement that otherwise does not occur in your program, for example, TB RT 200.0, so that your STEP 5 program appears as follows prior to conversion:

```
L  KF +200
TB RT 200.0
```

You now write the following instruction macro:

```
$MACRO: TB RT 200.0
T MD 250;
CALL SFC 32 (
 OB_NO   := 20,
 DTIME   := MD 250,
 SIGN    := W#16#0000,
 RET_VAL := MW 254);
$ENDMACRO
```

The STEP 5 statement TB RT 200.0 is replaced at conversion with the specified STEP 7 statement sequence. The delay time in ms is loaded into the (scratchpad) memory word MW 250 and then SFC 32 is called. In the dialog window before starting, the Converter suggests the number 20 instead of the number 6 for the delay OB.

*OB macros* replace an OB call (JU OB or JC OB) with the specified STEP 7 statement sequence. The general structure of an OB macro is as follows:

```
$OBCALL: <Number of the OB>
<Step 7 statement sequence>
$ENDMACRO
```

Example: In the STEP 5 program for the CPU 945, you use organization block OB 160 to start a waiting time. In STEP 7, a waiting time is implemented by system function SFC 47 WAIT. If you enter the macro

```
$OBCALL: 160
T MW 250;
CALL SFC 47 (WT := MW 250);
$ENDMACRO
```

the Converter replaces every OB 160 call (even a conditional call) with the specified statement sequence.

Input of the macros begins with EDIT → REPLACE MACRO. You enter the macros in the opened file S7U5CAPA.MAC and save the file with FILE → SAVE. Terminate macro input with FILE → EXIT.

### 27.3.2 Preparing the Conversion

If there is still no cross-reference list *name*XR.INI for your STEP 5 program, you must create one for conversion (under STEP 5 with MANAGEMENT → MAKE XREF).

You can now

- create your own working directory for the conversion and copy the required data into this directory or
- execute the conversion in the directory (folder) containing the STEP 5 files (if you have worked with the same programming device under STEP 5) or
- execute the conversion on diskette (if you have generated the STEP 5 files on another programming device).

The directory for the conversion must contain the files *name*ST.S5D and *name*XR.INI as well as *name*Z0.SEQ if appropriate. The Converter also puts the destination files *name*AC.AWL as well as *name*A0.SEQ and, if appropriate, *name*AF.SEQ and *name*S7.SEQ into this directory.

The file S7S5CAPx.MAC is stored in the Windows directory.

### 27.3.3 Starting the Converter

You call the S5/S7 Converter via the Windows 95/NT taskbar: START → SIMATIC → STEP 7 → S5 CONVERT FILE. With FILE → OPEN, you select the S5 program file you want to convert. If you click on "OK", the Converter displays the source and destination files as well as the assignment of the old blocks to the new. If necessary, you can change the names of the destination files in the text field. To change the suggested block numbers, doubleclick on the line and enter the new block number in the dialog field. The converter identifies standard blocks with a star (you must then copy these blocks from the block library into your off-line user program before compiling the S7 source file).

You start the Converter with the "Start" button. In the first run, it compiles the S5 program into an S5-ASCII text file (*name*A0.seq) and in the second run it compiles this into the S7 source file. The assignment list is compiled into the symbol table. The conversion is completed with the display of error messages and warnings. All errors and warnings are contained in the error file *name*AF.SEQ.

Error messages are output if parts of the S5 program are not convertible and can only be accommodated in the S7 program as comments. Warnings contain information on possible problems; they are output if the converted statements require to be checked again. The messages refer in part to the S5 program (for example, if an illegal MC 5 code is found) or to the S7 program (for example, if an unconvertible statement is found). If you click on a message, the Converter displays the environment of the message in a window.

It is advisable to print out the error list in order to process the error messages.

### 27.3.4 Convertible Functions

Table 27.2 lists the statements that are converted essentially unchanged. These also include statements with operands that are replaced in STEP 7 with others, such as the extended S memory bits that are replaced by the M memory bits from 256. Syntax changes can also occur (for example, +G becomes +R). You will normally not have to correct these statements.

The substitution statements (accesses to block parameters) are largely converted. Some editing is required with statements that access both timer and counter functions (for example SEC =*parname*) as well as in the processing of block parameters (DO =*parname*). Here, either code blocks or data blocks can be used actual operands and (important!): The block number can change as a result of the conversion.

Organization blocks contain the numbers used in STEP 7. All other blocks with user program become functions FC. The Converter converts data blocks DB to global data blocks with the same number. Data blocks DX are converted to data blocks DB from number 256 (DX 1 becomes DB 257, etc.). The Converter suggests numbers; you can change all the suggested block numbers in a window prior to the conversion run.

The converter takes over the library number of the blocks as AUTHOR in the block header. The name of a function block is taken over as NAME provided it does not contain any special characters (otherwise it is taken over without special characters with the original name as comment).

Special function calls are not converted (they must be replaced with system functions, for example).

The addresses of the inputs and outputs are taken over unchanged. In the case of load and transfer statements with operands from the P area, the Converter uses peripheral inputs PI and peripheral outputs PQ with unchanged addresses. Operands from the Q area are mapped to the peripheral I/O area (P) from address 256 (L QB 0 becomes L PIB 256, T QB 1 becomes T PQB 257, etc.).

The addresses of the memory bits F are taken over unchanged. The designation only contains the letter F (Flag) and not the letter M (Mem-

**Table 27.2** Conversion of the Operations

| Functions with STEP 5 | Functions with STEP 7 |
|---|---|
| Binary logic operations, memory functions | Binary logic operations, memory functions |
| Timer and counter functions | Timer and counter functions |
| Bit test functions | replaced with SET followed by check or with double negation Set/Reset |
| Load and transfer functions (without system data and absolute address) | Load and transfer functions |
| Comparison functions | Comparison functions |
| Calculation functions | Calculation functions |
| Digital logic operations | Word logic operations |
| Shift functions | Shift functions |
| Jump functions | Jump functions |
| Conversion functions | Conversion functions |
| Disable/enable interrupts | Replaced with SFC 41, SFC 42 |
| Stop functions | Replaced with SFC 46 |
| Null operations (NOP, ***, blank line) | NOP, NETWORK, // (blank line comment) |

ory). This also applies for the memory bits used as 'scratchpad memory' from memory byte FY 200 to FY 255. If you convert your STEP 5 program largely unchanged, you can retain the scratchpad memory bits as usual. If you want to continue to use the STEP 5 program or parts of it in a STEP 7 environment, I recommend that you store the 'scratchpad memory' blockwise in the temporary local data. This applies especially if you want to transfer your own program standards from STEP 5 to STEP 7. The extended S memory bits are mapped to the memory bits from address 256 (A S 0.0 becomes A M 256.0, L SY 2 becomes L MB 258, etc.)

Timer and counter functions are converted unchanged. Direct access to the individual bits of the timer or counter word is not longer possible under STEP 7. Influencing of the edge memory bits in these words with the bit test statements can be replaced with SET and CLR in conjunction with the relevant timer and counter operation.

Please note that in STEP 7 the data are address bytewise (in STEP 5, by contrast, wordwise). Thus, DL 0 becomes DBB 0, DR 0 becomes DBB 1; you can see the conversion for any addresses in Figure 27.3. With direct and indirect addressed operands, the Converter uses the correct S7 address; with data operands addressed via block parameters, you must make the conversion to bytewise addressing yourself.

Floating-point numbers are taken over unchanged, provided they are specified as constants in load statements or they have been used as actual parameters, and they are treated at conversion like STEP 7 floating-point numbers. The standard blocks supplied as a replacement for the STEP 5 standard function blocks also process floating-point numbers in the STEP 7 format (data type REAL). If you have put together floating-point yourself in your STEP 5 program or if you have taken them over from other devices via, for example, a computer link, you must adapt the STEP 5 representation of these floating-point numbers to the data type REAL. You will find a conversion example on the enclosed diskette in the "General examples" program (FC 45 GP_TO_REAL).

**Table 27.3**
Address Conversion for Data Operands

| STEP 5 | STEP 7 |
|---|---|
| DL [n] | DBB [2n] |
| DR [n] | DBB [2n+1] |
| DW [n] | DBW [2n] |
| DD [n] | DBD [2n] |
| D [(n).0..7] | DBX [(2n+1).0..7] |
| D [(n).8..15] | DBX [(2n).0..7] |

## 27.4 Post-Editing

### 27.4.1 Creating the STEP 7 Project

To complete the conversion, you create a STEP 7 project that corresponds in structure to your destination system (if you have not already created it in order to learn the S7 module addresses). If you want to change module addresses, parameterize modules or change the execution properties of the CPU, you require a hardware configuration (that is, a fully set up project). If the default settings of the module characteristics cannot be changed, it is enough to set up a module-independent program.

- You create a station (S7-300 or S7-400), open the object *Hardware* and configure the station. You also set the properties of the CPU with the Hardware Configuration (for example, numbers of the interrupt OBs). Together with the CPU, the SIMATIC Manager also sets up the lower-level object containers.
- With the object *Sources* marked, you fetch the generated file *name*AC.AWL into the source program container with INSERT → EXTERNAL SOURCE FILE...
- If your program uses S5 standard blocks, open the library "S5–S7 Converting Blocks" under "Standard Library V3.x" and copy the S7 standard blocks, indicated by the Converter in the block list with a star, into the off-line user program *Blocks* of your project. If you use S7 system blocks in the converted program (for example, SFC 20 BLKMOV), open the library "System Function Blocks" and copy the system blocks used into the off-line user program *Blocks*.
- If you have been working with symbolic programming, open the (empty) symbol table *Symbols* and fetch the converted symbols *name*S7.SEQ with SYMBOL TABLE → IMPORT...

**Table 27.4** Unconvertible Functions

| Functions in STEP 5 | Remarks |
|---|---|
| Load and transfer functions | |
| with system data | Replaced, for example, with system functions |
| with absolute addresses | Must be replaced with a new program |
| Register functions (LIR, TIR, LDI, TDI, MBA, MAB, MSA, MAS, MBA, MSB, MBR, ABR, ACR) | Must be replaced with a new program |
| Block transfer (TNB, TNW, TXB, TXW) | Replaced with SFC 20 BLKMOV |
| DO functions | |
| DO DW, DO FW | Converted |
| DO RS | Must be replaced with a new program |
| Calling special functions | Replace special functions with SFCs |
| LIM, SIM, IAE, RAE | Can be replaced with SFC 39 .. SFC 42 |
| Semaphore functions (SED, SEE, TSC, TSG) | No replacement |
| Other (IAI, RAI, ASM, UBE) | No replacement |

Following these preparations, you can now process the source file with the Editor before compiling it (you can reduce the number of error messages, if you carry out all corrections before compiling).

### 27.4.2 Unconvertible Functions

After conversion, you usually have to post-edit the source file. This affects all the statements listed in Table 27.4.

### 27.4.3 Address Changes

The address changes affect essentially the input and output modules. Under certain circumstances, you must adapt the accesses to the inputs and outputs as well as the direct peripheral I/O accesses to the (new) module addresses. You can carry out this adaptation before conversion in the STEP 5 file (if the address volume suitable for STEP 5) or you can swap the absolute addresses in the S7 source file with the help of the 'Replace' function of the Editor used (use caution if the old and new address areas overlap).

In the case of programming with symbolic addressing, you can also generate a source with symbol addresses, change the absolute addresses in the symbol table and then re-compile. Proceed as follows here:

- A requirement is that you have a symbol table with symbols for all the absolute addresses to be changed and a program compiled free of errors (the blocks in which the absolute addressed operands occur must be available in compiled form).
- You set the Editor to symbolic addressing: OPTIONS → CUSTOMIZE displays a dialog field; select the option SYMBOLIC REPRESENTATION in the 'Editor' tab.
- You generate a new source file using the Editor with FILE → GENERATE SOURCE FILE... After entering the file name, you select all blocks in the dialog window shown that you want as a source file with symbolic addressing. The new source file now contains the statements with symbolic addressing.
- Next, correct all absolute addresses in the symbol table from the (old) S5 version to the (new) S7 version.
- If you now compile the new source file, the new absolute addresses will be contained in the compiled blocks.

### 27.4.4 Indirect Addressing

The Converter can also understand indirect addressing with DO FW and DO DW with STEP 7 statements. However, it is necessary here to convert the pointer to the STEP 7 format, which, in conjunction with the buffering of accumulator contents and the status word, leads to an increased memory requirement. With suitable programming you can usually execute indirect addressing, whether it is memory-indirect or register-indirect, with fewer statements and a clearer program structure. If indirect addressing occurs frequently, STEP 7-adapted programming is certainly of advantage.

- Indirect addressing of timers, counters and blocks
  This is converted into memory-indirect addressing using a temporary local data word.
- Indirect addressing of blocks
  Allocation of the new block numbers cannot be taken into account (manual correction)
- Indirect addressing of operands
  Converted bitwise and wordwise using AR1, buffering of STW, Accum 1 and 2 in temporary local data (see below)
- Indirect addressing via the BR register
  No conversion possible, change manually via address registers
- Other indirect addressing
  Must be changed manually

| | |
|---|---|
| Jump functions | Replaced with jump distributor SPL |
| Shift functions | Replaced with shift functions with number of shifts in accum 2 |
| TNB, TNW | Replaced with SFC 20 BLKMOV with "variable" ANY pointer |
| LIR, TIR | No direct replacement available |
| Decrementing/incrementing | No direct replacement available |

The Converter changes indirect addressing with DO FW and DO DW of binary logic operations, memory functions, and load and transfer functions to a STEP 7 program. The STEP 5 pointer must be changed to the format of an area-internal STEP 7 pointer (with buffering of the accumulator contents and the status word). The result is a long sequence of statements (see example). If you have used a large number of indirect addresses in your program, manual conversion could be of advantage. As index register, you have unrestricted access to the two address registers AR1 and AR2 (in functions FCs). You can also address memory bits or data memory-indirect as in STEP 5, but you then require one doubleword per index register instead of one word.

The example in Table 27.5 shows in the first column a STEP 5 program which is compared with a data field with the bit pattern of an input word; if they are identical, a memory bit is set in each case. The second column contains the converted program. Using both address registers, you can write a directly comparable program requiring significantly fewer statements.

First, the address registers are loaded with the pointers (take account of bytewise addressing of the data!). Access to the data words and the memory bits is then register-indirect. After every comparison, address register AR1 is incremented by 2 bytes and address register AR2 is incremented by one bit (conversion to the byte address is omitted). In the example, the pointer to the data words is used as the break criterion, just as in STEP 5; at this point STEP 7 provides use of the loop jump LOOP.

### 27.4.5 Access to "Excessively Long" Data Blocks

Access to "excessively long" data blocks, that is access to data operands that had a byte address > 255, was carried out under STEP 5 with absolute addressing. The data block start address was calculated, the operand offset was added and the data operand was accessed either direct with LIR/TIR or via the BR register with LRW/TRW.

With STEP 7, you can address the data operands direct up to the permissible limit (8095 on the S7-300, 32767 on the S7-400). You can therefore replace access via the absolute address with a "normal" STL statement.

### 27.4.6 Working with Absolute Addresses

If is necessary to handle absolute memory addresses in STEP 5 if you address data operands in "excessively long" data blocks, or if you address indirectly with the BR register, or if you use the block transfer. Access to absolute memory addresses in no longer possible with STEP 7; the STEP address counter (with the associated operations) has been removed without replacement.

**Table 27.6** Converting the Special Function Organization Block

| Function | 115U | 135U | 155U | S7 Replacement |
|---|---|---|---|---|
| Process condition code byte | – | 110 | – | Statement sequence |
| Process accumulators | – | 111–113 | 131–133 | Statement sequence |
| Handle interrupts | | 120–123 | 122 | SFC 39 DIS_IRT, SFC 40 EN_IRT, |
| | | | 141–143 | SFC 41 DIS_AIRT, SFC 42 EN_AIRT |
| Activate a timer job | – | 151 | 151 | SFC 28 SET_TINT, SFC 29 CAN_TINT, SFC 30 ACT_TINT, SFC 31 QRY_TINT |
| Handle a delay interrupt | – | 153 | 153 | SFC 32 SRT_DINT, SFC 33 QRY_DINT, SFC 34 CAN_DINT |
| Variable waiting time | 160 | – | – | SFC 43 WAIT |
| Delete block | – | – | 124 | Data block: SFC 23 DEL_DB |
| Create block | 125 | – | 125 | Data block: SFC 22 CREAT_DB |
| Read block stack | – | 170 | – | – omitted – |
| Test data block | – | 181 | – | SFC 24 TEST_DB |
| Data block access | – | 180 | – | – omitted – |
| Copy data blocks | 183, 184 | 254, 255 | 254, 255 | SFC 20 BLKMOV (data areas) |
| Copy data areas | 182 | 182 | – | SFC 20 BLKMOV |
| | 190–193 | 190–193 | | |
| Set and read time-of-day | – | 150 | 121, 150 | SFC 0 SET_CLK, SFC 1 READ_CLK |
| Cycle statistics | – | 152 | – | Start information OB 1, SFC 6 RD_SINFO |
| Read status information | – | 228 | – | Start information, SFC 6 RD_SINFO |
| Multiprocessor communications | – | 200–205 | 200–205 | Replacement: GD communications |
| Compare restart types | – | 223 | 223 | – omitted – |
| Transfer interprocessor communication flags | – | 224 | – | GD communications |
| Set cycle time | – | 221 | – | CPU parameterization |
| Cycle time triggering | – | 222 | 31, 222 | SFC 43 RE_TRIGR |
| Transfer process images | 254, 255 | – | 126 | SFC 26 UPDAT_PI, SFC 27 UPDAT_PO |
| Counter loop | – | 160–163 | – | Statement sequence |
| Sign extension | 220 | 220 | – | Statement sequence |
| Page accesses | – | 216–218 | | – omitted – |
| System program access | – | 226, 227 | – | – omitted – |
| Process shift register | – | 240–248 | – | – omitted – |
| Handling blocks | – | 230–237 | – | SFB blocks for communications |
| PID algorithm | 251 | 250–251 | – | Standard blocks for PID control |
| Execute system service | 250 | – | – | (see above under 'Checking Program Execution Characteristics') |

**Table 27.7** Converting the Error Organization Blocks

| Function | S5-115 | S5-135 | S5-155 | S7 Replacement |
|---|---|---|---|---|
| Calling an unloaded block | 19 | 19 | 19 | OB 121 |
| Acknowledgment delay in the case of direct access to I/O modules | 23 | 23 | 23 | OB 122 |
| Acknowledgment in the case of updating the process image | 24 | 24 | 24 | OB 122 |
| Addressing errors | – | 25 | 25 | OB 122 |
| Cycle time exceeded | 26 | 26 | 26 | OB 80 |
| Substitution errors | 27 | 27 | 27 | – |
| Conditional Stop | – | 28 | – | – |
| Acknowledgment delay in the case of input byte IB 0 | – | – | 28 | OB 85 |
| Illegal operation code | – | 29 | – | STOP |
| Acknowledgment delay in the case of direct access in the extended I/O area | – | – | 29 | OB 122 |
| Illegal parameters | | 30 | – | – |
| Parity errors or acknowledgment in the case of access to the user memory | – | – | 30 | OB 122 |
| Special function group errors | – | 31 | – | – |
| Transfer errors in data blocks | 32 | 32 | 32 | OB 121 |
| Watchdog errors in the case of time-controlled execution | 33 | 33 | 33 | OB 80 |
| Battery failure | 34 | – | – | OB 81 |
| Controller errors | – | 34 | – | – |
| Error in creating a data block | – | – | 34 | (SFC) |
| I/O errors | 35 | – | – | OB 86 |
| Interface errors | – | 35 | – | OB 84 |
| Self-test errors | – | – | 36 | – |

# 28 Block Libraries

The Standard Library V3.x contains the following library programs and is included in the scope of supply of the STEP 7 Standard Software:

- Organization Blocks
- System Function Blocks
- IEC Converting Blocks
- S5-S7 Converting Blocks
- TI-S7 Converting Blocks
- PID Control Blocks
- Communication Blocks

You can copy blocks or interface descriptions into Version 3 projects from the library programs described. If you want to transfer blocks or interface descriptions into a Version 2 project, you must use the library "stdlibs".

## 28.1 Organization Blocks

(Prio = Default priority class)

| OB | Prio | Designation |
|---|---|---|
| 1 | 1 | Main program |
| 10 | 2 | Time-of-day interrupt 0 |
| 11 | 2 | Time-of-day interrupt 1 |
| 12 | 2 | Time-of-day interrupt 2 |
| 13 | 2 | Time-of-day interrupt 3 |
| 14 | 2 | Time-of-day interrupt 4 |
| 15 | 2 | Time-of-day interrupt 5 |
| 16 | 2 | Time-of-day interrupt 6 |
| 17 | 2 | Time-of-day interrupt 7 |
| 20 | 3 | Time-delay interrupt 0 |
| 21 | 4 | Time-delay interrupt 1 |
| 22 | 5 | Time-delay interrupt 2 |
| 23 | 6 | Time-delay interrupt 3 |
| 30 | 7 | Cyclic interrupt 0 (5 s) |
| 31 | 8 | Cyclic interrupt 1 (2 s) |
| 32 | 9 | Cyclic interrupt 2 (1 s) |

| OB | Prio | Designation |
|---|---|---|
| 33 | 10 | Cyclic interrupt 3 (500 ms) |
| 34 | 11 | Cyclic interrupt 4 (200 ms) |
| 35 | 12 | Cyclic interrupt 5 (100 ms) |
| 36 | 13 | Cyclic interrupt 6 (50 ms) |
| 37 | 14 | Cyclic interrupt 7 (20 ms) |
| 38 | 15 | Cyclic interrupt 8 (10 ms) |
| 40 | 16 | Hardware interrupt 0 |
| 41 | 17 | Hardware interrupt 1 |
| 42 | 18 | Hardware interrupt 2 |
| 43 | 19 | Hardware interrupt 3 |
| 44 | 20 | Hardware interrupt 4 |
| 45 | 21 | Hardware interrupt 5 |
| 46 | 22 | Hardware interrupt 6 |
| 47 | 23 | Hardware interrupt 7 |
| 60 | 25 | Multiprocessor interrupt |
| 80 | 26 | Time error |
| | 28 | Time error (startup) |
| 81 | 26 | Power supply fault |
| | 28 | Power supply fault (startup) |
| 82 | 26 | Diagnostics interrupt |
| | 28 | Diagnostics interrupt (startup) |
| 83 | 26 | Insert/remove interrupt |
| | 28 | Insert/remove interrupt (startup) |
| 84 | 26 | CPU hardware fault |
| | 28 | CPU hardware fault (startup) |
| 85 | 26 | Program sequence error |
| | 28 | Program sequence error (startup) |
| 86 | 26 | Rack failure |
| | 28 | Rack failure (startup) |
| 87 | 26 | Communication error |
| | 28 | Communication error (startup) |
| 90 | 29 | Background processing |
| 100 | 27 | Complete restart |
| 101 | 27 | Restart |
| 121 | – | Programming error |
| 122 | – | I/O access error |

## 28.2 System Function Blocks

### IEC timers and IEC counters

| SFB | NAME | Designation |
|---|---|---|
| 0 | CTU | Up counter |
| 1 | CTD | Down counter |
| 2 | CTUD | Up/down counter |
| 3 | TP | Pulse |
| 4 | TON | On delay |
| 5 | TOF | Off delay |

### Copy and block functions

| SFC | NAME | Designation |
|---|---|---|
| 20 | BLKMOV | Copy data area |
| 21 | FILL | Preassign data area |
| 22 | CREAT_DB | Create data block |
| 23 | DEL_DB | Delete data block |
| 24 | TEST_DB | Test data block |
| 25 | COMPRESS | Compress memory |
| 44 | REPL_VAL | Enter replacement value |

### Program control

| SFC | NAME | Designation |
|---|---|---|
| 43 | RE_TRIGR | Retrigger cycle time monitoring |
| 46 | STP | Change to STOP state |
| 47 | WAIT | Wait delay time |

### Drum

| SFB | NAME | Designation |
|---|---|---|
| 32 | DRUM | Drum |

### CPU clock and run-time meter

| SFC | NAME | Designation |
|---|---|---|
| 0 | SET_CLK | Set clock |
| 1 | READ_CLK | Read clock |
| 2 | SET_RTM | Set run-time meter |
| 3 | CTRL_RTM | Control run-time meter |
| 4 | READ_RTM | Read run-time meter |
| 48 | SNC_RTCB | Synchronize slave clocks |
| 64 | TIME_TCK | Read system time |

### Interrupt events

| SFC | NAME | Designation |
|---|---|---|
| 28 | SET_TINT | Set time-of-day interrupt |
| 29 | CAN_TINT | Cancel time-of-day interrupt |
| 30 | ACT_TINT | Activate time-of-day interrupt |
| 31 | QRY_TINT | Check time-of-day interrupt |
| 32 | SRT_DINT | Start time-delay interrupt |
| 33 | CAN_DINT | Cancel time-delay interrupt |
| 34 | QRY_DINT | Check time-delay interrupt |
| 35 | MP_ALM | Initiate multiprocessor interrupt |
| 36 | MSK_FLT | Mask synchronous error |
| 37 | DMSK_FLT | Demask synchronous error |
| 38 | READ_ERR | Read event status register |
| 39 | DIS_IRT | Disable asynchronous error |
| 40 | EN_IRT | Enable asynchronous error |
| 41 | DIS_AIRT | Delay asynchronous error |
| 42 | EN_AIRT | Enable asynchronous error |

### Data record transfer

| SFC | NAME | Designation |
|---|---|---|
| 54 | RD_DPARM | Read predefined parameter |
| 55 | WR_PARM | Write dynamic parameter |
| 56 | WR_DPARM | Write predefined parameter |
| 57 | PARM_MOD | Parameterize module |
| 58 | WR_REC | Write data record |
| 59 | RD_REC | Read data record |

### System diagnostics

| SFC | NAME | Designation |
|---|---|---|
| 6 | RD_SINFO | Read start information |
| 51 | RDSYSST | Read SYS ST sublist |
| 52 | WR_USMSG | Entry in the diagnostics buffer |

### Create block-related messages

| SFB | NAME | Designation |
|---|---|---|
| 33 | ALARM | Messages with acknowledgment display |
| 34 | ALARM_8 | Messages without accompanying values |
| 35 | ALARM_8P | Messages with accompanying values |
| 36 | NOTIFY | Messages without acknowledgment display |
| 37 | AR_SEND | Send archive data |

| SFC | NAME | Designation |
|---|---|---|
| 9 | EN_MSG | Enable messages |
| 10 | DIS_MSG | Disable messages |
| 17 | ALARM_SQ | Messages that can be acknowledged |
| 18 | ALARM_S | Messages that are always acknowledged |
| 19 | ALARM_SC | Determine acknowledgment status |

### Update the process image

| SFC | NAME | Designation |
|---|---|---|
| 26 | UPDAT_PI | Update process-image inputs |
| 27 | UPDAT_PO | Update process-image outputs |
| 79 | SET | Set I/O bit field |
| 80 | RSET | Reset I/O bit field |

### Address modules

| SFC | NAME | Designation |
|---|---|---|
| 5 | GADR_LGC | Determine logical address |
| 49 | LGC_GADR | Determine slot |
| 50 | RD_LGADR | Determine all logical addresses |

### Distributed I/O

| SFC | NAME | Designation |
|---|---|---|
| 7 | DP_PRAL | Initiate hardware interrupt |
| 11 | DPSYC_FR | SYNC/FREEZE |
| 13 | DPNRM_DG | Read diagnostics data |
| 14 | DPRD_DAT | Read slave data |
| 15 | DPWR_DAT | Write slave data |

### Global data communications

| SFC | NAME | Designation |
|---|---|---|
| 60 | GD_SND | Send GD packet |
| 61 | GD_RCV | Receive GD packet |

### Communications via unconfigured connections

| SFC | NAME | Designation |
|---|---|---|
| 65 | X_SEND | Send data externally |
| 66 | X_RCV | Receive data externally |
| 67 | X_GET | Read data externally |
| 68 | X_PUT | Write data externally |
| 69 | X_ABORT | Abort external connection |
| 72 | I_GET | Read data internally |
| 73 | I_PUT | Write data internally |
| 74 | I_ABORT | Abort internal connection |

### Communications via configured connections

| SFC | NAME | Designation |
|---|---|---|
| 62 | CONTROL | Check communications status |

| SFB | NAME | Designation |
|---|---|---|
| 8 | USEND | Uncoordinated send |
| 9 | URVC | Uncoordinated receive |
| 12 | BSEND | Block-oriented send |
| 13 | BRCV | Block-oriented receive |
| 14 | GET | Read data from partner |
| 15 | PUT | Write data to partner |
| 16 | PRINT | Write data to printer |
| 19 | START | Initiate complete restart in the partner |
| 20 | STOP | Set partner to STOP |
| 21 | RESUME | Initiate restart in the partner |
| 22 | STATUS | Check status of partner |
| 23 | USTATUS | Receive status of partner |

### Integrated functions CPU 312/314/614

| SFC | NAME | Designation |
|---|---|---|
| 63 | AB_CALL | Call assembler block |

| SFB | NAME | Designation |
|---|---|---|
| 29 | HS_COUNT | High-speed counter |
| 30 | FREQ_MES | Frequency meter |
| 38 | HSC_A_B | Control "Counter A/B" |
| 39 | POS | Control 'Positioning' |
| 41 | CONT_C | Continuous closed-loop control |
| 42 | CONT_S | Step-action control |
| 43 | PULSEGEN | Generate pulse |

## 28.3 IEC Converting Blocks

### Date and time functions

| FC | NAME | Designation |
|---|---|---|
| 3 | D_TOD_DT | Combine DATE and TOD to DT |
| 6 | DT_DATE | Extract DATE from DT |
| 7 | DT_DAY | Extract day-of-the-week from DT |
| 8 | DT_TOD | Extract TOD from DT |
| 33 | S5TI_TIM | Convert S5TIME to TIME |
| 40 | TIM_S5TI | Convert TIME to S5TIME |
| 1 | AD_DT_TM | Add TIME to DT |
| 35 | SB_DT_TM | Subtract TIME from DT |
| 34 | SB_DT_DT | Subtract DT from DT |

### Comparisons

| FC | NAME | Designation |
|---|---|---|
| 9 | EQ_DT | Compare DT for equal to |
| 28 | NE_DT | Compare DT for not equal to |
| 14 | GT_DT | Compare DT for greater than |
| 12 | GE_DT | Compare DT for greater than or equal to |
| 23 | LT_DT | Compare DT for less than |
| 18 | LE_DT | Compare DT for less than or equal to |
| 10 | EQ_STRNG | Compare STRING for equal to |
| 29 | NE_STRNG | Compare STRING for not equal to |
| 15 | GT_STRNG | Compare STRING for greater than |
| 13 | GE_STRNG | Compare STRING for greater than or equal to |
| 24 | LT_STRNG | Compare STRING for less than |
| 19 | LE_STRNG | Compare STRING for less than or equal to |

### String functions

| FC | NAME | Designation |
|---|---|---|
| 21 | LEN | Length of a STRING |
| 20 | LEFT | Left section of a STRING |
| 32 | RIGHT | Right section of a STRING |
| 26 | MID | Middle section of a STRING |
| 2 | CONCAT | Concatenate STRINGs |
| 17 | INSERT | Insert STRING |
| 4 | DELETE | Delete STRING |
| 31 | REPLACE | Replace STRING |
| 11 | FIND | Find STRING |
| 16 | I_STRNG | Convert INT to STRING |
| 5 | DI_STRNG | Convert DINT to STRING |
| 30 | R_STRNG | Convert REAL to STRING |
| 38 | STRNG_I | Convert STRING to INT |
| 37 | STRNG_DI | Convert STRING to DINT |
| 39 | STRNG_R | Convert STRING to REAL |

### Math functions

| FC | NAME | Designation |
|---|---|---|
| 22 | LIMIT | Limiter |
| 25 | MAX | Maximum selection |
| 27 | MIN | Minimum selection |
| 26 | SEL | Binary selection |

## 28.4 S5-S7 Converting Blocks

### Floating-point arithmetic

| FC | NAME | Designation |
|---|---|---|
| 61 | GP_FPGP | Convert fixed-point to floating-point |
| 62 | GP_GPFP | Convert floating-point to fixed-point |
| 63 | GP_ADD | Add floating-point numbers |
| 64 | GP_SUB | Subtract floating-point numbers |
| 65 | GP_MUL | Multiply floating-point numbers |
| 66 | GP_DIV | Divide floating-point numbers |
| 67 | GP_VGL | Compare floating-point numbers |
| 68 | GP_RAD | Find the square root of a floating-point number |

### Signal functions

| FC | NAME | Designation |
|---|---|---|
| 69 | MLD_TG | Clock pulse generator |
| 70 | MLD_TGZ | Clock pulse generator with timer function |
| 71 | MLD_EZW | Initial value single blinking wordwise |
| 72 | MLD_EDW | Initial value double blinking wordwise |
| 73 | MLD_SAMW | Group signal wordwise |
| 74 | MLD_SAM | Group signal |
| 75 | MLD_EZ | Initial value single blinking |
| 76 | MLD_ED | Initial value double blinking |
| 77 | MLD_EZWK | Initial value single blinking (wordwise) memory bit |
| 78 | MLD_EZDK | Initial value double blinking (wordwise) memory bit |
| 79 | MLD_EZK | Initial value single blinking memory bit |
| 80 | MLD_EDK | Initial value double blinking memory bit |

### Integrated functions

| FC | NAME | Designation |
|---|---|---|
| 81 | COD_B4 | BCD-binary conversion 4 decades |
| 82 | COD_16 | Binary-BCD conversion 4 decades |
| 83 | MUL_16 | 16-bit fixed-point multiplier |
| 84 | DIV_16 | 16-bit fixed-point divider |

### Basic functions

| FC | NAME | Designation |
|---|---|---|
| 85 | ADD_32 | 32-bit fixed-point adder |
| 86 | SUB_32 | 32-bit fixed-point subtractor |
| 87 | MUL_32 | 32-bit fixed-point multiplier |
| 88 | DIV_32 | 32-bit fixed-point divider |
| 89 | RAD_16 | 16-bit fixed-point square root extractor |
| 90 | REG_SCHB | Bitwise shift register |
| 91 | REG_SCHW | Wordwise shift register |
| 92 | REG_FIFO | Buffer (FIFO) |
| 93 | REG_LIFO | Stack (LIFO) |
| 94 | DB_COPY1 | Copy data area (direct) |
| 95 | DB_COPY2 | Copy data area (indirect) |
| 96 | RETTEN | Save scratchpad memory (S5 155U) |
| 97 | LADEN | Load scratchpad memory (S5 155U) |
| 98 | COD_B8 | BCD-binary conversion 8 decades |
| 99 | COD_32 | Binary-BCD conversion 8 decades |

### Analog functions

| FC | NAME | Designation |
|---|---|---|
| 100 | AE_460_1 | Analog input module 460 |
| 101 | AI_460_2 | Analog input module 460 |
| 102 | AI_463_1 | Analog input module 463 |
| 103 | AE_463_2 | Analog input module 463 |
| 104 | AE_464_1 | Analog input module 464 |
| 105 | AE_464_2 | Analog input module 464 |
| 106 | AE_466_1 | Analog input module 466 |
| 107 | AE_466_2 | Analog input module 466 |
| 108 | RLG_AA1 | Analog output module |
| 109 | RLG_AA2 | Analog output module |
| 110 | PER_ET1 | ET 100 distributed I/O |
| 111 | PER_ET2 | ET 100 distributed I/O |

**Math functions**

| FC | NAME | Designation |
|---|---|---|
| 112 | SINUS | Sine |
| 113 | COSINUS | Cosine |
| 114 | TANGENS | Tangent |
| 115 | COTANG | Cotangent |
| 116 | ARCSIN | Arc sine |
| 117 | ARCCOS | Arc cosine |
| 118 | ARCTAN | Arc tangent |
| 119 | ARCCOT | Arc cotangent |
| 120 | LN_X | Natural logarithm |
| 121 | LG_X | Logarithm to base 10 |
| 122 | B_LOG_X | Logarithm to any base |
| 123 | E_H_N | Exponential function with base e |
| 124 | ZEHN_H_N | Exponential function with base 10 |
| 125 | A2_H_A1 | Exponential function with any base |

## 28.5 TI-S7 Converting Blocks

| FB | NAME | Description |
|---|---|---|
| 80 | LEAD_LAG | Lead/lag algorithm |
| 81 | DCAT | Discrete control time interrupt |
| 82 | MCAT | Motor control time interrupt |
| 83 | IMC | Index matrix comparison |
| 84 | SMC | Matrix scanner |
| 85 | DRUM | Event maskable drum |
| 86 | PACK | Collect/distribute table data |

| FC | NAME | Designation |
|---|---|---|
| 80 | TONR | Retentive on delay |
| 81 | IBLKMOV | Transfer data area indirect |
| 82 | RSET | Reset process-image bitwise |
| 83 | SET | Set process-image bitwise |
| 84 | ATT | Enter value in table |
| 85 | FIFO | Output first value in table |
| 86 | TBL_FIND | Find value in table |
| 87 | LIFO | Output last table value |
| 88 | TBL | Execute table operation |
| 89 | TBL_WRD | Copy value from the table |
| 90 | WSR | Store data item |
| 91 | WRD_TBL | Combine table element |
| 92 | SHRB | Shift bit to bit-shift register |
| 93 | SEG | Bit pattern for 7-segment display |
| 94 | ATH | ASCII-hexadecimal conversion |
| 95 | HTA | Hexadecimal-ASCII conversion |
| 96 | ENCO | Least-significant set bit |
| 97 | DECO | Set bit in the word |
| 98 | BCDCPL | Generate ten's complement |
| 99 | BITSUM | Count set bits |
| 100 | RSETI | Reset PIQ bytewise |
| 101 | SETI | Set PI bytewise |
| 102 | DEV | Calculate standard deviation |
| 103 | CDT | Correlated data tables |
| 104 | TBL_TBL | Table combination |
| 105 | SCALE | Scale values |
| 106 | UNSCALE | Descale values |

## 28.6 PID Control Blocks

| FB | NAME | Designation |
|---|---|---|
| 41 | CONT_C | Continuous control |
| 42 | CONT_S | Step control |
| 43 | PULSGEN | Generate pulse |

## 28.7 Communication Blocks

| FC | NAME | Designation |
|---|---|---|
| 1 | DP_SEND | Send data |
| 2 | DP_RECV | Receive data |
| 3 | DP_DIAG | Diagnostics |
| 4 | DP_CTRL | Control |

# 29 Operation Set

The following overview lists the operations with absolute addressed operands. The following are also possible with the addressing types:

| | | |
|---|---|---|
| A I [doubleword] | memory-indirect with the doublewords<br>MD Memory doubleword<br>LD Local data doubleword<br>DBD Global data doubleword<br>DID Instance data doubleword | all operands |
| A I [AR1, P#offset]<br>A I [AR2, P#offset] | register-indirect area-internal with AR1<br>register-indirect area-internal with AR2 | no timer or counter functions and no blocks |
| A [AR1, P#offset]<br>A [AR2, P#offset] | register-indirect area-crossing with AR1<br>register-indirect area-crossing with AR2 | no timer or counter functions, and no blocks |
| A #*name* | parameter-indirect | all operands |

## 29.1 Basic Functions

### 29.1.1 Binary Logic Operations

| | | |
|---|---|---|
| A | – | AND with check for "1" |
| AN | – | AND with check for "0" |
| O | – | OR with check for "1" |
| ON | – | OR with check for "0" |
| X | – | Exclusive OR with check for "1" |
| XN | – | Exclusive OR with check for "0" |
| – | I | an input |
| – | Q | an output |
| – | M | a memory bit |
| – | L | a local data bit |
| – | T | a timer function |
| – | C | a counter function |
| – | DBX | a global data bit |
| – | DIX | an instance data bit |
| – | ==0 | Result equal to zero |
| – | <>0 | Result not equal to zero |
| – | >0 | Result greater than zero |
| – | >=0 | Result greater than or equal to zero |
| – | <0 | Result less than zero |
| – | <=0 | Result less than zero |
| – | UO | Result invalid |
| – | OV | Overlfow |
| – | OS | Stored overflow |
| – | BR | Binary result |

| | |
|---|---|
| A( | AND open bracket |
| AN( | AND NOT open bracket |
| O( | OR open bracket |
| ON( | OR NOT open bracket |
| X( | Exclusive OR open bracket |
| XN( | Exclusive OR NOT open bracket |
| ) | Close bracket |
| O | OR combination of AND |
| NOT | Negate RLO |
| SET | Set RLO |
| CLR | Reset RLO |
| SAVE | Save RLO to BR |

### 29.1.2 Memory Functions

| | | |
|---|---|---|
| = | – | Assign |
| S | – | Set |
| R | – | Reset |
| FP | – | Positive edge |
| FN | – | Negative edge |
| – | I | an input |
| – | Q | an output |
| – | M | a memory bit |
| – | L | a local data bit |
| – | DBX | a global data bit |
| – | DIX | an instance data bit |

### 29.1.3 Transfer Functions

| | | |
|---|---|---|
| L | – | Load |
| T | – | Transfer |
| – | IB | an input byte |
| – | IW | an input word |
| – | ID | an input doubleword |
| – | QB | an output byte |
| – | QW | an output word |
| – | QD | an output doubleword |
| – | MB | a memory byte |
| – | MW | a memory word |
| – | MD | a memory doubleword |
| – | LB | a local data byte |
| – | LW | a local data word |
| – | LD | a local data doubleword |
| – | DBB | a global data byte |
| – | DBW | a global data word |
| – | DBD | a global data doubleword |
| – | DIB | an instance data byte |
| – | DIW | an instance data word |
| – | DID | an instance data doubleword |
| – | STW | the status word |
| L | PIB | Load peripheral input byte |
| L | PIW | Load peripheral input word |
| L | PID | Load peripheral input doubleword |
| T | PQB | Transfer peripheral output byte |
| T | PQW | Transfer peripheral output word |
| T | PQD | Transfer peripheral output doubleword |
| L | T | Direct loading of a timer value |
| LC | T | Coded loading of a timer value |
| L | C | Direct loading of a counter value |
| LC | C | Coded loading of a counter value |
| L | *const* | Load a constant |
| L | P#.. | Load a pointer |
| L | P#var | Load a variable start address |
| PUSH | | Shift accums "forward" |
| POP | | Shift accums "back" |
| ENT | | Shift accums (without CC1) |
| LEAVE | | Shift accums (without CC1) |
| TAK | | Swap accum 1 and accum 2 |
| CAW | | Reverse bytes 0 and 1 in accum 1 |
| CAD | | Reverse all bytes in accum 1 |

### 29.1.4 Timer Functions

| | | |
|---|---|---|
| SP | T | Start timer as pulse |
| SE | T | Start as extended pulse |
| SD | T | Start as ON delay |
| SS | T | Start as retentive ON delay |
| SF | T | Start as OFF delay |
| R | T | Reset timer function |
| FR | T | Enable timer function |

### 29.1.5 Counter Functions

| | | |
|---|---|---|
| CU | C | Count up |
| CD | C | Count down |
| S | C | Set counter function |
| R | C | Reset counter function |
| FR | C | Enable counter function |

## 29.2 Digital Functions

### 29.2.1 Comparison Functions

| | |
|---|---|
| ==I | INT comparison for equal to |
| <>I | INT comparison for not equal to |
| >I | INT comparison for greater than |
| >=I | INT comparison for greater than or equal to |
| <I | INT comparison for less than |
| <=I | INT comparison for less than or equal to |
| ==D | DINT comparison for equal to |
| <>D | DINT comparison for not equal to |
| >D | DINT comparison for greater than |
| >=D | DINT comparison for greater than or equal |
| <D | DINT comparison for less than |
| <=D | DINT comparison for less than or equal to |

| | |
|---|---|
| ==R | REAL comparison for equal to |
| <>R | REAL comparison for not equal to |
| >R | REAL comparison for greater than |
| >=R | REAL comparison for greater than or equal to |
| <R | REAL comparison for less than |
| <=R | REAL comparison for less than or equal to |

### 29.2.2 Math Functions

| | |
|---|---|
| SIN | Sine |
| COS | Cosine |
| TAN | Tangent |
| ASIN | Arc sine |
| ACOS | Arc cosine |
| ATAN | Arc tangent |
| SQR | Finding the square |
| SQRT | Finding the square root |
| EXP | Exponent to base e |
| LN | Natural logarithm |

### 29.2.3 Arithmetic Functions

| | | |
|---|---|---|
| +I | | INT addition |
| –I | | INT subtraction |
| *I | | INT multiplication |
| /I | | INT division |
| +D | | DINT addition |
| –D | | DINT subtraction |
| *D | | DINT multiplication |
| /D | | DINT division (integer) |
| MOD | | DINT division (remainder) |
| +R | | REAL addition |
| –R | | REAL subtraction |
| *R | | REAL multiplication |
| /R | | REAL division |
| + | *const* | Adding a constant |
| + | P#.. | Adding a pointer |
| DEC | *n* | Decrementing |
| INC | *n* | Incrementing |

### 29.2.4 Conversion Functions

| | |
|---|---|
| ITD | Conversion of INT to DINT |
| ITB | Conversion of INT to BCD |
| DTB | Conversion of DINT to BCD |
| DTR | Conversion of DINT to REAL |
| BTI | Conversion of BCD to INT |
| BTD | Conversion of BCD to DINT |
| | Conversion of REAL to DINT |
| RND+ | Rounding to next higher number |
| RND– | Rounding to next lower number |
| RND | Rounding to next integer |
| TRUNC | Without rounding |
| INVI | INT one's complement |
| INVD | DINT one's complement |
| NEGI | INT negation |
| NEGD | DINT negation |
| NEGR | REAL negation |
| ABS | REAL absolute-value generation |

### 29.2.5 Shift Functions

| | | |
|---|---|---|
| SLW | – | Shift left wordwise |
| SLD | – | Shift left doublewordwise |
| SRW | – | Shift right wordwise |
| SRD | – | Shift right doublewordwise |
| SSI | – | Shift with sign wordwise |
| SSD | – | Shift with sign doublewordwise |
| RLD | – | Rotate left doublewordwise |
| RRD | – | Rotate right doublewordwise |
| – | *n* | by n positions |
| – | | with number of shifts in accum 2 |
| RLDA | | Rotate left through CC1 |
| RRDA | | Rotate right through CC1 |

### 29.2.6 Word Logic Operations

| | | |
|---|---|---|
| AW | – | AND wordwise |
| AD | – | AND doublewordwise |
| OW | – | OR wordwise |
| OD | – | OR doublewordwise |
| XOW | – | Exclusive OR wordwise |
| XOD | – | Exclusive OR doublewordwise |
| – | *const* | with a word/doubleword constant |
| – | | with the contents of accum 2 |

## 29.3 Program Flow Control

### 29.3.1 Jump Functions

| | | |
|---|---|---|
| JU | *label* | Unconditional jump |
| | | Jump if |
| JC | *label* | RLO = "1" |
| JCB | *label* | RLO = "1" store with RLO |
| JCN | *label* | RLO = "0" |
| JNB | *label* | RLO = "0" store with RLO |
| JBI | *label* | BR = "1" |
| JNBI | *label* | BR = "0" |
| | | Jump if result |
| JZ | *label* | zero |
| JN | *label* | not zero |
| JP | *label* | greater than zero |
| JPZ | *label* | greater than or equal to zero |
| JM | *label* | less than zero |
| JMZ | *label* | less than or equal to zero |
| JUO | *label* | invalid |
| JO | *label* | Jump on overflow |
| JOS | *label* | Jump on stored overflow |
| JL | *label* | Jump distributor |
| LOOP | *label* | Jump loop |

### 29.3.2 Master Control Relay

| | |
|---|---|
| MCRA | Activate MCR area |
| MCRD | Deactivate MCR area |
| MCR( | Open MCR zone |
| )MCR | Close MCR zone |

### 29.3.3 Block functions

| | | |
|---|---|---|
| CALL | FB | Call function block |
| CALL | FC | Call function |
| CALL | SFB | Call system function block |
| CALL | SFC | Call system function |
| UC | FB | Call function block unconditionally |
| CC | FB | Call function block conditionally |
| UC | FC | Call function unconditionally |
| CC | FC | Call function conditionally |
| BEU | | Unconditional block end |
| BEC | | Conditional block end |
| BE | | Block end |
| OPN | DB | Call global data block |
| OPN | DI | Call instance data block |
| CDB | | Swap data block registers |
| L | DBNO | Load global data block number |
| L | DINO | Load instance data block number |
| L | DBLG | Load global data block length |
| L | DILG | Load instance data block length |
| NOP | 0 | Null operation |
| NOP | 1 | Null operation |
| BLD | *n* | Program display instruction |

## 29.4 Indirect Addressing

| | | |
|---|---|---|
| LAR1 | – | Load AR1 with |
| LAR2 | – | Load AR2 with |
| – | MD | a memory doubleword |
| – | LD | a local data doubleword |
| – | DBD | a global data doubleword |
| – | DID | an instance data doubleword |
| LAR1 | | Load AR1 with accum 1 |
| LAR2 | | Load AR2 with accum 1 |
| LAR1 | AR2 | Load AR1 with AR2 |
| LAR1 | P#.. | Load AR1 with a pointer |
| LAR2 | P#.. | Load AR2 with a pointer |
| LAR1 | P#var | Load AR1 with a variable start address |
| LAR2 | P#var | Load AR2 with a variable start address |
| TAR1 | – | Transfer AR1 to |
| TAR2 | – | Transfer AR2 to |
| – | MD | a memory doubleword |
| – | LD | a local data doubleword |
| – | DBD | a global data doubleword |
| – | DID | an instance data doubleword |
| TAR1 | | Transfer AR1 to accum 1 |
| TAR2 | | Transfer AR2 to accum 1 |
| TAR1 | AR2 | Transfer AR1 to AR2 |
| CAR | | Swap AR1 and AR2 |
| +AR1 | | Add accum 1 to AR1 |
| +AR2 | | Add accum 1 to AR2 |
| +AR1 | P#.. | Add pointer to AR1 |
| +AR2 | P#.. | Add pointer to AR2 |

# Index


Absolute value generation 134
Accumulator function 97
Actual parameter 184
Address register 261
Addressing, absolute 57
Addressing, indirect, general 254
Addressing, indirect, special points 264
Addressing, memory-indirect 258
Addressing, register-indirect 259
Addressing, symbolic 58
AND function 75
ANY pointer, structure 256
ANY pointer, variable 274
Arc functions 128
Archiving a project 32
Area pointer 254
Arithmetic functions 121
ARRAY, data type 249
Assign function 82
Asynchronous errors 238
Authorization 27

BCD numbers 243
Binary flags (indicators) 144
Binary logic operation, processing 72
Binary result BR 146
Binary result, EN/ENO 150
Binary result, using 150
Binary scaler 87
Block call, CALL 163
Block call, UC, CC 164
Block calls, general 162
Block end functions 165
Block libraries 296
Block parameters, "passing on" 186
Block parameters, definition 180
Block parameters, storing in function blocks 272
Block parameters, storing in functions 271
Block, programming 56
Block, properties 54
Block, structure 54
Block, types 53
BOOL, data type 243
Bracket functions 78
BYTE, data type 243

Chain calculation 125
CHAR, data type 243
Check result 73
Clock memory 26
Comparison functions, general 117
Comparison functions, status bits 148
Complete restart 228
Compressing 200
Computing the logarithm 129
Configuration table 34
Configuring a station, configuring 33
Configuring a station, introduction 21
Connection table 39
Constant addition 126
Constants, introduction 67
Conversion functions, general 130
Conversion functions, status bits 146
Converter functions, library 300
Conveyor belt control system example 88
Conveyor example 187
Counter function 110
CPU information 44
Creating a project 31
Cycle statistics 198

Data block register 170
Data blocks, introduction 53
Data blocks, programming 62
Data operands, addressing special points 174
Data operands, addressing 171
Data types, complex (description) 247
Data types, complex (overview) 69
Data types, elementary (description) 243
Data types, elementary (overview) 68
Data types, introduction 68
Data types, user-defined 253
DATE, data type 246
DATE_AND_TIME, data type 248
DB pointer 255
Decrementing 126
Digital flags 144
Digital logic 141
DINT calculation, status bits 146
DINT, data type 245
Disable output modules 226
DP, configuring communication 36
DP, SFCs data interchange 201
DWORD, data type 243

Edge evaluation 85
Editor 41
Encoder type, signal state "0" 75
Error handling 234
Exclusive OR function 75
Exponentiation 128

Feed example 188
First check 73
First check, status bit /FC 144
Formal parameters 181
Function blocks, introduction 53
Function value, declaration 180
Functions, introduction 53

GD communications, introduction 20
GD, configuring communications 37
GD, SFCs data exchange 203
Global data blocks, data storage 267

Hardware Interrupts 214
HOLD mode 226

IEC functions, library 299
Incremental programming 59
Incrementing 126
Inputs I 25
Instance data blocks, data storage 269
INT calculation, status bits 146
INT, data type 244
Interrupt handling 213

Jump distributor 156
Jump functions, general 152

Libraries, Overview 296
Library, general 29
Load data block length 173
Load data block number 173
Load function 94
Load user program 45
Loading the variable address 266
Local data, data storage 269
Local data, static 167
Local data, temporary 165
Local instance, introduction 169
Logic step 273
Loop jump 156

Master Control Relay (MCR) 158
Math functions 127
Memory bits M 26
Memory reset 227
Message frame example 275
Modifying and forcing variables 47
MPI network, introduction 19
Multi-instance, see local instance
Multiprocessor interrupt 222
Multiprocessor mode 201

Negate RLO 77
Negation, conversion functions 133
Negation, result of logic operation 77
Network configuration 38
Null operations 177
Number range overflow, evaluation 149
Number representations 244

One's complement 133
Open data block 172
Operation overview 302
Operations, check 73
Operations, conditional 73
OR function 75
Organization blocks, asynchronous error 238
Organization blocks, interrupt 214
Organization blocks, introduction 53
Organization blocks, library 296
Organization blocks, startup 226
Organization blocks, synchronous error 234
Outputs Q 26
Overflow 145

Parameter types, overview 70
Parts counter example 113, 188
Peripheral inputs PIs 22
Peripheral outputs PQs 22
PLC (target system), switch on-line 44
Pointers, description 254
Priority classes 51
Process image, general 25
Process interrupts 214
Process-image updating 195
Program organization 194
Program processing methods 50
Program status 48
Program structure 193
Programming blocks 56
Programming data blocks 62
Project versions 32
Project, general 29

REAL calculation, status bits 146
REAL, data type 245
Real-time clock 198
Reference data 42
Reset function 82
Response time 197
Result of logic operation RLO 73, 144
Retentivity 228
Rewiring 43
Rotation operations 138
Rounding 132
RS flipflop function 84
Run-time meter 199

S5/S7 conversion 282
S5TIME, data type 246

Scan time monitoring 196
Set and reset RLO 148
Set function 82
Setting/resetting I/O bits 160
SFB 12 BSEND 207
SFB 13 BRCV 207
SFB 14 GET 208
SFB 15 PUT 208
SFB 16 PRINT 208
SFB 19 START 209
SFB 20 STOP 209
SFB 21 RESUME 209
SFB 22 STATUS 210
SFB 23 USTATUS 210
SFB 8 USEND 207
SFB 9 URCV 207
SFB, configuring communications 38
SFBs data interchange 206
SFC 0 SET_CLK 198
SFC 1 READ_CLK 199
SFC 11 DPSYN_FR 201
SFC 13 DPNRM_DG 201
SFC 14 DPRD_DAT 201
SFC 15 DPWR_DAT 201
SFC 2 SET_RTM 199
SFC 20 BLKMOV 98
SFC 21 FILL 98
SFC 22 CREAT_DB 176
SFC 23 DEL_DB 176
SFC 24 TEST_DB 176
SFC 25 COMPRESS 200
SFC 26 UPDAT_PI 195
SFC 27 UPDAT_PO 195
SFC 28 SET_TINT 219
SFC 29 CAN_TINT 219
SFC 3 CTRL_RTM 199
SFC 30 ACT_TINT 219
SFC 31 QRY_TINT 219
SFC 32 SRT_DINT 221
SFC 33 CAN_DINT 221
SFC 34 QRY_DINT 221
SFC 35 MP_ALM 222
SFC 36 MSK_FLT 235
SFC 37 DMSK_FLT 235
SFC 38 READ_ERR 235
SFC 39 DIS_IRT 222
SFC 4 READ_RTM 199
SFC 40 EN_IRT 222
SFC 41 DIS_AIRT 223
SFC 42 EN_AIRT 223
SFC 43 RE_TRIGR 196
SFC 44 REPL_VAL 238
SFC 46 STP 200
SFC 47 WAIT 200
SFC 48 SNC_RTCB 199
SFC 49 LGC_GADR 231
SFC 5 GADR_LGC 230
SFC 50 RD_LGADR 231
SFC 51 RDSYSST 241
SFC 52 WR_USMSG 240
SFC 54 RD_DPARM 231
SFC 55 WR_PARM 231
SFC 56 WR_DPARM 231
SFC 57 PARM_MOD 231
SFC 58 WR_REC 231
SFC 59 RD_REC 231
SFC 6 RD_SINFO 212
SFC 60 GD_SND 203
SFC 61 GD_RCV 203
SFC 62 CONTROL 210
SFC 64 TIME_TCK 199
SFC 65 X_SEND 205
SFC 66 X_RCV 205
SFC 67 X_GET 205
SFC 68 X_PUT 205
SFC 69 X_ABORT 205
SFC 72 I_GET 204
SFC 73 I_PUT 204
SFC 74 I_ABORT 204
SFC 79 SET 160
SFC 80 RSET 160
Shift functions, general 135
Shift functions, status bits 148
SIMATIC Manager 27
Source-oriented programming 62
Square root extraction 128
Squaring 128
Start information OB 1 211
Start information, interrupt handling 214
Start-up characteristics 225
STARTUP mode 228
Status bits, evaluation 148
Status bits, general 144
Status word 146
Status 72
STL statement 56
STOP mode 227
STRING, data type 248
STRUCT, data type 251
Subprocess images 195
Swap (exchange) data block registers 173
Symbol table 39
Synchronous error 234
System blocks 54
System diagnostics 239
System functions, library 297
System memory 25
System time 199

Temporary local data 165
TIME, data type 247
TIME_OF_DAY, data type 247
Time-delay interrupts 220
Time-of-day interrupts 217
Timer function 100

Transfer function 96
Transfer functions (move functions) 92
Trigonometric functions 127
Two's complement 133

**U**DT, user-defined data type 253
User blocks 53
User memory 24

**V**ariable declaration 60
Variable table 47

**W**arm restart 229
Watchdog interrupts 216
Word logic operations, general 139
Word logic operations, status bits 148
WORD, data type 243

# Abbreviations

| | |
|---|---|
| AI | Analog Input |
| AO | Analog Output |
| AS | Automation System |
| ASI | Actuator-Sensor Interface |
| BR | Binary Result |
| CFC | Continuous Function Chart |
| CP | Communications Processor |
| CPU | Central Processing Unit |
| DB | Data Block |
| DI | Digital Input |
| DO | Digital Output |
| DP | Distributed I/O |
| EPROM | Erasable Programmable Read Only Memory |
| FB | Function Block |
| FC | Function Call |
| FEPROM | Flash Erasable Programmable Read Only Memory |
| FM | Function Module |
| IM | Interface Module |
| LAD | Ladder Diagram |
| MCR | Master Control Relay |
| MPI | Multi Point Interface |
| OB | Organization Block |
| OP | Operator Panel |
| PG | Programming Device |
| PS | Power Supply |
| RAM | Random Access Memory |
| RLO | Result of Logic Operation |
| SCL | Structured Control Language |
| SDB | System Data Block |
| SFB | System Function Block |
| SFC | System Function Call |
| SM | Signal Module |
| STL | Statement List |
| SZL | System Status List |
| UDT | User Defined Data Type |
| VAT | Variable Table |